D0918239

INORGANIC SYNTHESES

Volume 30

Editors-in-Chief

DONALD W. MURPHY
AT&T Bell Laboratories

LEONARD V. INTERRANTE
Renselaer Polytechnic Institute

INORGANIC SYNTHESES

Volume 30

NONMOLECULAR SOLIDS

A Wiley-Interscience Publication
JOHN WILEY & SONS, INC.

New York   Chichester   Brisbane   Toronto   Singapore

This text is printed on acid-free paper.

Published by John Wiley & Sons, Inc.

Library of Congress Catalog Number: 39-23015

ISBN 0-471-30508-1

Printed in the United States of America

10 9 8 7 6 5 4 3 2 1

PREFACE

This special topical volume of *Inorganic Synthesis* focusses on the synthesis of nonmolecular inorganic solids. Like its predecessor, *Reagents for Transition Metal Complex and Organometallic Syntheses* (Vol. 28), edited by R. J. Angelici, it differs from the usual volumes by its focus and in the fact that it includes selected syntheses reprinted from earlier volumes in addition to new contributions. It is hoped that the volume will be of interest to a broad range of scientists and engineers whose need for inorganic solids goes beyond the range or forms of materials that are commercially available or who would like to adapt proven synthetic methods to other systems.

The choice of reprinted syntheses is selective rather than comprehensive. Our aim was to focus on syntheses of materials that are either not widely available commercially in high purity or that illustrate a generally useful methodology or approach to a specific type of inorganic solid. Previous syntheses are reprinted without change and the original reference is given. Syntheses of extended solids are found in many previous *Inorganic Syntheses* volumes (syntheses from six previous volumes are reprinted here), with Volumes 14 and 22 having the greatest concentration on solids.

The new syntheses in this volume include both solicited and contributed syntheses. An effort was made to fill gaps in materials types and synthetic approaches, as well as to include specific materials that are likely to be of broad current interest, such as oxide superconductors. As for all syntheses published in *Inorganic Syntheses*, these syntheses have been checked by independent investigators from another laboratory who have verified the reproducibility of the reported syntheses.

It is hoped that this volume will dispel the view that the synthesis of inorganic solids is simply "heat and beat." While it is true that many important inorganic solids can be prepared by applying sufficient heat to appropriate proportions of elements or binary compounds, a further level of sophistication is often needed to obtain materials of sufficient quality to determine intrinsic physical properties or prepare metastable materials. The chapters in this volume on growth of single crystals and on intercalation chemistry illustrate some of the routes by which these objectives can be accomplished. The volume illustrates a wide variety of techniques including electrochemical, hydrothermal, skull melting, flux growth, ion exchange, spray pyrolysis, and organometallic precursor routes. The marked increase in emphasis on materials as an important area of research makes a volume on compounds with extended structures particularly timely. We hope that the syntheses herein will encourage efforts to extend the range of new solid state materials and inspire new synthetic techniques.

We thank those who have submitted syntheses for the volume and especially the checkers who have provided independent verification. In addition to the checkers who verified included syntheses, we acknowledge J. Krajewski, J. Xu, and R. Kniep for efforts on syntheses that were not readily reproducible and thus not included in the volume.

We appreciate the conscientious effort of members of the *Inorganic Syntheses* board for passing along their editorial wisdom and reviewing each manuscript. Particularly, Duward Shriver's leadership in suggesting the volume is acknowledged as well as Thomas Sloan's vigilant oversight of nomenclature and his help in indexing. Lastly, we mourn the loss of John Bailar, one of the founders of *Inorganic Syntheses.* He continued to be an active member until his death in 1991, reviewing many of the manuscripts in this volume. His contributions to *Inorganic Syntheses* and to inorganic chemistry in general are given in more detail in the accompanying obituary.

Murray Hill, New Jersey
Troy, New York
November 1994

DONALD W. MURPHY
LEONARD V. INTERRANTE

JOHN C. BAILAR, JR. (1904–1991)

John Christian Bailar, Jr., Professor Emeritus of Inorganic Chemistry at the University of Illinois, who was born in Golden, Colorado on May 27, 1904, died of a heart attack in Urbana, Illinois on October 17, 1991 at the age of 87. President of the American Chemical Society (ACS) (1959), Priestley Medalist (1964), and recipient of the ACS Award for Distinguished Service in the Advancement of Inorganic Chemistry (1972), John spent his entire academic career (63 years) at Illinois—almost half the time that the university had been in existence. He educated several generations of chemists (90 doctoral candidates, 38 postdoctoral fellows, and scores of bachelor's and master's degree candidates) and published 338 papers and 58 book reviews, and he coauthored or edited 10 monographs, texts, or laboratory manuals.

When John began his career at Illinois in 1928 inorganic chemistry was languishing in the doldrums. In the United States there were very few inorganic chemists, and, like John, most were overburdened with teaching duties in general chemistry. Few inorganic courses existed beyond the freshman course, little inorganic research was being carried out, and avenues for publication were exceedingly limited.

At the 86th National American Chemical Society meeting (Chicago, September 11–15, 1933) five inorganic chemists—Harold S. Booth (1891–1950), Ludwig F. Audrieth (1901–1967), W. Conard Fernelius (1905–1986), Warren C. Johnson (1901–1983), and Raymond E. Kirk (1890–1957)—decided that there was a vital need for a series of volumes giving detailed, independently tested methods for the synthesis of inorganic compounds along the lines of *Organic Syntheses*, the series established by John's colleague at Illinois, Roger Adams. The five, soon joined by John, became the Editorial Board of the new series, *Inorganic Syntheses*, the first volume of which, under Booth's editorship, appeared in 1939. Since its beginning, John was an active participant and motivating force in its affairs, contributing 16 syntheses and checking five others, especially in the early years when the fledgling journal required considerable support. He served as Editor-in-Chief (Vol. IV, 1953), and eight of his former students (including LVI of this volume) and three of his academic grandchildren (students of his former students) later served in the same capacity.

Within the ACS, a division of Physical and Inorganic Chemistry had been founded in 1908. John was active in the division, serving as Secretary (1948), Councillor (1949–1950), member of the Executive Committee (1948–1954), and Chairman (1950). In view of the growing importance of inorganic chemistry, John and several colleagues thought that it merited separate divisional status, and they took steps to promote a Division of Inorganic Chemistry. When a physical chemist asked John why a separate division was necessary, he replied, "We need a forum for meetings, and, basically, we are just proud of our field." The physical chemist countered, "We already have many forums, and I don't really understand this 'pride thing,' even though I am a member of the most important field of chemistry." John responded, "I can see that you do feel some pride in your field."

In 1957 the new division was finally established largely through John's efforts, and he served as the first Divisional Chairman. Similarly, the first journal in the English language devoted exclusively to the field, *Inorganic Chemistry*, began publication in 1962, again, largely through John's efforts. In short, the resurgence in the field after World War II, which the late Sir Ronald S. Nyholm called the renaissance in inorganic chemistry, owed much to John's pioneering labors.

GEORGE B. KAUFFMAN

California State University, Fresno
Fresno, CA 93740

NOTICE TO CONTRIBUTORS AND CHECKERS

The *Inorganic Syntheses* series is published to provide all users of inorganic substances with detailed and foolproof procedures for the preparation of important and timely compounds. Thus the series is the concern of the entire scientific community. The Editorial Board hopes that all chemists will share in the responsibility of producing *Inorganic Syntheses* by offering their advice and assistance in both the formulation of and the laboratory evaluation of outstanding syntheses. Help of this kind will be invaluable in achieving excellence and pertinence to current scientific interests.

There is no rigid definition of what constitutes a suitable synthesis. The major criterion by which syntheses are judged is the potential value to the scientific community. An ideal synthesis is one that presents a new or revised experimental procedure applicable to a variety of related compounds, at least one of which is critically important in current research. However, syntheses of individual compounds that are of interest or importance are also acceptable. Syntheses of compounds that are readily available commercially at reasonable prices are not acceptable. Corrections and improvements of syntheses already appearing in *Inorganic Syntheses* are suitable for inclusion.

The Editorial Board lists the following criteria of content for submitted manuscripts. Style should conform with that of previous volumes of *Inorganic Syntheses*. The introductory section should include a concise and critical summary of the available procedures for synthesis of the product in question. It should also include an estimate of the time required for the synthesis, an indication of the importance and utility of the product, and an admonition if any potential hazards are associated with the procedure. The Procedure should present detailed and unambiguous laboratory directions and be written so that it anticipates possible mistakes and misunderstandings on the part of the person who attempts to duplicate the procedure. Any unusual equipment or procedure should be clearly described. Line drawings should be included when they can be helpful. All safety measures should be stated clearly. Sources of unusual starting materials must be given, and, if possible, minimal standards of purity of reagents and solvents should be stated. The scale should be reasonable for normal laboratory operation, and any problems involved in scaling the procedure either up or down should be discussed. The criteria for judging the purity of the final product should be delineated clearly. The Properties section should supply and discuss those physical and chemical characteristics that are relevant to judging the purity of the product and to permitting its handling and use in an intelligent manner. Under References, all pertinent literature citations should be listed in order. A style sheet is available from the Secretary of the Editorial Board.

The Editorial Board determines whether submitted syntheses meet the general specifications outlined above. Every procedure will be checked in an independent laboratory, and publication is contingent upon satisfactory duplication of the syntheses.

Each manuscript should be submitted in duplicate to the Secretary of the Editorial Board, Professor Jay H. Worrell, Department of Chemistry, University of South Florida, Tampa, FL 33620. The manuscript should be typewritten in English. Nomenclature should be consistent and should follow the recommendations presented in *Nomenclature of Inorganic Chemistry*, 2nd ed., Butterworths & Co, London, 1970 and in *Pure and Applied Chemistry*, Volume 28, No. 1 (1971). Abbreviations should conform to those used in publications of the American Chemical Society, particularly *Inorganic Chemistry.*

Chemists willing to check syntheses should contact the editor of a future volume or make this information known to Professor Worrell.

TOXIC SUBSTANCES AND LABORATORY HAZARDS

Chemicals and chemistry are by their very nature hazardous. Chemical reactivity implies that reagents have the ability to combine. This process can be sufficiently vigorous as to cause flame, an explosion, or, often less immediately obvious, a toxic reaction.

The obvious hazards in the syntheses reported in this volume are delineated, where appropriate, in the experimental procedure. It is impossible, however, to foresee every eventuality, such as a new biological effect of a common laboratory reagent. As a consequence, *all* chemicals used and *all* reactions described in this volume should be viewed as potentially hazardous. Care should be taken to avoid inhalation or other physical contact with all reagents and solvents used in this volume. In addition, particular attention should be paid to avoiding sparks, open flames, or other potential sources which could set fire to combustible vapors or gases.

A list of 400 toxic substances may be found in the *Federal Register*, Volume 40, No. 23072, May 28, 1975. An abbreviated list may be obtained from *Inorganic Syntheses*, Vol. 18, p. xv, 1978. A current assessment of the hazards associated with a particular chemical is available in the most recent edition of *Threshold Limit Values for Chemical Substances and Physical Agents in the Workroom Environment* published by the American Conference of Governmental Industrial Hygienists.

The drying of impure ethers can produce a violent explosion. Further information about this hazard may be found in *Inorganic Syntheses*, Volume 12, p. 317.

CONTENTS

Chapter One BINARY COMPOUNDS

Chapter Two TERNARY COMPOUNDS

Chapter Three CRYSTAL GROWTH

Chapter Four INTERCALATION COMPOUNDS

Chaper Five OXIDE SUPERCONDUCTORS AND RELATED COMPOUNDS

Chapter Six MISCELLANEOUS SOLID-STATE COMPOUNDS

INORGANIC SYNTHESES

Volume 30

Chapter One

BINARY COMPOUNDS

1. TUNGSTEN DICHLORIDE

$$WCl_6 + 2W \rightarrow 3WCl_2$$

Submitted by G. M. EHRLICH,* P. E. RAUCH,* and F. J. DISALVO*
Checked by ROBERT E. McCARLEY†

Tungsten(II) chloride has been used as a starting material in a variety of organometallic and inorganic syntheses.[1,2] In addition, there has been interest in its use as a chemical transport agent, such as in incandescent lamps or XeCl excimer lasers, which has resulted in several thermochemical and thermodynamic studies of the metal halide systems.

A large number of methods have been reported for the preparation of tungsten(II) chloride. Reduction of tungsten(VI) chloride with hydrogen,[3] sodium amalgam,[4] or aluminum in a sodium tetrachloroaluminate melt[5] and disproportionation or reduction of tungsten(IV) chloride[6,7] have been reported. Each of these methods has disadvantages. Hydrogen reductions are slow, requiring a large assembly of low efficiency.[3] Reactions that use a metal reducing agent can be hard to purify. Preparations from tungsten(IV) chloride are limited to small scales, owing to the flocculent nature of tungsten(IV) chloride, and can call for extended reaction times.[6,7] At present we know of no reaction that produces tungsten(II) chloride in 100% yield

* Department of Chemistry, Cornell University, Ithaca NY, 14853.
† Department of Chemistry, Iowa State University, Ames, IA 50011-3020.

because of the complex equilibria between tungsten(VI) chloride, tungsten(V) chloride, tungsten(IV) chloride, and tungsten(II) chloride.[8]

Because of the inconvenience of these preparations, and the increasing need for tungsten(II) chloride as a starting material, a convenient large-scale preparation was sought. This preparation, the reduction of tungsten(VI) chloride with tungsten metal in a single step, may be completed in 3 days from commercially available starting materials. It is a modification of an existing preparation.[9] Because tungsten is the reducing agent, no metal impurities are introduced. The yield of this preparation is only 35%, but the starting materials are relatively inexpensive and can usually be used as received without further purification. Although presented on a 50-g scale, it can easily be scaled down.

Starting materials, tungsten(VI) chloride [cat. # 74-2350] Strem Chemical Co., Newburyport, MA and 12-μm tungsten metal powder, 99.9% (cat. #26,751-1) Aldrich Chemical Co., Milwaukee, WI, were used as received. Both the tungsten metal and tungsten(VI) chloride are air-sensitive and should be manipulated under inert atmosphere.

The tungsten(VI) chloride should be purple and crystalline; the most common impurity is the red $WOCl_4$. If large amounts of the red impurity are apparent visually, the tungsten(VI) chloride may be purified by sublimation using an evacuated Pyrex sublimation tube with a frit (Fig. 1.) in a tube furnace. The sublimation tube is loaded with 30 g of tungsten(VI) chloride under inert atmosphere, and the end closed with a sealed ball joint. Evacuate the glass tube and seal it near the ball joint. Insert the tube, the sealed end first, into the furnace, leaving the 24/40 joint outside the furnace. Initially ramp the furnace from 120° to 190° to remove the more volatile $WOCl_4$, and then, after having removed the $WOCl_4$ in a dry box, reheat the tube to 240° to sublime the tungsten(VI) chloride away from the less volatile impurities.

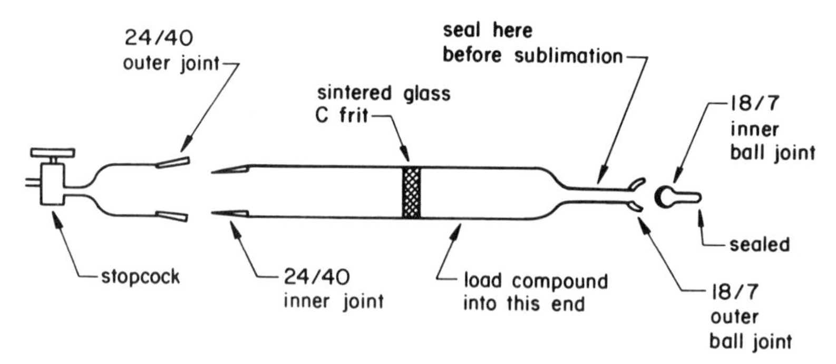

Figure 1. Sublimation tube with frit.

Procedure

The reduction of tungsten(VI) chloride is accomplished in a sealed tube. A 33-cm-long, 26-mm-i.d., 29-mm-o.d. quartz tube is attached to an 8-cm-long, 18-mm-i.d., 21-mm-o.d. quartz tube to simplify sealing (Fig. 2). If one wishes to run the reaction on a 10-g scale, a 30-cm-long, 10-mm-i.d., 12-mm-o.d. tube is more appropriate. A slight bend, $\sim 30°$, is placed in the tube 20 cm from end *D* in Fig. 2 by clamping the tube in a horizontal position and heating the tube about the bend with a torch. Heat the underside of the tube first, until it is white-hot, before heating the topside. Allow gravity to bend the tube. Check the apparatus for leaks and that it fits in the dry-box antechamber and furnace.

In a dry box, a funnel is placed into the small attached tube to keep the tube walls clean, and the tube is loaded with tungsten(VI) chloride (24 g, 0.06 mole) and an excess of 12-μm tungsten metal powder (25 g, 0.14 mole). A $\frac{1}{2}$-in. stainless-steel Ultra-Torr Cajon union (part number SS8UT-6) is attached to a stopcock and to the small tube on the sample container. The gastight assembly is brought out of the dry box and the tube evacuated to < 10 mtorr and sealed. If tungsten powder is on the tube walls at the point of sealing, sealing may be difficult or the tube may fracture on heating or cooling. After the seal has cooled, the tube is inverted so that the reagents fall to the longer end.

■ **Caution.** *Wear a face shield, apron, and heat-resistant gloves when heating volatile compounds in sealed tubes. A cracked tube of tungsten(VI) chloride will evolve irritating fumes, including HCl. Further, precautions should always be undertaken to avoid injury when handling volatile species that are heated in closed systems. There is the possibility of explosion if the temperature is inadvertently increased beyond those recommended here. When not attending to the reaction, enclose it with a blast shield.*

The tungsten(VI) chloride is sublimed away from the tungsten to the short end by inserting the tube, long end first, into a 20-in. tube furnace with a 2.5-in. bore, leaving 6 cm of the quartz tube outside the furnace as in Fig. 3.

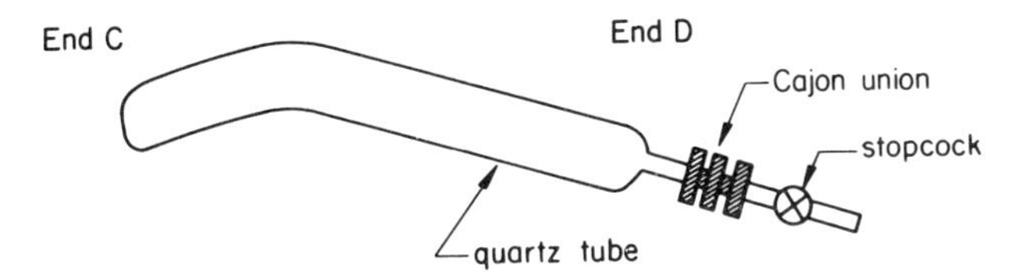

Figure 2. Assembled bent fused silica tube with Cajon union and stopcock.

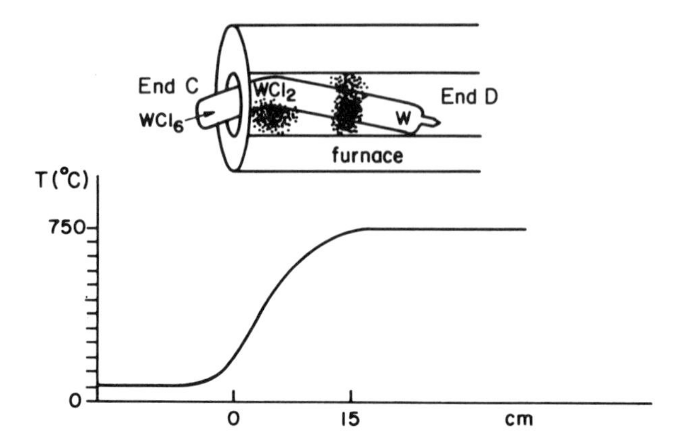

Figure 3. Sealed tube in furnace, showing refractory wool and temperature profile.

Support the tube with a few tufts of refractory wool. Raise the temperature of the furnace to 270° and hold at that temperature until the tungsten(VI) chloride has distilled to the cold end (end *C* in Fig. 3); this should take about 30 min. Remove the tube from the furnace.

Establish a shallow temperature gradient in the furnace from 200° to 750° by leaving the end of the furnace open and placing a plug of refractory wool 15 cm into the furnace, as in Fig. 3. Beginning with all the tungsten(VI) chloride well outside the furnace, slowly insert the tube into the furnace 1 cm every 15 min until the tungsten(VI) chloride melts and begins to reflux. A shallow temperature gradient will prevent the formation of a plug of tungsten(II) chloride that would block the tube and prevent the tungsten(VI) chloride vapor from reaching the hot tungsten. The cooler end containing the tungsten(VI) chloride should be at about 200°, depending on the degree of convection. Do not allow the liquid tungsten(VI) chloride to flow onto the hot tungsten powder because the tungsten(VI) chloride vapor pressure may cause the tube to explode.

Alternatively, if the temperature gradient over the length of the furnace is carefully mapped beforehand when the center of furnace is at 750°, a fused-silica tube may be constructed so that the cool end will be within the furnace and at 200°. With this arrangement movement of the tube during the reaction is unnecessary.

After about 6 h, when enough tungsten(VI) chloride has been consumed, the remaining tungsten(VI) chloride will cool and become solid. Remove the tube and inspect it. If any black plugs of tungsten(II) chloride have formed, they may be dislodged by carefully rapping the tube with a rubber mallet.

Slowly reinsert the tube into the temperature gradient until the tungsten(VI) chloride melts, repeating the above procedure until the cold end of the tube is in the furnace at 200°. Leave it there for one day. Total reaction time will be about 36 h, depending on the purity of the tungsten.

After completion of the reaction with tungsten, sublime the impurities away from the tungsten(II) chloride [mostly red $WOCl_4$ or black WCl_4 and unreacted tungsten(VI) chloride] by partially inserting the end of the tube containing the WCl_2 (*C* in Fig. 2) into a 450° furnace for 2 h, leaving 2 in. of end *D* outside the furnace. The impurities will sublime into end *D*.

Carefully break open the tube in air, taking care that the impurities do not mix with the tungsten(II) chloride, as they drastically reduce the yield. This may be done by first scoring the tube about the bend with a carbide blade or diamond pencil and wetting the tube with water. Insert the tube into a length of thin-walled, oversized PVC or Tygon tubing, and, while holding one end in a vise, jerk the other end to break the tube.

Dissolve the black impure tungsten(II) chloride in 100 mL of hot 6*M* HCl and filter over a coarse frit using a filter aid such as Celite. Wash the frit 3 times with boiling 6*M* HCl and slowly evaporate the filtrate until the product just begins to crystallize. Triple the volume with 12*M* HCl and cool the filtrate to −10°. Large yellow-green needles of $(H_3O)_2W_6Cl_{14}*6H_2O$ will form. Filter off the crystals and wash with cold (−10°) 6*M* HCl.

To dry the hydrate, place the crystals in a fritted Pyrex tube and heat under a dynamic vacuum; ramp at no more than 100°/h to 300° and hold at 300° for one hour to yield the yellow-brown tungsten(II) chloride. To prevent the finely powdered product from blowing into the vacuum system, use a fritted tube. To obtain a more crystalline product heat to 450° for a few hours. Yield 16.5 g.

Properties

X-ray powder diffraction (Scintag XDS 2000 using CuKα radiation): $(H_3O)_2W_6Cl_{14}*6H_2O$ d(Å) = 8.573, 8.359, 7.222, 6.497, 5.501. WCl_2 dried to 300° is poorly crystallized, giving broad peaks by powder diffraction. Broad peaks were observed at d(Å) = 6.94, 5.63, 2.88, and 2.61 Å. When heated to 450° the crystallinity improves considerably. The diffraction pattern is given in the JCPDS file, card number 17-261.

Acknowledgment

This work has supported by the Department of Energy, Division of Basic Energy Sciences, Grant DE-FG02-87ER45298.

References

1. T. Saito, A. Yoshikawa, T. Yamagata, H. Imoto, and K. Unoura, *Inorg. Chem.*, **28**, 3588–3592 (1989).
2. G. Natta, G. Dall'Asta, G. Mazzanti, U.S. Patent 3476728, 4 Nov. 1969; *Chem. Abstr.*, 72(8): 32460c.
3. G. I. Novikov, N. V. Andreeva, and O. G. Polyanchenok, *Russ. J. Inorg. Chem.*, **6**, 1019 (1961).
4. J. B. Hill, *J. Am. Chem. Soc.*, **38**, 2383 (1916).
5. W. C. Dorman, *Nucl. Sci. Abstr.*, **38**, 43612 (1972).
6. R. E. McCarley and T. M. Brown, *Inorg. Chem.*, **3**, 1232 (1964).
7. R. E. McCarley, personal communication on the use of Fe as a reducing agent with tungsten(IV) chloride.
8. J. H. Canterford and R. Colton, *Halides of the Second and Third Row Transition Metals*, Wiley, London, 1968.
9. H. Schäfer, M. Trenkel, and C. Brendel, *Monatsh. Chem.*, **102**, 1293 (1971).

2. ZIRCONIUM MONOCHLORIDE AND MONOBROMIDE

$$3\text{Zr (sheet, turnings)} + \text{ZrX}_4\,(\text{l, g}) \xrightarrow[\text{Ta}]{350,\ 600,\ 800^\circ} \text{ZrX (s)}$$

Submitted by RICHARD L. DAAKE* and JOHN D. CORBETT†
Checked by DONALD W. MURPHY‡

Reprinted from *Inorg. Synth.* **22**, 26 (1983)

The two zirconium monohalides represent the first examples of metallic salts with double metal layers sandwiched between two halogen layers in a close-packed array.[1-3] At the present time, they are also the most stable of these monohalides and the easiest to prepare in microcrystalline form.

Direct high-temperature reactions of Zr and the corresponding ZrX_4 are at present the only practical route to these phases, but these processes are characteristically slow and incomplete, giving mixtures with higher halides. In one of the procedures to be described, the zirconium is finely divided. Even so, the product is impure and requires reequilibration with fresh metal. A second approach utilizes thin turnings in a stoichiometric reaction that is carried out at increasing temperatures as the reduction proceeds through the less volatile, intermediate phases. Both procedures require the use of

* Bartlesville Wesleyan College, Bartlesville, OK 74003.

† Ames Laboratory and Department of Chemistry, Iowa State University, Ames, IA 50011. The Ames Laboratory is operated by the U.S. Department of Energy by Iowa State University under contract No. W-7405-Eng-82.

‡ Bell Laboratories, 600 Mountain Avenue, Murray Hill, NJ 07974.

tantalum containers, not only to contain the sizable pressures involved but also to provide the necessary inertness to the products, a property that glass does not possess. Techniques and equipment necessary for effective use of tantalum as a container material are described elsewhere in this volume (Chapter 2, Section 15).

Procedure

Because zirconium halides are air- and moisture-sensitive, handling of these compounds must be done in an inert atmosphere dry box equipped to maintain both oxygen and water at below the 10-ppm level. Prepurified nitrogen low in oxygen serves as an adequate working atmosphere. The nitrogen is passed over activated Linde molecular sieves to maintain an acceptably low moisture level. Generally, the reactivity of the zirconium halides to air and moisture decreases with the decreasing oxidation state.

The $ZrCl_4$ and $ZrBr_4$ starting materials should be of high purity, particularly with respect to other nonmetallic components, for these are apt to be carried into the final monohalide product. Commercial products* should be sublimed under high vacuum (at $\sim 300°$), preferably through an intervening, coarse-grade glass frit to reduce entrainment of oxyhalide and other involatile impurities.

The metal for the reduction needs to be low in nonmetallic impurities in order to have adequate ductility for rolling. Reactor-grade (crystal bar) quality or equivalent is recommended.† Thin metal sheet 0.25–0.5 mm thick or turnings are made using standard vacuum melting, cold-rolling, and machining techniques. Thin uniform turnings ($\leqslant 0.1$ mm) are highly desirable for the second procedure described below.

Because of the relatively high intermediate pressure (20–40 atm) generated in the reactions, tantalum end caps with thickness 30–50% greater than that of the tubing itself are recommended.

■ **Caution.** *Crimp welds cannot dependably withstand the necessary pressures. A steel or Inconel tube furnace liner is recommended as partial protection against possible explosion hazard. It also serves to smooth out temperature gradients.*

Thermal cycling will be lessened if the control (but not the measuring) thermocouple is placed between the liner and the furnace wall. It is highly

* Available from Materials Research Corp., Orangeburg, NY; Alfa Division, Ventron Corp., Danvers, MA; Cerac, Butler, WI.

† Available from Materials Research Corp., Orangeburg, NY; Goodfellow Metals, Cambridge, England; Atomergic Chemetals Corp., Plainview, NY.

advisable that the fused-silica jacket about the welded container be sufficiently long to extend from the end of the furnace to permit condensation of the tetrahalide, should a leak develop. Unless a thermal gradient is required, the tantalum tube itself should be centered in the furnace. Although the explosion hazard is very slight if these procedures are followed, the open ends of the furnace should still be shielded from workers.

A. SYNTHESIS WITH ZIRCONIUM FOIL

The original two-step method described required zirconium sheet or foil in large excess to compensate for a surface-limited reaction. The sheet form also allows the ready separation of the product from unreacted metal. Although the literature describes this method only for ZrCl (and HfCl),[1] the technique works equally well for ZrBr.

A 3–8 g quantity of tetrahalide together with a fourfold (by weight) excess of zirconium sheet is contained in a length of 9–19-mm-diameter tantalum tubing with at least at 0.4-mm wall which is capped as described in the welding section (Chapter 2, Section 15). This is sealed in an evacuated fused-silica tube and heated in a gradient with the Zr at the hot end. The gradient serves mainly to provide a cool reservoir for volatile phases (ZrX_4 and ZrX_3) while also keeping the metal foil at a sufficient temperature to maintain an acceptable reaction rate. The reaction temperature is raised slowly over 2–3 days from a 350°/450° gradient to a 500°/600° gradient where it is held for 4–6 days. (The gradient is measured with thermocouples strapped to the outside of the silica jacket and the real gradient is doubtless less.)

This first-stage reaction gives a nonequilibrium mixture of reduced phases, owing to physical blockage of the metal surface to further reaction. As expected, the somewhat volatile trichloride is found at the cooler end of the reaction tube, often as a plug adjacent to a band of well-formed $ZrCl_3$ needles up to 1 cm or more in length. The metal foil is found coated with a relatively thick (up to 1 mm) layer of strongly adhering blue-black product of variable composition ($1.2 \leqslant Cl{:}Zr \leqslant 1.5$) containing ZrCl and $ZrCl_2$ (according to X-ray powder pattern analysis).

In the second stage, the $ZrCl/ZrCl_2$ mixture, scraped from the foil in the dry box, is equilibrated isothermally with a 10-fold excess of *fresh* foil in a new reaction tube of the same type. The temperature is initially set at 600° and slowly raised to 800° over a period of 4–6 days, resulting in an essentially quantitative conversion to ZrCl, which can be recovered easily from the metal substrate in excellent purity.

If only a small amount (<1 g) of ZrCl or ZrBr is desired, it can be prepared directly in a single step with a 15-fold excess (by weight) of metal foil, the temperature being gradually raised from 400° to 800° over 10–14 days.

B. SYNTHESIS USING ZIRCONIUM TURNINGS

Thin turnings of metal can also be caused to react with a stoichiometric amount of tetrahalide to achieve 100% conversion to ZrCl or ZrBr. The method is also suitable for much larger quantities (50–60 g) in a single batch and is the preferred route if thin, uniform turnings can be obtained. The turning process is believed to produce cracks along grain boundaries, thereby making the bulk of the metal more accessible to the gaseous ZrX_4. (Very thin foil has not been tried.) The monohalides are essentially line compounds, and a homogeneous product is obtained without taking special precautions regarding moderate thermal gradients along the reaction tube.

The weighed tetrahalide is first transferred into the cap-welded end of a tantalum tube of size noted above (tube volume $\simeq 1.5\ cm^3/g$ ZrX). The appropriate weight of turnings ($\pm 0.1\%$) is then packed into the tube. Reactants are loaded in this order to prevent welding difficulties from the volatile tetrahalides.

After the second cap is welded in placed, the tube is jacketed in fused silica, and the assembly is placed in a tube furnace with an Inconel liner. The temperature is then slowly raised from 350° to 850° over a period of 7–10 days and maintained there for 3–5 days.

Both ZrCl and ZrBr are stoichiometric line compounds as far as can be judged. Although standard chemical analyses for metal and halogen[1] may be useful in assaying the monohalides produced from reactions using excess metal foil, X-ray diffraction patterns are suitable indicators of phase purity for most purposes. If normal-abundance zirconium is used, there will be some fractionation of the fraction of hafnium contained therein, particularly in the first procedure. This must be taken into account in gravimetric procedures, but it will affect X-ray results only slightly. The values (Å) and intensities (10 maximum) reported below were obtained with a precision flat-plate Guinier camera and NBS silicon powder as an internal standard. The platelet morphology of ZrCl and ZrBr causes preferred orientation in the flat-plate technique, so that Debye–Scherrer intensities may deviate substantially from those reported here.

For ZrCl, the 14 most intense lines in the Guinier pattern ($\theta < 45°$) are 8.90 Å(5), 2.943(2), 2.706(10), 2.224(2), 2.215(3), 1.710(10), 1.679(6), 1.596(3), 1.481(3), 1.445(3), 1.354(3), 1.120(2), 1.118(5), 1.104(4).[3] (Areas of the pattern are usually banded together because of grinding damage.) Some characteristic lines of potential impurity phases occur at 5.53, 2.832, 2.682, and 2.212 Å for $ZrCl_3$; 6.46, 2.507, 2.338, and 1.691 Å for 3*R*-$ZrCl_2$,[4] and 6.92, 6.49, 2.541, and 1.797 (doublet) Å for the Zr_6Cl_{12} phase.[5] A line at 2.461 Å is the best measure of α-Zr in the presence of any of these halides.

For the monobromide, the 11 most intense lines in the Guinier pattern[2] are: 9.36 Å(2), 3.016(2), 2.965(2), 2.669(10), 2.419(2), 2.060(3), 1.752(5), 1.464(4),

1.402(2), 1.279(2), 1.124(3). Characteristic lines of potential impurity phases occur at 5.84, 3.157, 2.976, and 2.305 Å for $ZrBr_3$; 6.86, 3.053, 2.540, and 1.763 Å for one form of $ZrBr_2$;[6] and 7.29, 3.077, 2.669, 1.892, and 1.883 Å for Zr_6Br_{12}.[5]

Properties

Zirconium monochloride and monobromide are black powders or highly reflective microcrystals. Although the monohalides appear to be stable in air for days to weeks, they should be kept and handled under inert atmosphere if high purity is required, owing to a probable slow reduction of water vapor to form the hydrides (see below) and ZrO_2. These monohalides possess a rhombohedral ($R\bar{3}m$), three-slab structure in which each slap consists of four tightly bound, close-packed layers sequenced X–Zr–Zr–X.[2,3] The two monohalides actually possess slightly different structures that are interrelated through interchange of two of the four-layered slabs. Very strong Zr–Zr bonding is evident between the pairs of metal layers. In ZrCl, for example, each metal atom has six like neighbors in the same layer at 3.43 Å plus three in the adjoining layer at 3.09 Å. For comparison, the average internuclear distance in the 12-coordinate metal is 3.20 Å. Both phases are Pauli paramagnetic. As is typical of most layered compounds, the weak van der Waals interactions between two neighboring chlorine layers in two adjacent slabs allow easy cleavage.[3] The zirconium monohalide structures, formally d^3 for zirconium, have also been found for a number of d^2 examples, namely in ScCl[7] and many of the rare earth elements.[9].

The zirconium monohalides are essentially two-dimensional metals with good delocalization and conduction within the double metal sheets. The metallic conductivity has been established by both X-ray photoelectron spectroscopy[2] and conductivity measurements where single crystal data (obtained by an unspecified method) indicated a conductivity of $55\ \Omega^{-1}\ cm^{-1}$ parallel to the plates and about $10^{-3}\ \Omega^{-1}\ cm^{-1}$ normal to them.[9]

Both compounds are sufficiently ductile that they may be pressed at room temperature to 97% of theoretical density.[2] The ZrCl particularly shows extensive damage on grinding, analogous to that in graphite, which is evident in the powder pattern through banding of certain classes of reflections. In contrast the bromide shows this only to a small extent, evidently because of different bonding interactions with second nearest neighbors.[2] It has not proved possible to intercalate these phases. However, both monohalides do take up hydrogen readily just above room temperature to form discrete hemihydride and monohydride phases.[10] The hydrogen atoms therein are mobile above $\sim 80°$, occupying primarily the tetrahedral holes.[11] The monohydride phases appear to be metastable and to disproportionate above 600°

into ZrH_2 and the corresponding Zr_6X_{12} phases, thereby providing a better route to these compounds than achieved by direct reactions.[5]

The monohalides are useful reducing agents for the preparation of a number of intermediate zirconium chloride and bromide phases since their reactivities are considerably greater than those of the refractory metal. Thus, ZrX–ZrX_4 reactions have been used to obtain the trihalides in 100% yields[12] as well as $ZrCl_2$ (3*R*–NbS_2 type)[4] and a different polytype of $ZrBr_2$.[6]

References

1. A. W. Struss and J. D. Corbett, *Inorg. Chem.*, **9**, 1373 (1970).
2. R. L. Daake and J. D. Corbett, *Inorg. Chem.*, **16**, 2029 (1977).
3. D. G. Adolphson and J. D. Corbett, *Inorg. Chem.*, **15**, 1820 (1976).
4. A. Cisar, J. D. Corbett, and R. L. Daake, *Inorg. Chem.*, **18**, 836 (1979).
5. H. Imoto, J. D. Corbett, and A. Cisar, *Inorg. Chem.*, **20**, 145 (1981).
6. R. L. Daake, Ph.D. thesis, Iowa State University, 1976.
7. K. R. Poeppelmeier and J. D. Corbett, *Inorg. Chem.*, **16**, 294 (1977).
8. Hj. Mattausch, A. Simon, N. Holzer, and R. Eger, *Z. Anorg. Allgem. Chem.*, **466**, 7 (1980).
9. S. I. Troyanov, *Vestn. Mosk. Univ., Khim.*, **28**, 369 (1973).
10. A. W. Struss and J. D. Corbett, *Inorg. Chem.*, **16**, 360 (1977).
11. T. Y. Hwang, R. G. Barnes, and D. R. Torgeson, *Phys. Lett.*, **66A**, 137 (1978).
12. R. L. Daake and J. D. Corbett, *Inorg. Chem.*, **17**, 1192 (1978).

3. LANTHANUM TRIIODIDE (AND OTHER RARE-EARTH METAL TRIIODIDES)

$$2La(s) + 3HgI_2(l) \xrightarrow{330^\circ} 2LaI_3(s) + 3Hg(l)^1$$

$$2La(s) + 3I_2(g) \xrightarrow{800^\circ} 2LaI_3(l)^2$$

Submitted by JOHN D. CORBETT*
Checked by ARNDT SIMON†

Reprinted from *Inorg. Synth.* **22**, 31 (1983)

Lanthanum triiodide may be obtained from the oxide by a number of routes, such as dissolution of the oxide in HI(aq) followed by precipitation of the

* Ames Laboratory and Department of Chemistry, Iowa State University, Ames, IA 50011. The Ames Laboratory is operated by the U.S. Department of Energy by Iowa State University under contract No. W-7405-Eng-82.

† Max-Planck Institut für Festkörperforschung, Heisenbergstrasse 1, 7000 Stuttgart 80, Germany.

hydrated iodide and vacuum dehydration (20–350°), perhaps in the presence excess NH_4I.[3] However, the yield will be low because of significant hydrolysis to LaOI, and one or two vacuum sublimations or distillations of the LaI_3 product will be necessary to obtain a usable material. On the other hand, the ready availability of lanthanum (and other rare-earth elements) as good-quality metal makes direct reactions considerably more attractive. A simple route suitable for small quantities is the reduction of excess liquid HgI_2 by lanthanum in a sealed Pyrex container.[1] The direct combination of the elements is less expensive and much better for larger quantities; however, some attention must be paid to the choice of container. The metal must be maintained above 734° (the LaI_2–LaI_3 eutectic[4]) so that a layer of solid iodides does not block the surface of the metal to further oxidation, but the reducing melt above that temperature acts as a flux for the spontaneous reduction of SiO_2 by La. Therefore, both the metal and the reduced melt must be kept out of contact with silica (and most other ceramic materials). Among the available metal containers that are suitable for the rare-earth elements only tungsten is also sufficiently inert to iodine at about 0.1–1 atm and 800° to give minimal side reactions.[2] The reaction also illustrates the safe application of a hot–cold, sealed reaction tube for the reaction of materials with very different volatilities.

The reactions described below have been applied to the syntheses of the triiodides of the rare-earth elements as well as those of yttrium and scandium.

A. SYNTHESIS USING HgI_2

Lanthanum metal* in the form of sheet, small lumps, or turnings should be stored under vacuum or inert gas and protected from the atmosphere as much as possible. (The reactivity of the rare-earth metal decreases across the series so that the heavier members are relatively inert to air and moisture at room temperature.) These metals all form very stable carbides and hydrides, and degreasing may be necessary on the materials as received.

Preparation

The container for the HgI_2–La reaction consists of a Pyrex tube 15–25 mm in diameter and 6–10 times as long. This is closed at one end, and the other end is connected to a ground-glass joint through a section of Pyrex tubing of 1.5–2 mm wall thickness, 6–10 mm i.d., suitable for seal-off under vacuum.

* See Chapter 2, Section 16 for list of suppliers.

The main tube is either bent 20–30° near the middle to form a flat inverted V or a dike is inblown there to allow separation of the products. A weighed quantity of La (1–5 g) and at least a three fold excess of HgI_2 are placed in the reaction container, and the tube is evacuated under high vacuum ($<10^{-5}$ torr) and sealed off at the thick-walled tubing. (The seal-off should be practiced ahead of time if this is a new technique. Go slowly and do not attempt to reheat the seal-off. Sealing off silica glass rather than Pyrex requires much more heat but is less troublesome.)

The container is placed in the furnace slightly inclined so that the melt will cover the metal in the lower closed end, and the tube is heated for 12–48 hr at 300–330°.

■ **Caution.** *The reaction should be run in a good hood in case of unexpected container failure and release of toxic mercury vapor. A reliable automatic controller should also be used to avoid accidental overheating and dangerous pressures of HgI_2 and Hg. (The total pressure is 1.95 atm at 354° neglecting dilution of HgI_2 by Hg_2I_2 and LaI_3.)*

After the reaction is complete, the container is partly withdrawn from the furnace to expose the empty end, and heating is continued until the HgI_2 and Hg are distilled from the LaI_3 product. A final heating of the LaI_3 to 600° in high vacuum (preferably in a tantalum crucible) will remove the last traces of mercury compounds.

B. SYNTHESIS FROM THE ELEMENTS

Preparation

As noted above, a temperature in excess of 734° is necessary to obtain an adequate rate of reaction of the metal with I_2, either as an exposed metal surface or with reduced iodide melts. (The metal readily dissolves in liquid LaI_3, giving solid LaI_2 only at high concentrations or lower temperatures.) Iodine is highly volatile (boiling point 184°) and must be maintained at a relatively low temperature to avoid dangerous pressures. Though this difference has been accommodated through the use of rather complex flow systems,[3] reactions run in sealed hot–cold tubes are in general simple, direct, and safe if attention is paid to one condition.

■ **Caution.** *A portion of the reaction tube must be kept at low enough temperature that the volatile component will not exert an unsafe pressure. Use of an evacuated container permits rapid volatilization of that species in and out of the cooler zone.*

Two designs that can be used to accomplish this synthesis are shown in Fig. 1. Both are constructed from fused silica. One furnace is used to maintain iodine either in the bottom of design *a* or in the side arm of design *b* at a temperature T_2 sufficient to generate a pressure of 0.2–1 atm, while a second furnace is used to heat the metal and its salt products in a tungsten crucible to the higher temperature T_1 necessary to give an adequate reaction rate. The separation of the two arms in design *b* then must be sufficient to accommodate the combined wall thickness of two tubular furnaces.

The tungsten crucible, about 3.2 cm diameter ×5 cm high,* is sealed within the apparatus, taking care that it rests either on a fairly flat tube end or on a silica support so that greater thermal expansion of tungsten will not lead to

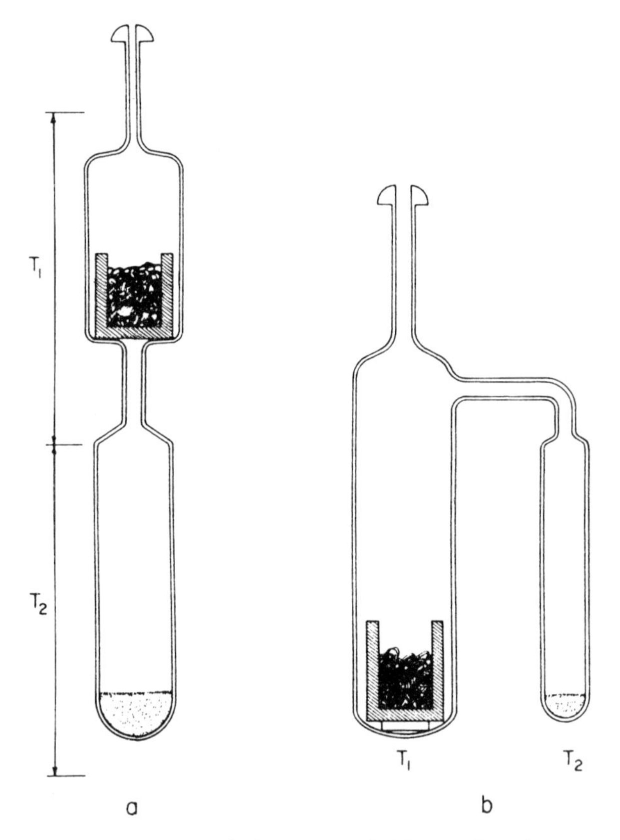

Figure 1. Two designs for sealed hot–cold tubes.

* Available from Kulite Tungsten Co., Ridgefield, NJ, and Ultramet, Pacoima, CA.

cracking of the glass vessel during heating. The crucible is fairly porous and so should be strongly heated (>500°) in high vacuum before loading. Lanthanum chunks or turnings (not powder) (5–20 g) and reagent-grade I_2 are quickly loaded in air into the crucible with the aid of a small funnel that fits through the joint and neck. The apparatus is evacuated (through a trap cooled with dry ice to protect the pumps), and then the portion containing iodine is also cooled to −80° once the ambient moisture has been pumped out. The apparatus is then evacuated below 10^{-3} torr and sealed off below the joint.

■ **Caution.** *Welder's glasses are required for eye protection.*

The iodine is then sublimed to the lower or right-hand portion of apparatus *a* or *b*, respectively, and two tubular furnaces arranged as to provide the temperature profiles shown. Insulation (Fiberfrax, Carborundum Co.) is used to cover the tops of the furnaces and especially the smaller horizontal tube in *b*. The hotter furnace is positioned so that sublimed or molten LaI_3, which may overflow the crucible, will not block the narrow tubes. The temperature of the metal end is run right up to ~800°, with the I_2 reservoir held at 110–130°.

■ **Caution.** *A reliable automatic temperature controller should always be used on the lower temperature furnace to avoid dangerous I_2 pressures.*

After some salt has formed and covered the metal, the I_2 furnace temperature is increased to about 180°. The reaction will take 4–12 h, depending on quantities and temperatures. The furnace around the I_2 can be lowered occasionally to judge progress. If excess I_2 has been used, the iodine reservoir is cooled to −80° after reaction is complete while the furnace around the salt is still at 200–300° and the I_2 reservoir sealed off. If a slight deficiency of I_2 has been used (or the reaction has been incomplete), the excess metal can be separated (and intermediate iodides allowed to disproportionate) through sublimation of the crude product in high vacuum (below).

Purification

The LaI_3 product obtained from either of the above routes is sufficiently pure for some purposes. However, some impurities will have been introduced by handling of the reactants, La especially, in the air. One may also choose to use a relatively poor grade of metal (with respect to nonmetals) for the synthesis, anticipating the purification afforded by the sublimation. Also, some reaction of LaI_3 with fused silica to give SiI_4 plus LaOI or a lanthanum silicate will occur[5] in the high-temperature reaction B. Although the amount would not seem serious in a sealed vessel, some SiI_4 may diffuse back to the metal, producing silicide. In any event the quality of either product, especially as judged by color and melting point, will be definitely improved by vacuum sublimation or distillation.[6] This should be carried out at about 900°, either in a sealed Ta tube designed so as to avoid refluxing or under a high vacuum using something like the two-piece apparatus shown in Fig. 2F of the earlier synthesis on tantalum welding (J. D. Corbett, *Inorg. Synth.*, **22**, 15 (1983)). A second sublimation may be necessary if entrainment is large in the first. Given adequate starting materials and the absence of adventitious contamination during the process, a sublimed product of high purity is assured, $\geqslant$99.9%. The yield is nearly quantative if the initial reactions go to completion but is reduced by up to 5% per sublimation through recovery losses.

The most likely impurity is LaOI, which has the PbFCl structure as do many of the other rare-earth metal oxyiodides.[3] The stronger 25% of the powder diffraction lines for LaOI (Å), with intensities in parentheses, are: 3.05(10), 2.92(8), 2.06(4), 1.72(5), 1.71(5). The powder pattern of LaI_3 is not especially useful for establishing its purity unless all the lines from a high-resolution (Guinier) pattern are compared.

Properties

The light yellow LaI_3, melting point 778–779°, exhibits the $PuBr_3$-type structure. The material is very sensitive to traces of both moisture and O_2. The absence of a cloudy appearance on the dissolution of LaI_3 (and other rare-earth metal trihalides) in absolute ethanol is *not* a good assurance of purity unless the material has been heated strongly to ensure the formation of crystalline LaOI (or other MOX phases) from absorbed moisture, hydroxide, or other agents. The same applies to the appearance of MOI in the powder pattern.

References

1. F. L. Carter and J. F. Murray, *Mat. Res. Bull.*, **7**, 519 (1972).
2. L. F. Durding and J. D. Corbett, *J. Am. Chem. Soc.*, **83**, 2462 (1961).
3. D. Brown, *Halides of the Lanthanides and Actinides*, Wiley-Interscience, New York, 1968, p. 219.
4. J. D. Corbett, L. F. Druding, W. J. Burkhard, and C. B. Lindahl, *Discuss. Farad. Soc.*, **32**, 79 (1962).
5. J. D. Corbett, *Inorg. Nucl. Chem. Lett.*, **8**, 337 (1972).
6. J. D. Corbett, R. A. Sallach, and D. A. Lokken, *Adv. Chem. Ser.*, **71**, 56 (1967).

4. LANTHANUM DIIODIDE

$$2LaI_3(l) + La(s) \xrightarrow[\text{Ta}]{870°} 3LaI_2(l)^1$$

Submitted by JOHN D. CORBETT*
Checked by ARNDT SIMON†

Reprinted from *Inorg. Synth.*, **22**, 36 (1983)

This straightforward reaction provides a method of preparing a simple metallic salt, $La^{3+}(I^-)_2e^-$. The principal requirements for the synthesis are high-purity reactants, an inert container, and a good dry box (not a glove-bag). A small amount of H_2O in the LaI_3 can have a large effect on the purity of the LaI_2 product owing to the formation of LaOI and LaH_2.[2]

Procedure

The synthesis involves the direct reaction of LaI_3 with a modest excess of La. A Ta (Nb or Mo) container is essential (see J. D. Corbett, *Inorg. Synth.*, **22**, 15 (1983)) and sublimed LaI_3 is preferable.[3] An open crucible can be used under a noble-gas atmosphere with only a small amount of sublimation of LaI_3 during the reaction (and accompanying reaction with the silica walls), but a sealed container gives better control of stoichiometry and purity. Lanthanum metal is somewhat reactive toward air and the best product is obtained by minimizing atmospheric exposure of the lanthanum metal used in the reduction.

The LaI_3 (5–25 g) and somewhat more than the stoichiometric amount of La‡ (13.46% of LaI_3 by weight) are weighed in the dry box and transferred into the crucible. If an open crucible is used, this is placed in the closed end of a fused-silica tube long enough so that the upper end, which is equipped with a standard taper joint, will extend 10–15 cm from the (vertical) furnace. This end is capped with a Pyrex top equipped with a standard taper joint and stopcocks to allow evacuation and introduction of a noble gas. The joint

* Ames Laboratory and Department of Chemistry, Iowa State University, Ames, IA 50011. The Ames Laboratory is operated by the U.S. Department of Energy by Iowa State University under contract No. W-7405-Eng-82.

† Max-Planck Institut für Festkörperforschung, Heisenbergstrasse 1, 7000 Stuttgart 80, West Germany.

‡ For list of suppliers see Chapter 2, Section 16.

should be greased with a high-temperature grease (e.g., Apiezon T) and cooled with a small air blower. A closed crucible is crimped or capped in the dry box, welded, and jacketed as described. The reactants are heated to 840–900° (melting point LaI_2 = 830°) for 3–4 h, after which the product is cooled by turning off power to the furnace.

According to the phase diagram,[1] LaI_2 melts congruently in equilibrium with La. The reaction can thus be carried out using excess metal if a bulk form is used so that the excess can be readily removed after grinding the product. Alternatively, the sample may be heated above the melting point of La (845°) in this system to aggregate the excess metal.

Samples of LaI_2 equilibrated for weeks near 850° with a large excess of metal, preferably with high surface area, have been found to contain also the cluster phase $La(La_7I_{12})$[4] (isostructural with Sc_7Cl_{12}[5]). The phase relationships of this with the other phases are not known, but the yield is negligible when the excess metal has limited surface area.

The surest identification of LaI_2 is accomplished by the determination of melting point and powder pattern. The latter may be calculated from the known crystal structure.[6] The stronger lines (Å) are at 6.98(m), 3.00(vs), 2.773(s), 2.275(m), 1.961(m), and 1.641(m). A search for the neighboring phases by the same means is a good check of purity; the stronger lines (Å) are, for La_7I_{12}, 3.117(vs), 2.209(mw), and 2.197(mw),[4] and for $LaI_{2.5}$,[1,7] 3.46(vs), 3.08(vs), 3.015(s), 2.838(s), and 2.202(vs). The dissolution of LaI_2 in water for analytical purposes should be performed in a closed container to avoid loss of HI (and perhaps I_2) during the vigorous reaction. A small amount of acetic acid is added afterward to dissolve the hydrolysis products. Standard methods for determining La^{3+} and I^- are used.[1] With sufficient care regarding purity of the reactants and execution of the reaction, a substantially quantitative yield of the single phase is obtained, and wet analyses should give $LaI_{2.00 \pm .02}$ with 100.0 ± 0.3% of recovery of La plus I.

Properties

The blue-black lanthanum diiodide is a very good reducing agent and should be stored and handled in the absence of O_2, H_2O, or other reducible series. It is one of the better characterized members of a small group of metallic diiodides and is best formulated $La^{3+}(I^-)_2e^-$ so as to emphasize delocalization of the differentiating electron. The phase exhibits a conductivity comparable to that of lanthanum metal and a small (Pauli) paramagnetism appropriate to the metallic state and diamagnetic cores. The above representation probably overstates the ionic character of the bonding, however; covalency and iodine participation are thought to be important in the band formation. Support for this comes from the fact that none of the elements that form metallic diiodides (La, Ce, Pr, Gd, Th) yield similar chlorides or bromides in spite of the closer approach of the cations that would be possible with smaller anions.[3]

The structure of LaI_2 and of the isostructural CeI_2 and PrI_2 is of the $MoSi_2$ (or $CuTi_2$) type and gives an almost alloy-like impression with its square layers. For LaI_2 the lattice constants are $a = 3.922$, $c = 13.97$ Å, space group $I4/mmm$, $d = 5.46$ g/cm^3.[6]

The synthetic procedure described for LaI_2 should also work for CeI_2 and PrI_2, subject to two conditions. First, the compounds melt increasingly incongruently in the order listed, meaning that complete conversion is best accomplished by quenching the equilibrated melt (to avoid segregation) followed by equilibration with excess metal below the peritectic melting point.[1] Small amounts of the more reduced M_6I_{12} are formed only on extended reaction with excess metal.[4] In addition, PrI_2 and CeI_2, to a limited extent, have been shown to form a variety of diiodide polymorphs at the higher temperatures, namely $CdCl_2$, $2H_1$-MoS_2, 3R-MoS_2, and a $M_4I_4I_4$ cluster type.[6] The metallic $MoSi_2$-type material appears to be more stable at lower temperatures.[7] These same three MI_3-M systems also contain an isostructural series of M_2I_5 phases of known structure, now melting congruently only for M = Pr, and techniques similar to those described here can also be used for their preparation.[1,7]

References

1. J. D. Corbett, L. F. Druding, W. J. Burkhard, and C. B. Lindahl, *Discuss. Farad. Soc.*, **32**, 79 (1962).
2. J. D. Corbett, *Prep. Inorg. React.*, **3**, 10 (1966).
3. J. D. Corbett, R. A. Sallach, and D. A. Lokken, *Adv. Chem. Ser.*, **71**, 56 (1967).
4. E. Warkentin and A. Simon, private communication.
5. J. D. Corbett, K. R. Poeppelmeier and R. L. Daake, *Z. Anorg. Allgem. Chem.*, **491**, 51 (1982).
6. E. Warkentin and H. Bärnighausen, *Z. Anorg. Allgem. Chem.*, **459**, 187 (1979).
7. E. Warkentin and H. Bärnighausen, private communication.

5. RARE-EARTH SESQUISULFIDES, Ln_2S_3

$$2Ln + 3S \rightarrow Ln_2S_3$$

$$3Ln + 3S + 3I \rightarrow 3LnSI \rightarrow Ln_2S_3 + LnI_3$$

Submitted by A. W. SLEIGHT* and D. P. KELLY*
Checked by R. KERSHAW† and A. WOLD†

Reprinted from *Inorg. Synth.*, **14**, 152 (1973)

Rare-earth sesquisulfides have generally been prepared by the reaction of hydrogen sulfide with rare-earth oxides. However, such a procedure frequently gives oxysulfides or nonstoichiometric sulfides. Direct reaction of the elements readily gives pure stoichiometric sesquisulfides, and this procedure is easily modified to give crystals.[1]

Procedure

High-purity sulfur, iodine, and rare-earth metals are commercially available. The sponge or powder form of the rare-earth metal is easiest to work with, but such forms are not satisfactory (and usually not available) for the more active rare-earth metals. Thus, although the examples below assume that Gd (gadolinium) sponge or powder is used, ingot forms are used for La, Pr, Nd, Tb, Tm, Yb, and Lu. Such ingots must be freshly cut or filed.

Polycrystalline gadolinium sesquisulfide can be prepared from 1.5725 g (0.01 mol) of gadolinium and 0.4810 g (0.015 mol) of sulfur. The reactants are not ground together or in any way intimately mixed. They are simply placed together in a 10-mm silica tube presealed at one end. The tube is evacuated and sealed to produce a 20-cm-long ampule. The ampule is placed in a two-zone or gradient furnace with all reactants initially in the hot end. The hot-zone temperature is raised to 400°, while the other zone is maintained at about 100°. The sulfur quickly moves to the cooler zone, and thereafter vapors of sulfur will react with the gadolinium in a controlled fashion.

■ **Caution.** *If the entire ampule is heated to 400°, a violent reaction will occur, with attack on the ampule, which will probably break.*

* Central Research Department, E. I. du Pont de Nemours & Company, Wilmington, DE 19898.
† Brown University, Province, RI 02912.

When all the sulfur is consumed, the ampule is transferred to a muffle furnace and heated at 1000° for about 10 h.

Single crystals of gadolinium sesquisulfide may be prepared from 1.5725 g (0.01 mol) of gadolinium, 0.3206 g (0.01 mol) of sulfur, and 1.2690 g (0.01 mol) of iodine. The reactants are sealed in an evacuated silica tube as described above. The vacuum should be applied no longer than necessary before sealing because of the volatility of iodine. The ampule is then heated in a two-zone or gradient furnace exactly as described above. When all the sulfur and iodine are consumed, the ampule is removed from the furnace. At this point the sample will be basically GdSI. However, this compound will decompose when heated at higher temperatures. Crystals of gadolinium sesquisulfide will grow in a gadolinium iodide melt if the ampule is held at 1100–1200° for 20 h or longer. The crystals may be washed free of gadolinium iodide by using alcohol or water–alcohol mixtures. Red rods several millimeters in length are obtained.

Polycrystalline rare-earth sesquisulfides have been prepared[1] by this method for La, Ce, Pr, Nd, Sm, Gd, Tb, Dy, Ho, Er, Tm, Yb, Lu, and Y. Europium sesquisulfide does not exist. For the more reactive rare-earth metals (La to Sm), the silica ampule will be severely attacked unless protected. This may also be a problem for other rare earths if high temperatures and long heating times are employed. Carbon is the most suitable material for protecting the silica in these syntheses. A graphite crucible may be used, but it is generally satisfactory simply to coat the inside of the silica tube with carbon by the pyrolysis of benzene. Benzene is poured into the silica tube, which is closed at one end; it is then poured back out with the residue left clinging to the tube. The tube is placed in a furnace at 800° for a few minutes.

Crystals of rare-earth sesquisulfides have been prepared by this method except when Ln is La, Er, Tm, or Y. In these cases the LnSI compounds are stable even at 1250°C.

Anal. Calcd. for Gd_2S_3: Gd, 76.58; S, 23.42. Found: Gd, 76.4; S, 22.2.

Properties

The rare-earth sesquisulfides are reasonably stable when exposed to air at room temperature, although a weak odor of hydrogen sulfide is frequently present. The orthorhombic A structure is found for La, Ce, Pr, Nd, Sm, Gd, Tb, and Dy. The monoclinic D structure is found for Dy, Ho, Er, and Tm (Dy_2S_3 is dimorphic), and the rhombohedral E structure is found for Yb and Lu. The B type of "rare-earth sesquisulfide" has been shown to be an oxysulfide,[2] and it is not obtained by this synthesis. The A or TH_3P_4 type of structure is also not obtained by this method. This form is usually (perhaps always) stabilized by impurities or nonstoichiometry.[1]

All crystals of the A and D structure types grow as rods (up to 1 cm in length), where the rod axis corresponds to the short crystallographic axis. Crystals of the E structure type grow as hexagonal plates which are generally twinned.

The crystals are electrically semiconducting.[1]

References

1. A. W. Sleight and C. T. Prewitt, *Inorg. Chem.*, **7**, 2282 (1968).
2. D. Carre, P. Laruelle, and P. Besançon, *C. R. Acad. Sci. Paris*, **270**, 537 (1970).

6. GROUP IV SULFIDES

$$M + 2S \rightarrow MS_2$$

$$MS_2 + 2I_2 \xrightarrow{900^\circ C} MI_4 + 2S$$

$$MI_4 + 2S \xrightarrow{800^\circ C} MS_2 + 2I_2$$

Submitted by LAWRENCE E. CONROY*
Checked by R. J. BOUCHARD†

Reprinted from *Inorg. Synth.*, **12**, 158 (1970)

Titanium(IV) sulfide, zirconium(IV) sulfide, hafnium(IV) sulfide, and tin(IV) sulfide constitute a group of isostructural compounds that can be prepared by similar methods. A procedure for preparing polycrystalline titanium(IV) sulfide has appeared in this series.[1] A review of methods, covering the literature to 1956, is contained in that article. Titanium(IV) sulfide can also be prepared directly from the elements[2] and by using the chemical transport technique[3] described below. Zirconium(IV) sulfide can be prepared from reaction of the elements[4] at 1400°; by heating zirconium(IV) chloride, hydrogen, and sulfur vapor;[5] by heating Zr_3S_5 in hydrogen sulfide gas[6] at 900–1300°; and by treating zirconium(IV) chloride with hydrogen sulfide[7] at 500°. A procedure utilizing reaction of the elements and chemical transport[3]

* Department of Chemistry, University of Minnesota, Minneapolis, MN 55455.
† E. I. du Pont de Nemours & Company, Wilmington, DE 19898.

is described here. Hafnium(IV) sulfide can be prepared from reaction of the elements alone[2] or with chemical transport.[3]

Tin(IV) sulfide can be prepared by hydrogen sulfide precipitation of Sn(IV) from solution, to produce a microcrystalline material that is contaminated with oxide. *Mosaic gold* is a crystalline form of tin(IV) sulfide prepared by high-temperature sublimation procedures. Mosaic gold is the reported product of heating mixtures of (1) tin and sulfur;[8] (2) tin, sulfur, and ammonium chloride;[8,9] (3) tin, sulfur, mercury, and ammonium chloride;[9] (4) tin(II) oxide, sulfur, and ammonium chloride;[9] (5) tin(II) chloride and sulfur;[9] (6) tin(II) sulfide, tin(II) chloride, and sulfur.[9]

Many of the transition-metal chalcogenides can be prepared as large crystals of high purity by means of high-temperature chemical transport reactions.[10] The syntheses of TiS_2, ZrS_2, HfS_2, and SnS_2 that are described below adapt the procedure of Greenaway and Nitsche.[3] The polycrystalline disulfide is synthesized from the elements and transported through a temperature gradient in the presence of iodine vapor. Both the synthesis and transport reaction are carried out in the same evacuated ampule. On heating, the polycrystalline disulfide and iodine equilibrate with the volatile metal tetraiodide and sulfur. The equilibria favor the disulfide and iodine more in the cooler zone than in the hotter zone. The disulfide may thus be transported from a hotter to a cooler zone under conditions that yield very pure, and often rather large, crystals. Because the iodine is regenerated in the cooler zone, the ratio of iodine to disulfide that is required is quite small.

A. TITANIUM(IV) SULFIDE

$$Ti + 2S \rightarrow TiS_2$$

$$TiS_2 + 2I_2 \underset{800^\circ}{\overset{900^\circ}{\rightleftarrows}} TiI_4 + 2S$$

Procedure

The synthesis and transport reactions are carried out in silica or Vycor ampules (Fig. 1). Convenient dimensions are 1.5–2.5 cm i.d. and 10–15 cm long. Titanium sponge* (0.96 g, 0.020 mol) and resublimed sulfur (1.34 g, slight excess over 0.04 mol) are placed in the tube along with a weighed

* The sponge form of titanium, zirconium, or hafnium is preferable in these syntheses because the tendency for explosive reaction is much less than with the powdered metals. More massive forms, such as wire or shot, react too slowly and less completely. Suitable sources are City Chemical Corp., 132 W. 22d St., New York, NY 10011, and Electronic Space Products, Inc., 854 S. Robertson Blvd., Los Angeles, CA 90035.

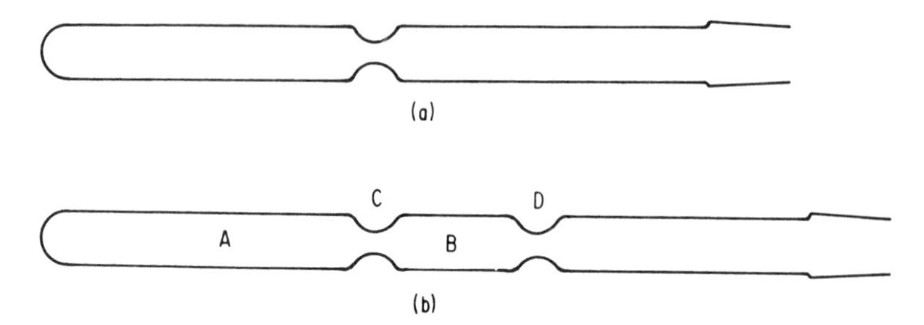

Figure 1. Silica ampule for transport reactions.

quantity of the iodine transporting agent corresponding to 5 mg/mL of the tube volume. Two relatively simple procedures for introducing iodine into the ampule are available:

1. The reactants are added to the transport tube through a long-stemmed funnel, and the tube is attached to a vacuum line and evacuated at 10^{-5} torr. A tube containing the calculated weight of purified iodine is attached to the vacuum line and evacuated. The iodine is then sublimed over into the transport tube, which is then sealed off.*
2. A transport tube of type *b* (Fig. 1) is prepared. The required weight of iodine (for the volume of zone *A*) is loaded into a *soft-glass* ampule (5 mm o.d.) that is evacuated and sealed off at a length of approximately 3 cm. The reactants are loaded into zone *A* of the silica tube, and the iodine ampule is inserted into zone *B*. The silica tube is then evacuated, outgassed, and sealed off at *D*. The iodine ampule is opened by clamping the silica tube in a horizontal position and heating zone *B* at one point with a fine oxygen-gas flame until the iodine ampule melts at the hot spot and liberates the iodine. The iodine is then sublimed over into zone *A*, and the silica ampule is sealed off at *C*. Other more elaborate procedures for adding the transport agent have been described.[12, 13]

The transport ampule is placed in a horizontal two-zone furnace in which different temperatures may be maintained at either end of the ampule. Such a furnace may be constructed by bolting together two laboratory tube furnaces end to end or by winding a center-tapped heating element on a refractory

* The checkers pointed out that commercially available titanium, zirconium, and hafnium (even the ultrapure variety) are usually contaminated with oxides, in which case the high-vacuum procedure suggested here is not justified.

tube–furnace core. Commercial two-zone furnaces are available. For short ampules the normal center-to-end gradient of an ordinary tube furnace may be used. Because the reaction of titanium, zirconium, or hafnium with sulfur is highly exothermic, slow heating of the reactants is necessary to avoid explosion. After the ampule is inserted in the cold furnace, the furnace is *slowly* heated to transport temperatures, 900° at the reactant end and 800° at the growth end, over a period of 5 h.*

■ **Caution.** *Do not insert the ampule into a hot furnace!*

Crystals are visible in the growth zone within 6 h after the reactant zone reaches 900°. The 900 → 800° gradient is maintained for at least 70 h or until no further transport is evident. The ampule is allowed to cool and is cracked open.

■ **Caution.** *Wrap the tube in several layers of cloth before attempting to open.*

The crystals of TiS_2 are washed successively with CCl_4 and CS_2 to remove any surface iodine and sulfur.

If large crystals are desired, some modifications will promote the growth of a smaller number of large crystals. An ampule of at least 2.5 cm i.d. is desirable. Before the reaction zone is heated above 800°, the reaction zone is heated to 900°, and transport is allowed to proceed at 900°800° for 70–80 h. The yield of transported crystals depends on tube dimensions, purity of reactants, and duration of the transport process.

B. ZIRCONIUM(IV) SULFIDE

$$Zr + 2S \rightarrow ZrS_2$$

$$ZrS_2 + 2I_2 \underset{800^\circ}{\overset{900^\circ}{\rightleftarrows}} ZrI_4 + 2S$$

* The checkers suggested an alternative heating procedure to avoid explosions. The reactant end of the ampule is inserted into the cold furnace, leaving the empty growth end projecting. (■ **Caution.** *Provide a protective shield.*) The furnace is slowly heated to transport temperatures over a period of at least an hour, and the ampule is then slowly moved, over a period of 4–5 h, into the heated zones. During this procedure most of the sulfur will sublime into the cool projecting portion of the ampule and will be removed from the reaction zone. As its vapor pressure increases with the temperature, reaction with the metal can proceed gradually.

Procedure

Zirconium sponge (1.82 g, 0.020 mol) and resublimed sulfur (1.34 g, a slight excess over 0.040 mol) are charged into the transport tube. The procedures for adding iodine (5 mg/mL of ampule volume) and heating that are described in the TiS_2 synthesis (Section A above) are followed. Transport is carried out at 900 → 800° for at least 70 h. The yield of transported crystals depends on tube dimensions, purity of reactants, and duration of the transport process.

C. HAFNIUM(IV) SULFIDE

If hafnium sponge (3.57 g, 0.020 mol) and resublimed sulfur (1.34 g, a slight excess over 0.040 mol) are used in the above procedure in place of zirconium sponge and sulfur, HfS_2 is obtained.

D. TIN(IV) SULFIDE

$$Sn + 2S \rightarrow SnS_2$$

$$SnS_2 + 2I_2 \underset{600^\circ}{\overset{700^\circ}{\rightleftarrows}} SnI_4 + 2S$$

Procedure

Tin metal (2.37 g, 0.020 mol) and resublimed sulfur (1.34 g, a slight excess over 0.040 mol) are charged into the transport tube. The procedures for adding iodine (5 mg/mL of ampule volume) and heating that are described in the TiS_2 synthesis (Section A above) are followed. Because of the low melting point of tin, this reaction proceeds more smoothly at lower temperature with less danger of explosions than in the case of the transition metals. Transport is carried out at 700 → 600° for 25–30 h. An alternative procedure is to prepare polycrystalline SnS_2 by the usual solution precipitation procedures. This material is also available commercially. Precipitated SnS_2 contains a considerable fraction of SnO_2, but it may be purified by the same transport procedure described above. Precipitated SnS_2 should be dried at 350° for at least a day before sealing in a transport ampule. Crystalline SnS_2 (mosaic gold) is usually prepared from tin amalgam and contains appreciable mercury contamination. This reagent is to be avoided in high-purity preparations. The yield depends on tube dimensions, purity of reactants, and duration of the transport process.

Properties

The group IV disulfides form platelike crystals that are stable in air* and are insoluble in water, dilute acids and bases, and most organic solvents. Titanium(IV) sulfide is gold-colored, zirconium(IV) sulfide is brown-violet, hafnium(IV) sulfide is violet, and tin(IV) sulfide is golden yellow. The transition-metal compounds all exhibit a high metallic luster even in very thin sections. The tin(IV) sulfide appears metallic in thick section, but thin crystals are a transparent yellow color. All four compounds crystallize in the hexagonal cadmium iodide structure (C6 in the Strukturbericht classification) in which the sulfur atoms are arranged in a hexagonal close-packed array and the metal atoms reside in octahedral holes between alternate sulfur layers. The lattice parameters are: TiS_2[3], $a = 3.405$ Å, $c = 5.687$ Å; ZrS_2,[11] $a = 3.661$ Å, $c = 5.825$ Å; HfS_2,[11] $a = 3.625$ Å, $c = 5.846$ Å; SnS_2,[3] $a = 3.639$ Å, $c = 5.884$ Å. All four compounds decompose or sublime, without melting, at temperatures above 800°. They are all *n*-type semiconductors* when prepared by the known procedures, indicating a slight excess of metal atoms in the lattice[11] (of the order of 0.1% excess in TiS_2, 0.01% in ZrS_2, but less than 10^{-6}% in HfS_2 or SnS_2 when prepared by the procedure given here).

References

1. R. C. Hall and J. P. Mickel, *Inorg. Synth.*, **5**, 82 (1957).
2. F. K. McTaggart and A. D. Wadsley, *Austral. J. Chem.*, **11**, 445 (1958).
3. D. L. Greenaway and R. Nitsche, *J. Phys. Chem. Solids*, **26**, 1445 (1965).
4. R. Vogel and A. Hartung, *Arch. Eisenhuettenw.*, **15**, 413 (1941–1942).
5. A. E. van Arkel, *Physica*, **4**, 286 (1924).
6. M. Picon, *Compt. Rend.*, **196**, 2003 (1933).
7. E. F. Strotzer, W. Biltz, and K. Meisel, *Z. Anorg. Allgem. Chem.*, **242**, 249 (1939).
8. P. Woulfe, *Phil. Trans.*, **61**, 114 (1771).
9. J. W. Mellor, *A Comprehensive Treatise on Inorganic and Theoretical Chemistry*, Vol. 7, Longmans, Green, London, 1927, p. 469 contains an extensive list of early references
10. H. Schafer, *Chemical Transport Reactions*, Academic Press, New York, 1961.
11. L. E. Conroy and K. C. Park, *Inorg. Chem.*, **7**, 459 (1968).
12. A. G. Karipides and A. V. Cafiero, *Inorg. Synth.*, **11**, 5 (1968).
13. R. Kershaw, M. Vlasse, and A. Wold, *Inorg. Chem.*, **6**, 1599 (1967).

* *Notes added in proof:* Checker reports TiS_2 and ZrS_2 changed color on standing in closed vials at 25°. Authors report that TiS_2 is a metallic conductor down to 1 K.

7. TITANIUM DISULFIDE

$$Ti + 2S \rightarrow TiS_2$$

Submitted by M. J. McKELVY* and W. S. GLAUNSINGER*†
Checked by G. OUVRARD‡

Titanium disulfide was shown to exist through the Ti–S phase diagram studies undertaken by Biltz et al. in 1937.[1] Subsequently, Bernard and Jeannin concluded that stoichiometric TiS_2 cannot be prepared by direct reaction of the elements at high temperatures (800 and 1000°), with the resulting disulfide being metal-rich ($Ti_{1+x}S_2$).[2] However, further studies carried out at lower temperatures, 600–640°, have shown that well-characterized, nearly stoichiometric TiS_2 ($Ti_{1.002\pm0.001}S_2$) can be prepared by direct reaction of the elements.[3–6] TiS_2 is important as both a host for a wide range of intercalation reactions and as a cathode for high-energy-density batteries.[7–10] Several aspects of the synthetic procedure used to prepare TiS_2 for these purposes need to be considered. The formation of essentially stoichiometric TiS_2 is of particular importance, as any excess Ti residing in the van der Waals (vdW) gap appears to "pin" the host layers together, inhibiting, if not preventing, intercalation.[2,3] Similarly, it is desirable to avoid Ti vacancy-interstitial pair formation which occurs for higher temperature syntheses (1000°), as this also results in the presence of Ti in the vdW gap.[11] TiS_2 prepared at lower temperatures (~600°) shows no evidence of these defects.[12] Care must also be taken to avoid TiS_3 formation, which can occur at the elevated sulfur pressures necessary to minimize the presence of excess Ti. TiS_2 combined with liquid sulfur cannot form TiS_3 above 630–632°.[13–15] Thus, quenching of TiS_2 in the presence of liquid sulfur from temperatures above this range effectively avoids TiS_3 formation.

TiS_2 can also be prepared by reaction of titanium halides with hydrogen sulfide, halogen-associated transport coupled with the reaction of the elements and various solution reactions.[15–17] However, the disulfides resulting from these methods have not been as throughly characterized and have the disadvantage of significant impurity incorporation, such as halogen and oxygen containing impurities.

The following procedure describes the synthesis of the most stoichiometric and thoroughly characterized TiS_2 prepared to date by direct reaction of the

* Center for Solid State Science, Arizona State University, Tempe, AZ 85287-1704.
† Department of Chemistry and Biochemistry, Arizona State University, Tempe, AZ 85287-1604.
‡ Institut des Matériaux de Nantes, CNRS-UMR 110, Nantes, France.

elements ($Ti_{1.002 \pm 0.001}S_2$).[5,6] The procedure takes 2 weeks to complete. However, only about 8–10 hours of actual lab time are required.

Procedure

All glassware and fused silica ampules associated with the synthesis are cleaned using a cleaning solution consisting of 45 mL of 48% HF, 165 mL of concentrated HNO_3, 200 mL of H_2O, and 10 g of Alconox detergent.

■ **Caution.** *Aqueous hydrogen fluoride solutions are highly corrosive and cause painful, long-lasting burns. Gloves, safety glasses and a lab coat should be worn, and the solution should be handled in a well-ventilated hood.*

After a 3-min exposure to the cleaning solution, the glassware or fused silica ampules are rinsed at least 25 times with distilled water and subsequently rinsed five additional times with MΩ·cm water that has been filtered to remove organic contaminants. The glassware or fused silica ampules are then dried at 110°.

Polycrystalline TiS_2 (trigonal 1T form) is prepared by direct reaction of the elements in two steps. 0.020″ diameter Marz-grade Ti wire (Materials Research Corp.) is used with a stated purity of 99.93%, with O (500 ppm), Fe (50 ppm), Ni (30 ppm), C (30 ppm), Cr (20 ppm), Pb (20 ppm), Sn (20 ppm), Al (15 ppm), N (10 ppm), Ag ($<$10 ppm), Cu ($<$10 ppm), Mn ($<$10 ppm), V ($<$ 5 ppm), Ca ($<$5 ppm), Mg ($<$5 ppm), Si ($<$5 ppm), and H (3 ppm) being the measured impurities by weight. Substitution of powder or sponge forms of Ti may result in a runaway reaction and will generally yield higher TiS_2 impurity levels because of their greater surface area combined with the higher concentrations of impurities found near the surface.[15] Substitution of lower-surface-area forms of Ti may result in significantly longer reaction times. The Ti is degreased prior to use by sonicating twice in trichloroethylene (Baker AR grade, evaporation residue $<$0.0001%) followed by sonicating three times in toluene (Baker AR grade, evaporation residue 0.00001%). Each sonication has about a 15-min duration.

■ **Caution.** *Trichloroethylene is a known carcinogen and toluene may cause anemia. These solvents should be handled and the sonication procedure carried out in a well-ventilated hood. Protective gloves, safety glasses and a lab coat should be worn.*

The first reaction step involves adding stoichiometric amounts of Ti and S (99.9995%) (Alfa Products), with an extra 5 mg/cm^3 S, to a 12-mm-o.d. fused silica ampule having a 1-mm wall thickness. (Care should be taken to avoid using scratched ampules or ampules with significantly larger diameters as they are less likely to be able to withstand the sulfur pressures associated with the synthesis.) This amounts to combining Ti (400 mg, 8.356 mmole) and S (580 mg, 18.09 mmole) in a sealed ampule volume of 9 cm^3 for the preparation of slightly less than a gram of TiS_2. The ampule is then evacuated to $<10^{-4}$ torr, with the upper empty two-thirds of the reaction volume subsequently heated to 200–500° to remove adsorbed water. A wet tissue around the lower section of the ampule containing the reactants is sufficient to avoid significant heating of the reactants and substantial sulfur volatilization. The ampule is then sealed at $\leq 10^{-4}$ torr, with a sealed ampule length of about 13 cm being convenient for handling. The ampule is heated to 450° for one day, with the temperature subsequently raised about 50°/day in two 25° steps to a final temperature of 640°. After holding the ampule at 640° for 3 days, the ampule is pulled out of the furnace to quickly air-cool the product to ambient temperature.

■ **Caution.** *Because of the potentially explosive high sulfur pressure in the ampules, appropriate protection, including a lab coat, heavy gloves and a face shield, should be worn when inserting the ampules into or removing them from a hot furnace.*

The excess S is vapor transported to the opposite end of the ampule from the disulfide using a 325 ± 25°-to-ambient temperature gradient. The ampule is left in the gradient for 1–2 h, removed, allowed to cool, and its TiS_2 is gently agitated. The ampule is then replaced in the gradient. After an additional 1–2 h, the ampule is removed and allowed to cool to ambient temperature.

In order to ensure complete reaction and optimum stoichiometry, a second reaction step is necessary. As TiS_2 is mildly air-sensitive, it is exclusively handled in a glovebox (<1 ppm total H_2O and O_2) (Vacuum Atmospheres Corp.). The ampule from the first step is scratched in the middle with a glass knife and wrapped with parafilm. It is then carefully snapped in the glovebox to avoid sudden implosion and/or the incorporation of quartz-chip impurities. The disulfide is then crushed with a clean agate mortar and pestle and transferred along with 40 mg/cm^3 of excess S to a 12-mm-o.d. (1 mm wall thickness) fused-silica ampule previously evacuated to $\leq 10^{-4}$ torr and flamed to remove adsorbed moisture. The ampule is then sealed at $\leq 10^{-4}$ torr (with a wet tissue around the sample area), placed in a furnace at 640 ± 2° for 3 days and water-quenched to avoid the formation of TiS_3, a fibrous black compound. The excess S is again vapor-transported to the opposite end of the ampule from the TiS_2 using a 325 ± 25°-to-ambient temperature gradient. The ampule is left in the gradient for 1–2 h, then removed to agitate the TiS_2 to break up any clumps and reinserted into the gradient. After repeating this procedure twice, the ampule is inserted into the furnace to a depth about 1–2 cm less than before. If no additional sulfur transport is observed, vapor transport is complete. The ampule is then opened in the glovebox, as before, to obtain the final product, which is a free-flowing golden powder.

Properties

Thermogravimetric analysis (TGA) of the heating of the resulting disulfide to 900° in pure oxygen, as shown in Fig. 1, yields a residual weight percent of $71.45 \pm 0.04\%$ for the resulting TiO_2.[6] The TiO_2 (rutile form) produced by this procedure was analyzed for residual sulfur by Guelph Chemical Laboratories and found to contain 800 ± 30 ppm by weight. Further taking into account the oxygen impurity in the Ti wire yields a composition of $Ti_{1.002 \pm 0.001}S_2$.[6] X-ray powder diffraction (XPD) of the resulting material yields trigonal-cell constants of $a = 3.407 \pm 0.002$ Å and $c = 5.695 \pm 0.003$ Å, which are in good agreement with the most stoichiometric TiS_2 previously prepared.[3] As mentioned earlier, TiS_2 is mildly air-sensitive and should be

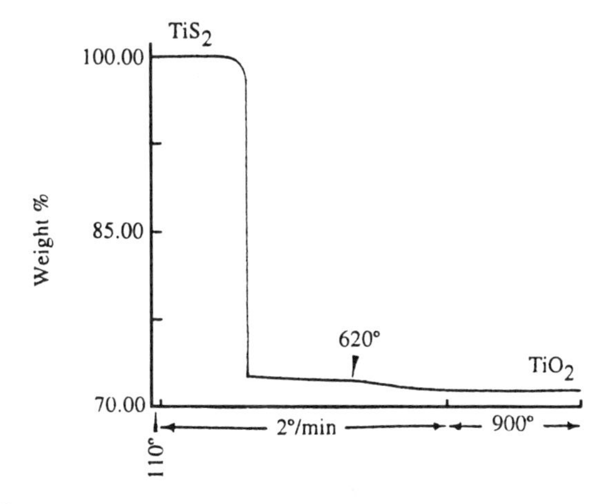

Figure 1. TGA curve of the oxidation of TiS_2 to TiO_2. The unusual inflection at $620 \pm 6°$ is apparently associated with the anatase-to-rutile (TiO_2) phase transition.[5] XPD combined with TGA suggests the rapid oxidation at about 350° results in the presence of similar amounts of anatase and rutile, with the residual sulfur "trapped" in amorphous grain boundaries. The enhanced oxidation of this residual sulfur at about 620° may be associated with interfacial changes initiated by the onset of the anatase-to-rutile transition, which is complete by the end of the analysis.[5]

handled under glovebox conditions. However, it can be quickly transferred in air for both TGA and XPD investigation without measurably affecting the results.

The product can be further characterized by magnetic susceptibility studies, as previously described.[6] Magnetic susceptibility is sensitive to TiS_2 stoichiometry,[3,6] with $Ti_{1.002 \pm 0.001}S_2$ having an intrinsic temperature-independent paramagnetic susceptibility of $9 \pm 2 \times 10^{-6}$ emu/mol.[6] This contrasts with the susceptibility of $Ti_{1.01}S_2$, which is about 6 times greater.[3]

The product composition, $Ti_{1.002 \pm 0.001}S_2$, is also consistent with the carrier concentration of 1.4×10^{20} electrons/cm^3 determined by Hall-coefficient measurements for similarly-prepared, highly stoichiometric TiS_2.[6,18,19] The carrier concentration for this extrinsic semiconductor is consistent with the contribution of four electrons per excess Ti to the host conduction band to produce its observed extrinsic semiconducting properties.[6]

Acknowledgment

This work was supported by National Science Foundation Grants DMR-8605937 and DMR-8801169.

References

1. W. Biltz, P. Ehrlich, and K. Meisel, *Z. Anorg. Allg. Chem.* **234**, 97 (1937).
2. J. Benard and Y. Jeannin, *Adv. Chem.*, **39**, 191 (1963).
3. A. H. Thompson, F. R. Gamble, and C. R. Symon, *Mater. Res. Bull.*, **10**, 915 (1975).
4. D. A. Winn and B. C. H. Steele, *Mater. Res. Bull.*, **11** 551 (1976).
5. M. J. MacKelvy and W. S. Glaunsinger, *Mater. Res. Bull.*, **21**, 835 (1986).
6. M. J. McKelvy and W. S. Glaunsinger, *J. Solid State Chem.*, **66**, 181 (1987).
7. F. R. Gamble, *Ann. N. Y. Acad. Sci.*, **313**, 86 (1978).
8. F. Levy (ed.), *Intercalated Layered Materials*, Reidel, Dordrecht, 1979.
9. M. S. Whittingham, *Science*, **192**, 112 (1976).
10. M. S. Whittingham, *J. Solid State Chem.* **29**, 303 (1979).
11. S. Takeuchi and H. Katsuta, *J. Japan Inst. Metals*, **34**, 758 (1970).
12. R. R. Chianelli, J. C. Scanlon, and A. H. Thompson, *Mater. Res. Bull.*, **10**, 1379 (1975).
13. J. C. Mikkelson, *Nuovo Cimento.*, **38B**, 378 (1977).
14. A. H. Thompson, U.S. Patent 4,069,301 (1978).
15. M. S. Whittingham and J. A. Panella, *Mater. Res. Bull.*, **16**, 37 (1981).
16. A. H. Thompson and F. R. Gamble, U.S. Patent 3,980,761.
17. M. J. Martin, G. Qiang, and D. M. Schleich, *Inorg. Chem.*, **27**, 2804 (1988).
18. C. A. Kukkonen, W. J. Kaiser, E. M. Logothetis, B. J. Blumenstock, P. A. Schroeder, S. P. Faile, R. Colella, and J. Gambold, *Phys. Rev.*, **B24**, 1691 (1981).
19. P. C. Klipstein and R. H. Friend, *J. Phys. C*, **17**, 2713 (1984).

8. METATHETICAL PRECURSOR ROUTE TO MOLYBDENUM DISULFIDE

$$MoCl_5 + 5/2Na_2S \rightarrow MoS_2 + 5NaCl + \tfrac{1}{2}S$$

Submitted by PHILIPPE R. BONNEAU,* JOHN B. WILEY,† and RICHARD B. KANER*
Checked by MICHAEL F. MANSUETTO‡

The layered transition-metal dichalcogenide, molybdenum disulfide, has been investigated for its properties as a lubricant,[1,2] battery cathode material,[3–5] and catalyst.[6–8] This compound is most often synthesized through reaction of the elements for several days at elevated temperatures (>900°).[9] Other synthetic routes are also known.[10] Methods based on metathesis (exchange) reactions have been reported where transition-metal halides and alkali-metal chalcogenides[11] or covalent sulfiding agents[12] are combined in nonaqueous solvents. The synthesis described here involves the solid-state metathesis reaction of molybdenum(V) chloride and sodium sulfide. This route is effective for the preparation of molybdenum disulfide in that it produces good-quality, easily isolable, and highly crystalline material. Once the precursors are prepared, the final product can be synthesized in a matter of seconds.[13] An additional feature of this synthetic method is that it can be extended to other dichalcogenides such as WS_2, $MoSe_2$, and WSe_2.

Procedure

Precursors should be stored and handled in a helium- or argon-filled dry box.

■ **Caution.** *$MoCl_5$ will evolve corrosive HCl gas on exposure to moisture. Also, heating volatile materials in a sealed tube can create an explosion hazard. Before carrying out such procedures, ideal-gas-law calculations should be performed to verify that if all the reactants go into the gas phase, the reaction vessel pressure does not exceed 5 atm.*

$MoCl_5$. Solid-state metathesis reactions are best performed with high-quality molybdenum(V) chloride. $MoCl_5$ can be obtained commercially (Aesar)

* Department of Chemistry and Biochemistry and Solid State Science Center, University of California, Los Angeles, CA 90024.
† Department of Chemistry, University of New Orleans, New Orleans, LA 70148.
‡ Department of Chemistry, Northwestern University, Evanston, IL 60208.

but, depending on its history, may contain undesired lower halides (e.g., $MoCl_3$) and oxychlorides. Transport methods are useful for the purification of this material and have been described in detail.[14] $MoCl_5$ (8.0 g, 0.029 mole) is loaded in a dry box through a long-stem funnel into the bottom of a flame-dried Pyrex tube that is closed at one end (2.5 cm o.d. × 25 cm length). The open end of the tube is fitted with a closed stopcock and the entire reaction vessel is transferred to a vacuum line. The tube is then sealed under vacuum with a oxygen-gas hand torch. [It is often desirable to have the open end of the tube fitted with a smaller-bore Pyrex tubing (1.2 cm o.d.) so that it can be more easily sealed.] The transport reaction is carried out over a temperature gradient in a tube furnace. The tube is horizontally positioned with the $MoCl_5$ end at 250° and the other end, protruding out of the furnace, at approximately room temperature. Typically the lower boiling oxychlorides (green-colored) transport farthest from the heat source while the $MoCl_5$ (black) deposits at an intermediate temperature just inside the opening of the furnace. Lower chlorides do not transport well under these conditions and essentially remain in the hot zone. Fifty minutes is often a sufficient amount of time to complete the process. After cooling, the reaction tube is returned to the dry box and carefully opened so as not to remix the $MoCl_5$ with the residual material.

■ **Caution.** *Inhalation of ammonia can cause respiratory problems. Liquid ammonia should be handled in a fume hood. The synthesis of* Na_2S *in liquid ammonia must be performed at low temperatures. Elevated temperatures will produce excessive ammonia pressure that could result in an explosion.* Na_2S *is air-sensitive and will give off toxic* H_2S *gas on exposure to water.*

Na_2S. Na_2S is made from the reaction of sodium metal and elemental sulfur in liquid ammonia at low temperature (ca. −30°).[15–16] High-purity sodium metal (5.89 g, 0.256 mole) is freshly cut in the dry box and added to sulfur (4.11 g, 0.128 mole) in a flame-dried, thick-walled Pyrex tube (500 mL) equipped with a side arm and a Teflon stopcock fitted with an O-ring. The tube is evacuated and then cooled in liquid nitrogen. Approximately 50 mL of liquid ammonia are condensed into the tube through the side arm. The tube is then removed from the liquid nitrogen to allow the ammonia to melt. At this point, the blue color characteristic of alkali-metal ammonia solutions is observable. Special attention is given so as not to allow the reaction temperature to rise too far above the melting point of ammonia (−78°). The temperature can be easily regulated by frequently returning the reaction vessel to the liquid nitrogen bath or by the use of a dry ice/acetone bath. The reaction mixture is gently shaken until the blue color disappears and the material appears homogeneous (ca. 20 min). The disappearance of the blue color indicates that

all of the sodium metal has been converted to Na_2S and that the reaction is complete. The side arm of the reaction vessel is then attached to a mercury bubbler, the stopcock is carefully opened, and the ammonia is slowly allowed to evaporate. Not all the ammonia is removed by this process, therefore the reaction vessel is heated to 300° under dynamic vacuum on a vacuum line. A liquid nitrogen trap is used to collect any residual ammonia. Finally, under static vacuum, the sample is transferred back to the dry box. Pure Na_2S is white, although the product can appear light yellow due to a small amount of polysulfide. This synthetic method produces finely divided product that can spontaneously react with $MoCl_5$ on light stirring with a spatula. To reduce its surface area and consequently allow better control of the metathesis reaction, the Na_2S reagent is annealed in a dry, evacuated, sealed Pyrex tube at ~500° for several hours.

■ **Caution.** *The reaction between $MoCl_5$ and Na_2S is very exothermic (− 210 kcal/mole) and can be explosive. Reactions will produce a bright flash of light and a mushroom cloud of sulfur. Additionally, hot, sometimes molten, reaction products can be expelled from the fulminating mixture. Reactions should be contained in a stainless steel bomb and carried out on a small scale (<1 g total reactants). Once starting materials are mixed together, they can spontaneously react. This possibility can be lessened by minimizing the surface area of the starting materials (i.e., annealing Na_2S) and by minimizing the time between the mixing of the reactants and the controlled initiation of the reaction. The precursors are themselves potentially hazardous. On exposure to air or on washing with water, the product mixtures can evolve toxic gases from some of the unreacted starting materials: H_2S and HCl from Na_2S and $MoCl_5$, respectively. Workups should at a minimum be done in a fume hood. Incomplete reactions can sometimes also be a problem. If unreacted starting materials are present in the reaction mixture the addition of solvent can promote further reaction. This can be quite hazardous since the heat of the reaction is often enough to ignite flammable solvents such as methanol.*

MoS_2. The reaction of $MoCl_5$ and Na_2S should be carried out in a bomb similar to those used in calorimetry. A schematic of the bomb used in this laboratory is shown in Fig. 1. This bomb is fitted with a nichrome or iron initiation wire (26-gauge) that can be heated by the passage of current. $MoCl_5$ (0.50 g, 0.0018 mol) and Na_2S (0.36 g, 0.0046 mol) are lightly stirred with a metal spatula. The resulting mixture need not be completely homogeneous. The mixture is placed in the sample cup, which in turn is placed in the bomb. The bomb is closed (the bomb does not have to have an airtight seal), and the reaction is initiated with the hot filament. The reaction will go to completion in less than two seconds. The dark MoS_2 product mixture is

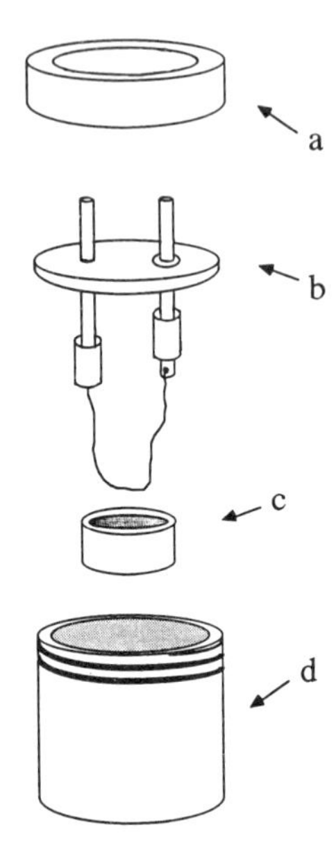

Figure 1. Reaction bomb. (*a*) Threaded screw-cap ring to secure bomb cover. (*b*) Bomb cover with two posts. The bottom of each post has a small hole to connect the nichrome or iron detonator wire and a metal sleeve that slides down to hold the wire in place. The right post is insulated from the rest of the bomb with a Teflon ring. Leads from the electric detonator are attached to the posts – when the detonator is activated, a current is passed and the detonator wire is heated resistively. (*c*) Sample cup. (*d*) Stainless-steel reaction bomb (48 mm i.d., 35 mm o.d. × 50 mm in height).

removed from the bomb and ground with a mortar and pestle. MoS_2 is not air or moisture sensitive, so the reaction mixture can now be removed from the dry box and taken to a fume hood. The MoS_2 product is isolated by washing with methanol to remove any unreacted $MoCl_5$, with water to remove the sodium chloride byproduct and any unreacted Na_2S, and then with carbon disulfide or carbon tetrachloride to remove any elemental sulfur byproduct that has not sublimed from the product. The typical yield of MoS_2 is 80%.

Properties

The MoS_2 product is a gray, highly crystalline material in which shiny crystallites can often be observed. The product contains both the 2*H*- and 3*R*-MoS_2 polytypes. The powder X-ray diffraction pattern has major reflections at $d = 6.16$ (100% relative intensity), 2.28 (58%), and 1.83 (29%) for the 2*H*-

polytype, and 6.15 (100%), 2.36 (27%), and 2.20 (27%) for the 3*R*-polytype.[16] The sulfur content of the sample (typically 2.00(1)) can be determined gravimetrically by heating in flowing dilute hydrogen (5% in nitrogen) at 900–1000° for ~40 h. Under these conditions, MoS_2 is completely reduced to molybdenum metal as indicated by X-ray diffraction.

References

1. F. L. Claus, *Solid Lubricants and Self-Lubricating Solids*, Academic Press, New York, 1972.
2. J. R. Lince and P. D. Fleischauer, *J. Mater. Res.*, **2**, 827 (1987).
3. J. Rouxel and R. Brec, *A. Rev. Mater. Sci.*, **16**, 137 (1986).
4. J. J. Auborn, Y. L. Baberio, K. J. Hansen, D. M. Schleich, and M. J. Martin, *J. Electrochem. Soc.*, **134**, 580 (1987).
5. A. J. Jacobson, R. R. Chianelli, and M. S. Whittingham, *J. Electrochem. Soc.*, **126**, 2277 (1979).
6. R. R. Chianelli, *Catal. Rev. Sci. Engng.*, **26**, 361 (1984).
7. T. A. Pecoraro and R. R. Chianelli, *J. Catal.*, **67**, 430 (1981).
8. S. Gobolos, Q. Wu, F. Delannay, P. Grange, and B. Delmon, *Polyhedron*, **5**, 219 (1986).
9. R. M. A. Leith and J. C. J. M. Terhell, in *Preparation and Crystal Growth of Materials with Layered Structures*, R. M. A. Leith, (ed.) Reidel, Dordrecht, The Netherlands, 1977, pp. 200–204. For related syntheses, see also L. E. Conroy, *Inorg. Synth.* **12**, 158 (1970).
10. (a) J. C. Wildervanck and F. Jellinck, *Z. Anorgan. Allg. Chem.*, **328**, 309 (1964); (b) W. Kwestroo, in *Preparative Methods in Solid State Chemistry*, P. Hagenmuller (ed.), Academic Press, New York, 1972, pp. 563–574; (c) S. K. Srivastava and B. N. Avasthi, *J. Less-Common Met.* **124**, 85 (1986).
11. R. R. Chianelli and M. B. Dines, *Inorg. Chem.*, **17**, 2758 (1978).
12. (a) D. M. Schliech and M. J. Martin, *J. Solid State Chem.*, **64**, 359 (1986); (b) M. J. Martin, G. Qiang, and D. M. Schliech, *Inorg. Chem.*, **27**, 2804 (1988).
13. P. R. Bonneau, R. F. Jarvis, Jr., and R. B. Kaner, *Nature*, **349**, 510 (1991).
14. H. Schafer, *Chemical Transport Reactions*, Academic Press, New York, 1964.
15. W. Klemm, H. Sodomann, and P. Langmesser, *Z. Anorg. Allg. Chem.*, **241**, 281 (1939).
16. G. Brauer, *Handbook of Preparative Inorganic Chemistry*, vol. 1, 2nd ed., Academic Press, New York, 1963, p. 358.

9. LITHIUM NITRIDE, Li_3N

$$6Li + N_2 \rightarrow 2Li_3N$$

Submitted by E. SCHÖNHERR,* A. KÖHLER,* and G. PFROMMER*
Checked by Kj. Fl. NIELSEN†

Reprinted from *Inorg. Synth.*, **22**, 48 (1983)

In the last several years, lithium nitride has excited much interest as a solid ionic conductor.[1–7] Lithium nitride is formed by direct reaction between lithium metal and nitrogen gas. The reaction goes to completion if the temperature of the lithium and the ambient pressure of the nitrogen are sufficiently high. The synthesis of Li_3N ("$AzLi_3$") was first described by Ouvrard,[8] who passed nitrogen over glowing lithium in an iron boat. Several other groups have investigated the applicability of various materials as containers for lithium during the reaction,[9,10] Additional information on this synthesis was given by Zintl and Brauer,[11] who used a LiF-coated ZrO_2 boat between 400 and 800°. A detailed procedure for Li_3N synthesis is described by Masdupuy and Gallais.[12] These authors used Fe boats for temperatures between 370 and 450°. The application of elevated N_2 pressures in order to effect the low-temperature synthesis of Li_3N has been described by Neumann et al.,[13] Yonco et al.,[14] and Kutolin and Vulikh.[15]

The phase diagram of the Li–Li_3N system was first described by Bol'shakov,[16] while exact thermodynamic data for Li_3N are reported by Yonco et al.[14] Lithium nitride forms as a simple eutectic with Li with the eutectic point near 0.05 mol% Li_3N at 180.3°. The melting point of Li_3N is $813 \pm 1°$.

A. POLYCRYSTALLINE MATERIAL

Procedure

The reactants used throughout this synthesis are N_2 gas (99.996%) and Li metal rods (99.9%)‡ that are about 12 mm in diameter and up to 200 mm in length and are packed in argon-filled cans. Opening of the cans and handling

* Max-Planck-Institut für Festköperforschung, Heisenbergstrasse 1, 7000 Stuttgart 80, Germany.

† Physics Laboratory III, The Technical University of Denmark, DK-2800, Lyngby, Denmark.

‡ Alfa Products, Danvers, MA 01923.

of the Li rods is carried out in a glovebox that contains high-purity argon (flow rate 50 m^3/h, $H_2O \approx 1$ ppm). The impurity layer on the rods is removed with a knife. The cleaned rods are placed into a conical crucible of tungsten or molybdenum,* and the remaining rods are stored in an evacuated container of stainless steel mounted within the glovebox. The crucible is about 50 mm in diameter and 70 mm in height. The crucible is filled with about 20 g Li, closed with a rubber stopper, and transferred to the external synthesis equipment.

■ **Caution.** *Lithium reacts vigorously with water, liberating explosive H_2 gas. It will cause severe burns if allowed to contact bare flesh. Appropriate precautions should be taken.*

The synthesis equipment consists of a stainless-steel cylinder with an inside diameter of 100 mm and a height of 265 mm and is shown schematically in Fig. 1. The wall of the cylinder is 3 mm and the top plate 10 mm in thickness. The cylinder is mounted with 12 MX10 stainless-steel screws on a baseplate made of stainless steel. The baseplate is furnished with a rabbet that contains the O-ring Teflon seal between the cylinder flange and base. In addition, three 6-mm-o.d. stainless-steel lines are welded to the base. One line leads to a vacuum pump, another line to the gas bottle, and the third to a mechanical safety valve that opens for pressure higher than 75 atm. The inlet line has connections to a manometer and a second electrochemical safety valve that can be controlled by the manometer. A Pt/Pt 10% Rh, stainless-steel shielded thermocouple allows the temperature to be recorded during the procedure. The rubber stopper of the crucible is removed after the crucible has been placed on the pedestal of the synthesis equipment. The equipment is then evacuated to a pressure of about 10^{-2} torr, flushed two times with N_2 at about 5 atm pressure, and loaded with 10 atm of N_2.

■ **Caution.** *Since the gas pressure within the vessel ($\sim$2 L in volume) is higher than 1 atm, the precautions for operating containers with compressed gases must be followed. The equipment described in this synthesis may not be safe for applied pressures higher than 90 atm.*

The crucible is heated with a heating strip, mounted at the upper half of the vessel, until the reaction starts. The temperature of the strip should be limited to 250°. The increase of N_2 pressure during heating is about 20% (i.e., from 10 to 12 atm). The reaction between Li and N_2 goes to completion spontaneously, owing to the high heat of formation, which produces a reaction

* Available from Metallwerk Plansee GmbH, A-6600 Reute, Austria.

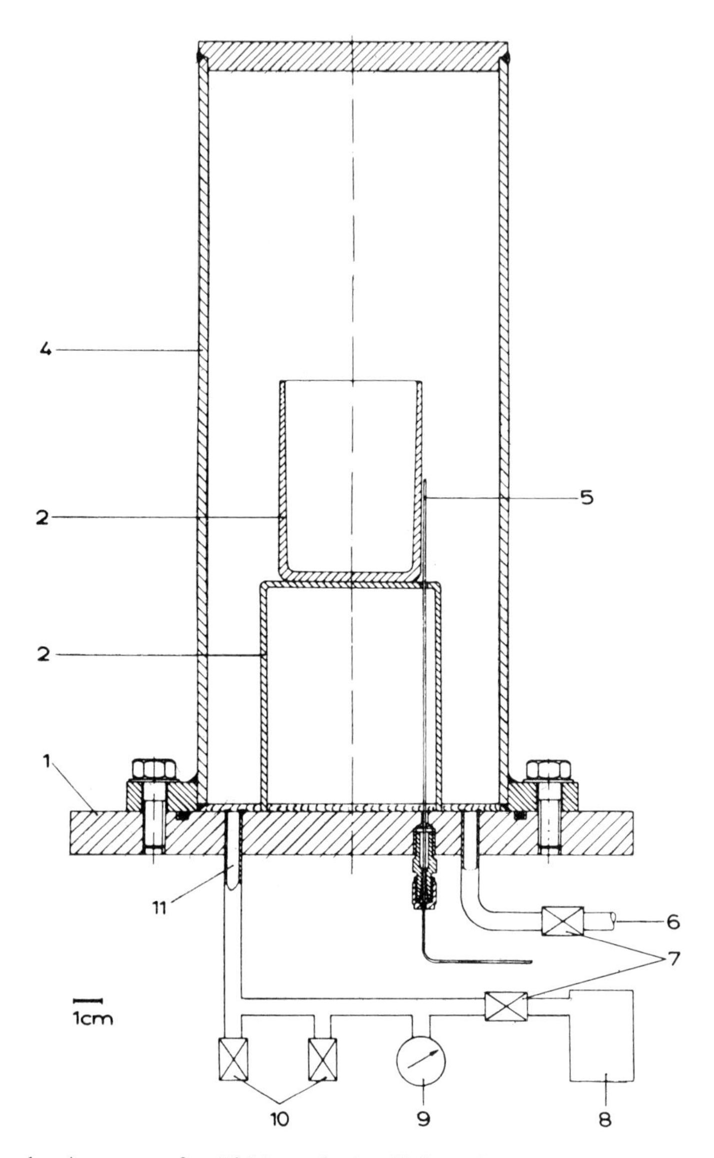

Figure 1. Apparatus for Li_3N synthesis: (1) baseplate; (2) pedestal; (3) crucible; (4) vessel; (5) thermocouple; (6) to vacuum pump; (7) valves; (8) N_2 gas bottle; (9) manometer; (10) safety valves; (11) N_2 gas inlet.

temperature of approximately 800°. To initiate the reaction, the rate of heating has to be sufficiently high, that is, higher than 10°/min; otherwise the Li_3N layer that is formed near the melting point of Li can prevent further nitridation. The reaction can start at temperatures as low as 100°, however, if the lithium is contaminated with LiOH. If the nitridation has not commenced by the time the temperature has reached 230°, cooling the container to room temperature and reheating will in general initiate the reaction. When a rapid temperature increase during the synthesis is noticed, heating is stopped and the sample allowed to cool. The remaining nitrogen is vented, lowering the pressure to 1 atm before opening. The final product tends to stick to the walls of the crucible, depending on the degree of surface oxidation of the crucible. Therefore, a crucible having a conical shape is preferred. Since the crucible is in contact with Li_3N at high temperatures for only a short time, even a Ni crucible can be used without serious contamination of the product.

Properties

The Li_3N is brown red in color in reflected light and consists of thin shells and compact material. The shape of the thin shells corresponds partly with the initial nitridated surface of the Li rods. The compact material consists of agglomerated thin plates up to 3 mm in diameter. The plates are intensely red in transmitted light. The color of the surface changes to dark blue and violet on exposure to air. The compound forms NH_3 in humid air.

Lithium nitride cyrstallizes with a hexagonal point symmetry. The lattice constants are $a = 3.648$ Å and $c = 3.875$ Å,[11,17] and the space group is *P6/mm*.[11,17,18] Along the c axis, Li_3N forms an alternating structure of Li and Li_2N layers. The movement of Li^+ ions may be preferred within the Li_2N layers.[18]

B. SINGLE CRYSTALS OF Li_3N

Procedure

Small crystals in the millimeter range can be grown from solution,[11,17,19] or from a mixture of lithium and sodium as a solvent.[20] Large crystals can be grown by the Czochralski technique similar to that described in Reference 21. The growth equipment is schematically shown in Fig. 2. The vessel (1) is made from stainless steel and furnished with four slots at the top. The center slot carries the seed holder (2), which consists of a nickel clamp welded to a stainless-steel tube. An inner-tube supplies the clamp with water for cooling. One slot (3) is used as the inlet for N_2 gas. The crucible is illuminated with a 150-W halide bulb through a small-diameter slot (20 mm i.d.) behind the center slot (not drawn in the Fig. 2). The fourth slot (4) with an inside diameter of 32 mm is used for the observation of the crystal growth. At least three thin stainless-steel tubes with an inside diameter of 8 mm and 50 mm in length are mounted along the slot to prevent the contamination of the

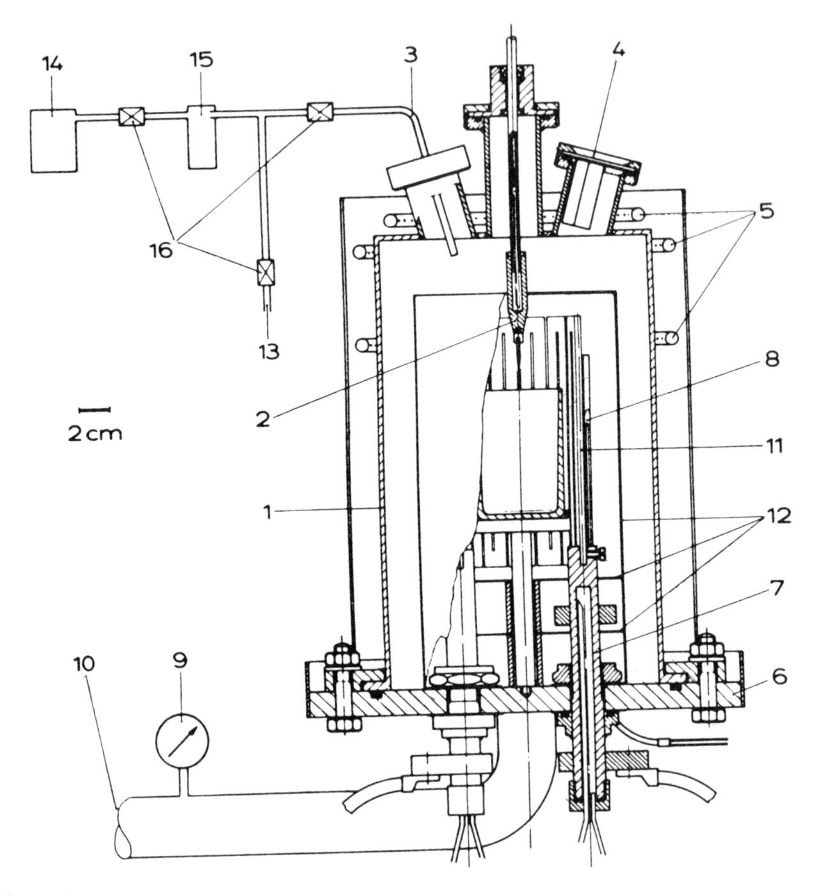

Figure 2. Apparatus for Li_3N crystal growth: (1) vessel (ss); (2) seed holder; (3) N_2 gas inlet; (4) observation window; (5) cooling tubes; (6) baseplate (Ni); (7) electrodes (Ni); (8) thermocouple (Pt/Pt 10% Rh, Inconel shielded); (9) Inconel manometer; (10) to vacuum pump; (11) furnace (ss); (12) heat shields (Mo); (13) N_2 gas outlet; (14) N_2 gas bottle; (15) flow meter; (16) valves.

window by Li_3N deposition. Three copper tubes (5) with an array of fine holes supply the outside of the vessel with water for cooling.

The nickel baseplate (6) supports water-cooled electrodes made of nickel (7), a thermocouple (8), and a pipe leading to a manometer (9) and to a vacuum pump system (10).

A resistance furnace (11) with sufficient mechanical stability and chemical resistance is machined from stainless steel 150 mm in height, 70 mm in outside diameter, and 2 mm in wall thickness by cutting it lengthwise into 24 elements. The clearance between elements is 2 mm. A power supply with 30 V and 200 A is sufficient. The lifetime of the furnace is about 80 h in the presence of molten Li_3N. A graphite heater cannot be used because of the formation of CO_2, which reacts with the Li_3N crystals. The temperature is controlled by a Pt/Pt 10% Rh thermocouple, which is encapsulated in an Inconel tube with 1 mm outside diameter. The thermocouple is mounted on the outside of the furnace. The heat loss and material loss by convection are reduced by several heat shields made of molybdenum (12).

For crystal growth, a slightly conical crucible of tungsten with an outside diameter of 50–49 mm, a height of 70 mm, and a wall thickness of 2 mm is used. Tungsten has been found to be the most suitable material, showing less corrosion by the Li_3N melt than either tantalum or zirconium. The Li_3N melt reacts with W to form what appears to be Li_6WN_4, which is not noticeably soluble in the Li_3N melt and solid. The Li–W–N compound is formed at the crucible walls and sticks to neither the tungsten nor the Li_3N cast. The average weight loss of the crucible is 0.6 ± 0.2 g/h. The crucibles can be used, for example, either 20 times for 11 h or 13 times for 22 h, before they become leaky. Molybdenum crucibles are also fairly compatible with the Li_3N melt. (Use of them, however, reduces the maximum stable growth rate.) Crucibles of iron or vitreous carbon cannot be used for crystal growth.

For crystal growth the crucible can be loaded with presynthesized Li_3N material, but to prevent additional contact of Li_3N with air, the crucible can be loaded with Li metal, and the synthesis of Li_3N carried out in the growth equipment. For this purpose the crucible is filled with Li rods and heated to about 300° in the glove box (Section A) under an Ar atmosphere. The floating impurities are removed from the molten Li metal with a stainless-steel spoon. When the purified Li has cooled, the crucible is closed with a rubber stopper and transferred to the growth equipment.

After loading, the growth chamber is evacuated to 10^{-5} torr. The lithium is then heated to 350–400° to remove hydrogen. After a period of 15 h, the chamber is filled with N_2 until the pressure is approximately 600 torr. The temperature is increased until the Li metal begins to glow and the N_2 pressure drops, an indication that the reaction has started. Immediately, N_2 is added to restore the original pressure and the crucible is heated in less than 5 min to the melting point of the Li_3N. If longer times are used, the synthesized material may solidify. This will lead to bubbling when it is remelted. When the temperature of about 830° is reached, the N_2 pressure is reduced to atmospheric pressure by venting any excess N_2 through the outlet (13).

The seeding can be controlled either with a small Li_3N crystal with dimensions of $4 \times 4 \times 25$ mm or more conveniently with a tungsten pin imbedded in a copper rod. Details of the seeding device are shown in Fig. 3. If the tapered end is not inserted too far into the melt, only a few seeds are

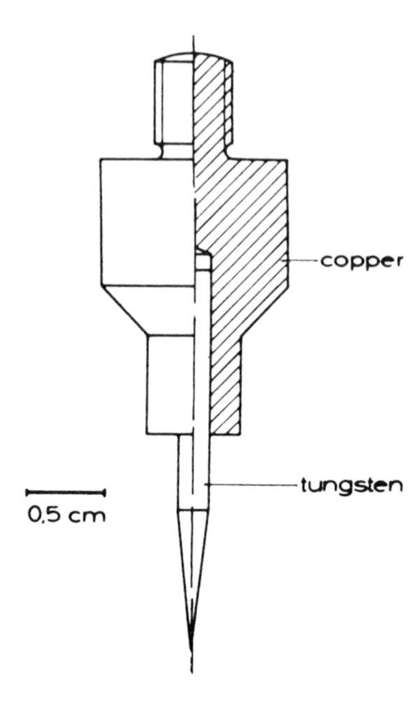

Figure 3. Schematic sketch of the seeding tip.

formed, fewer than in the case when a Li_3N seed crystal is used. Material is deposited from the vapor phase at the initial grains so that the grains grow toward the seed holder. After about an hour, the deposited material reaches the copper rod, which then serves for removing the heat of fusion.

The growth of the boules is achieved with a pulling rate of 0.7–3 mm/h and a rotation rate of 30 rpm. The resulting growth rate is 3.3 ± 0.8 mm/h. The effective time of growth is 5–20 h. The growth is terminated by withdrawing the boules from the melt and the cooling to room temperature at a rate of 40°/h.

■ **Caution.** *Finely crystalline Li_3N is deposited at all inner parts of the equipment which have been cooler than the furnace. When the vessel is opened, the deposited Li_3N can ignite. Burning melt droplets can be formed which drop from the walls of the vessel or seed holder. The finely divided Li_3N also ignites on contact with H_2O.*

■ **Caution.** *The equipment should be opened in a hood since NH_3 gas and Li_2O smog may be formed.*

The equipment is cleaned in a hood with water and then evacuated until a pressure of 10^{-5} torr is reached with the furnace at 600°.

Properties

The shiny red-brown boules are up to 50 mm in length and 35 mm in diameter. When the boules are stored in air, only the surface becomes dark. The boules contain along the long-axis single crystalline grains up to 30 mm in length and up to 15 mm in diameter. Those grains are dark red in transmission. The density is determined to be 1.294 ± 0.006 g/cm^3. Although the impurities in the Li metal are Na $\approx$ 70, Mg $\approx$ 10, and Cu $\approx$ 20 ppm, the main impurities in the Li_3N are Na $\approx$ 50, K $\approx$ 50, Ca $\approx$ 25, Mg $\approx$ 50, and Cu $\approx$ 100 ppm.

References

1. F. Gallais and E. Masdupuy, *C. R. Hebd. Seances Acad. Sci.*, **227**, 635 (1948).
2. B. A. Boukamp and R. A. Huggins, *Phys. Lett.*, **58A**, 231 (1976).
3. U. v. Alpen, A. Rabenau, and G. H. Talat, *Appl. Phys. Lett.*, **30**, 621 (1977).
4. B. A. Boukamp and R. A. Huggins, *Mater. Res. Bull.*, **13**, 23 (1978).
5. S. G. Bishop, P. J. Ring, and P. J. Bray, *J. Chem. Phys.*, **45**, 1525 (1966).
6. U. v. Alpen, *J. Solid State Chem.*, **29**, 379 (1979).
7. A. Rabenau, in *Festkörperprobleme* (*Adv. Solid State Phys.*), Vol. 18, J. Treusch (ed.), Vieweg, Braunschweig, 1978, p. 77.
8. L. Ouvrard, *C. R. Hebd. Seances Acad. Sci.*, **114**, 120 (1892).
9. M. Guntz, *C. R. Hebd. Seances Acad. Sci.*, **123**, 995 (1896).
10. W. Lenz-Steglitz, *Ber. Deut. Pharm. Ges.*, **20**, 227 (1910).
11. E. Zintl and G. Brauer, *Z. Elektrochem.*, **41**, 102 (1935).
12. E. Masdupuy and F. Gallais, *Inorg. Syntheses*, **4**, 1 (1953).
13. B. Neumann, C. Kröger, and H. Haebler, *Z. Anorg. Allgem. Chem.*, **204**, 81 (1932).
14. R. M. Yonco, E. Veleckis, and V. A. Maroni, *J. Nuclear Mater.*, **57**, 317 (1975).
15. S. A. Kutolin and A. I. Vulikh, *Zh. Prikl. Khim*, **41**, 2529 (1968).
16. K. A. Bol'shakov, P. I. Fedevov, and L. A. Stepina, *Izv. Vysshikh. Ucheb. Zaved, Tsve t. Met.*, **2**, 52 (1959).
17. A. Rabenau and H. Schulz, *J. Less Common Metals*, **50**, 155 (1976).
18. H. Schulz and K. Schwarz, *Acta Cryst.*, **A34**, 999 (1978).
19. M. D. Lyutaya and T. S. Bartnitskaya, *Inorg. Mater.*, **6**, 1544 (1970).
20. M. G. Down and R. J. Pulham, *J. Crystal Growth*, **47**, 133 (1979).
21. E. Schönherr, G. Müller, and E. Winckler, *J. Crystal Growth*, **43**, 469 (1978).

10. ALUMINUM NITRIDE

$$(1)\ (Et_3Al)_2 + 2NH_3 \xrightarrow[n\text{-hexane}]{50^\circ} 2[Et_3AlNH_3] \rightarrow$$

$$\tfrac{2}{3}(Et_2AlNH_2)_3 + 2C_2H_6$$

$$(2)\ \tfrac{1}{3}(Et_2AlNH_2)_3 \xrightarrow{145^\circ} (EtAlNH)_n + C_2H_6$$

$$(3)\ (EtAlNH)_n \xrightarrow[NH_3]{\Delta} AlN + C_2H_6$$

Submitted by F. C. SAULS,† L. V. INTERRANTE,* S. N. SHAIKH,* and L. CARPENTER, JR.*
Checked by WAYNE L. GLADFELTER‡

Aluminum nitride is hard, has a high thermal[1] and low electrical conductivity, a wide bandgap, and a coefficient of expansion closely matched to that of silicon. It is transparent throughout the visible and near-infrared regions of the spectrum.[2,3] It is thus an attractive material for a variety of structural, electronic, and optical applications, including substrates for Si-based circuits. Unfortunately, small amounts of impurities and lattice defects seriously degrade these properties.[3-6] A simple preparation of AlN free of metal, oxygen, carbon, and halogen impurities using simple starting materials is therefore important. The ability to control particle size is critical for sintering studies.

Several approaches to this preparation have been previously employed. Direct reaction of aluminum with N_2 or NH_3, carbothermal reduction of Al_2O_3 in a nitrogen atmosphere, and nitrogenation of Al_4C_3 can give products of off-stoichiometry or incorporating carbon and/or oxygen.[2,7-10] Pyrolysis of precursors derived from $AlCl_3$ and amines (or silylamines) leads to chlorine (and silicon) impurities.[11-13] Aluminum amide–imide polymers may also be pyrolyzed to AlN, but solvent or electrolyte residues from the polymer preparation cause carbon or halogen contamination of the AlN produced.[14-18] An aluminum amide–imide polymer free of supporting electrolyte and solvent can be prepared by condensing atomic Al with NH_3, but his approach is not easily scaled up.[15,19] Current methods for preparing

* Department of Chemistry, Rensselaer Polytechnic Institute, Troy, NY 12180.
† Department of Chemistry, King's College, Wilkes-Barre, PA 18711.
‡ Department of Chemistry, University of Minnesota, Minneapolis, MN 55455.

high-purity AlN are based on the pyrolysis of AlF_3 or $(NH_4)AlF_6$ with ammonia.[3,20] This procedure is expensive and gives HF as a byproduct.

The preparation given here is based on chemistry first discovered by Wiberg.[21] It provides a high-purity AlN powder at relatively low materials cost. The mechanism of the process has been studied.[22-27]

General Remarks

■ **Caution.** *$(Et_3Al)_2$ and, to a lesser extent the nitrogen-containing intermediates described below, react extremely vigorously with oxygen and water. The procedure must be carried out in carefully dried glassware purged with purified nitrogen. Reactions 1 and 2 are subject to thermal runaway. It is safer if they are carried out in an inert-atmosphere box.*

$(Et_3Al)_2$ (~93% purity) can be purchased from Aldrich or Texas Alkyls. The principal impurities are organoaluminum species with butyl or hexyl groups in place of some of the ethyl. There is no need to remove these, as the higher alklyaluminum compounds undergo reactions analogous to $(Et_3Al)_2$. *n*-Hexane may be dried by distillation from sodium under nitrogen, using benzophenone as indicator. Gaseous ammonia may be dried by passage through a column filled with KOH; however, the use of anhydrous, high-purity ammonia is recommended to ensure a low oxygen level in the final AlN product. The glovebox used for steps 1 and 2 was equipped with a recirculating heptane system for cooling the condenser, as well as gas inlet and outlet ports.

Procedure

Within the glovebox, the reaction is carried out in a 500-mL, three-necked round-bottomed flask equipped with reflux condenser (gas outlet at the top) and silicone rubber septa. Teflon sleeves are used to avoid silicone grease contamination. A solution of $(Et_3Al)_2$ (85 mL, 0.62 mol Al) in 150 mL of *n*-hexane is placed in the flask and heated in a sandbath to 50°, with continuous stirring. A flow of N_2 (25 mL/min) through a stainless-steel needle is bubbled into the solution to establish a positive flow and prevent back-diffusion of air; ammonia gas (75 mL/min) is then added to this stream. A test of the effluent gas with wet litmus paper shows no ammonia present.

■ **Caution.** *The solution should be warmed before beginning ammonia addition and must remain warm during the reaction, or what we believe to be the unstable Et_3AlNH_3 may accumulate in the flask. The sudden exothermic decomposition of this intermediate, accompanied by ethane evolution, can cause* *an explosion and scatter the highly reactive reaction mixture. Containment of the apparatus in a glovebox is recommended as a safety precaution during the entire Et_3Al loading, handling, and reaction procedure. In addition, this facilitates the recovery of the "EtAlNH" powder eventually obtained without risk of exposure to air or moisture.*

After 12 h, a similar litmus test shows NH_3 in the effluent. The ammonia and nitrogen are shut off. The resulting solution of $(Et_2AlNH_2)_3$ is then cooled and the *n*-hexane removed under reduced pressure at ambient temperature.

■ **Caution.** *Heating concentrated solutions containing* $(Et_2AlNH_2)_3$ *during solvent removal can result in runaway exothermic decomposition (see below).*

The liquid $(Et_2AlNH_2)_3$ is transferred to a Schlenk tube fitted with a 19-cm air condenser. The tube is gradually heated to 145° so that a slow evolution of ethane gas occurs, then held at 145° for 12 h, yielding a solid of approximate composition $(EtAlNH)_n$.

■ **Caution.** *This reaction is exothermic and evolves a gas. Heating too rapidly can lead to runaway decomposition. Heating beyond 150° is not recommended.*

In the glovebox, the white chunky solid is pulverized with a spatula and transferred to a molybdenum or tungsten boat. This is then placed in a fused-silica tube equipped with inlet and outlet valves (Fig. 1). The tube is inserted into a furnace and heated under flowing ammonia (10 mL/min) to 1000° over a period of 6 h, held at this temperature 8 h, then cooled. Overall yield is 17.8 g (70%).

Anal. The structure of $(Et_2AlNH_2)_3$ was confirmed by mass-spectroscopic and ^1H NMR and ^{13}C NMR analyses. Reproducible elemental analyses were not obtained because of decomposition during analysis. $(EtAlNH)_n$: % found (calculated): C = 30.84 (33.8); H = 7.80 (8.51); N = 17.31 (19.71); Al = 38.46 (37.97). AlN: Al = 65.48 (65.82); N = 34.03 (34.17); C = 0.06; O < 0.3; Si = 20 ppm; Cu = 0.5 ppm; other metals (Fe, Cr, Ni, Mg, Sn, W, B, Sc, Ti, V, Mn, Ca, Li, Be) less than the limit of detection by spark emission analysis.

Properties

The white to slightly gray powder is shown to be AlN by X-ray powder diffraction, and has broad IR absorption at 700 cm^{-1}, in agreement with the

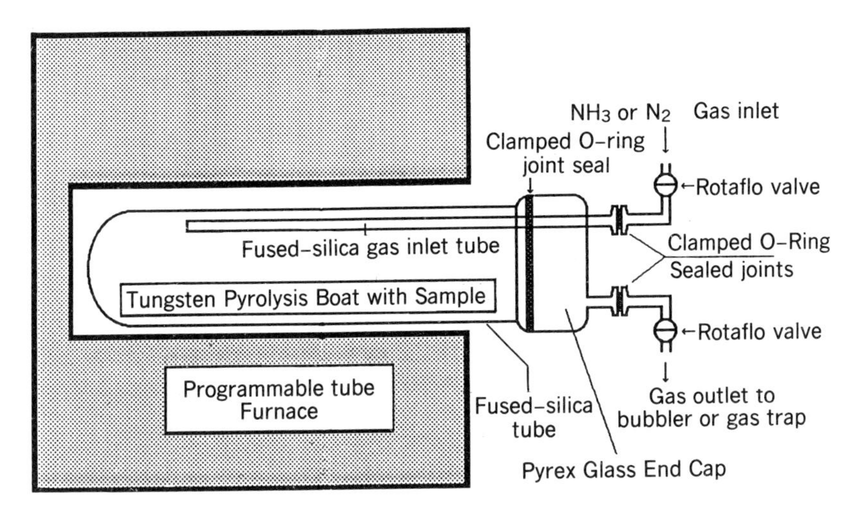

Figure 1. Apparatus used to pyrolyze the $(EtAlNH)_n$ precursor to AlN.

literature.[28,29] It has a high surface area [BET, 40–250 m^2/g] and picks up oxygen (due to surface hydrolysis) readily; after 24 h of atmospheric exposure the powder contains $> 5\%$ oxygen. The crystallinity may be increased and grains coarsened by heating in a nitrogen atmosphere, which reduces the sensitivity to the atmosphere. After 16 h at 1605°, the surface area is reduced to 10 m^2/g; 24-h exposure of this material to the atmosphere gives 0.54% oxygen in the powder. The unexposed product may be hot-pressed [1740°, 4400 psi (lb/in^2)] to yield translucent pellets.

References

1. G. A. Slack, R. A. Tanzilli, R. O. Pohl, and J. W. Vandersande, *J. Phys. Chem. Solids*, **48**, 641 (1987).
2. N. Kuramoto and H. Taniguchi, *J. Mats. Sci. Lett.*, **3**, 471 (1984).
3. G. A. Slack and T. F. McNelly, *J. Cryst. Growth*, **34**, 263 (1976).
4. G. A. Slack, *J. Cryst. Growth*, **42**, 560 (1977).
5. G. A. Slack, *J. Phys. Chem. Solids*, **34**, 321 (1973).
6. G. A. Slack and S. F. Bartram, *J. Appl. Phys.*, **46**, 39 (1975).
7. A. Rabineau, in *Compound Semiconductors*, Vol. 1, R. K. Willardson and H. L. Goering (eds.), Reinhold, New York, 1962, pp. 174–176.
8. *Gmelin's Handbuch der anorganischen Chemie*, No. 35, Verlag Chemie GmbH., Berlin, 1993.
9. G. Brauer (ed.), *Handbook of Preparative Inorganic Chemistry*, Vol. 1, 2nd ed., Academic Press, London, 1962.
10. T. Maeda and K. Harada, Sumitomo Chemical Co. Ltd., Japan Kokai 78 68,700 19 June 1978, appl. 76/145,137 01 Dec. 1976 (CA 89:165623f).
11. D. C. Lewis, *J. Electrochem. Soc.*, **117**, 978 (1970).

12. R. Riedel, G. Petzow, and U. Klingiebel, *J. Mater. Sci. Lett.*, **9**, 222 (1990).
13. D. M. Schleich, U.S. Patent 4,767,607 (1978).
14. E. Wiberg and A. May, *Z. Naturforsch. B.*, **10**, 229 (1955).
15. L. Maya, *Adv. Cer. Mat.*, **1**, 150 (1986).
16. M. Sebold and C. Rüssel, *Mater. Res. Soc. Symp. Proc.*, **121**, 477 (1988).
17. M. Sebold and C. Rüssel, *J. Am. Ceram. Soc.*, **72**, 1503 (1989).
18. A. Oichi, H. K. Bowen, and W. E. Rhine, *Mater. Res. Soc. Symp. Proc.*, **121**, 663 (1988).
19. P. S. Skell and L. R. Wolf, *J. Am. Chem. Soc.*, **94**, 7919 (1972).
20. I. C. Huseby, *J. Am. Ceram. Soc.*, **66**, 217 (1983).
21. Work of E. Wiberg, reported in G. Bahr, Inorganic Chemistry; W. Klemm (ed.), *FIAT Review of German Science 1939–1946*, Vol. 24; Dieterichsche Verlagsbuchhandlung, Wiesbaden, 1948; Part 2, p. 155.
22. H. M. Manasevit, F. M. Erdman, and W. I. Simpson, *J. Electrochem. Soc.*, **118**, 1864 (1971).
23. L. V. Interrante, L. Carpenter Jr., C. Whitmarsh, W. Lee, M. Garbauskas, and G. A. Slack, *Mater. Res. Soc. Symp. Proc.*, **73**, 359 (1986).
24. F. C. Sauls, L. V. Interrante, and Z. Jiang, *Inorg. Chem.*, **29**, 2989 (1990).
25. Z. Jiang and L. V. Interrante, *Chem. Mater.*, **2**, 439 (1990).
26. W. Hurley Jr., F. C. Sauls, P. Marchetti, L. V. Interrante, and Maciel, G. E. in preparation.
27. M. Cohen, J. D. Gilbert, and J. D. Smith, *J. Chem. Soc. Dalton*, 1092 (1965).
28. M. Hoch, T. Vernadakis, and K. M. Nair, *Sci. Ceram.*, **10**, 227 (1980).
29. G. K. Makarenko, D. P. Zyatkevich, and K. I. Arsenin, *Inorg. Mater.*, **15**, 535 (1979).

11. ELECTROCHEMICAL SYNTHESES OF BINARY SILVER OXIDES

Submitted by PETER FISCHER* and MARTIN JANSEN*
Checked by S. M. ZAHURAK†

Anodic oxidation of aqueous silver(I) solutions is a convenient and efficient route for the syntheses of highly oxidized binary silver oxides in an extremely pure and at the same time well-defined crystalline state. Applying this technique the previously unknown oxides Ag_2O_3[1,2] and Ag_3O_4[3,4] are obtained as single crystals with metallic luster. The maximum size achieved is 2 mm. Silver oxide, as prepared by conventional chemical oxidation, is a nonstoichiometric black power. The procedure presented here yields macroscopic crystals of this oxide with stoichiometric composition.[5]. All silver oxides with silver in an oxidation state >1 are of some technical interest as components for the design of primary power cells. The preparation takes one day per charge, in general.

* Institut für Anorganische Chemie, Universität Bonn, 53121 Bonn, Gerhard-Domagh-Strasse 1, Germany.

† AT&T Bell Laboratories, 600 Mountain Ave., Murray Hill, NJ 07974.

■ **Caution.** *During electrolysis the pH of the electrolyte decreases. So, if the electrolyte used is based on fluorides, free hydrofluoric acid will form.*

Procedure

The electrolytic syntheses of the oxides Ag_2O_3 and Ag_3O_4 are carried out in a simple beaker containing the electrolyte. Only for the synthesis of AgO a closed cell is required, because in the latter case electrolysis has to be performed at a temperature of about 100°, which is close to the boiling point of the electrolyte.

The cathode material can be either silver or platinum, whereas for the anode platinum is the best material because of its stability against corrosion and high overpotential with respect to oxygen evolution. It should be mentioned that the state of the surface of the platinum anode has some effect on the deposition of the respective product; for instance, a polished surface will cause a very irregular deposition (absence of crystallization nuclei). Applying time-dependent instead of constant current offers the advantage of adjusting the current to the increasing surface of the growing crystals. In this way a nearly constant current density is maintained during the experiment, which is crucial for the formation of single-phase product.

The stock solution is prepared as follows. A solution containing 1.00 ± 0.05 mol/L Ag^+ is prepared by dissolving silver perchlorate monohydrate (225.3 g, 1.0 mol) in one liter of doubly distilled water. The solid salt and the aqueous solution are both stable and easy to handle. Alternatively, the stock solution may be prepared by adding an excess of silver(I) carbonate to aqueous perchloric acid. Silver hexafluorophosphate or tetrafluoroborate, which may also be used for the synthesis of Ag_2O_3, are difficult to prepare in sufficient purity.

■ **Caution.** *Avoid contact of any perchlorates with organic material. The silver oxides prepared here are highly oxidizing, and any contact with organic material should be avoided.*

The products are isolated as follows. After electrolysis the anode covered with product crystals is removed cautiously together with the cap of the cell. Both are transferred to a 20-mL test tube filled with cold, doubly distilled water. The crystals are removed mechanically from the electrode and sampled using a small filter crucible. After washing with cold, doubly distilled water the crystals are dried in vacuum (– 1 mbar) for at least 10 min, and, depending on the total mass of the product, up to several hours.

For collecting and drying the samples we have used a special apparatus as shown in Fig. 1. The crystals can be transferred through the open ground

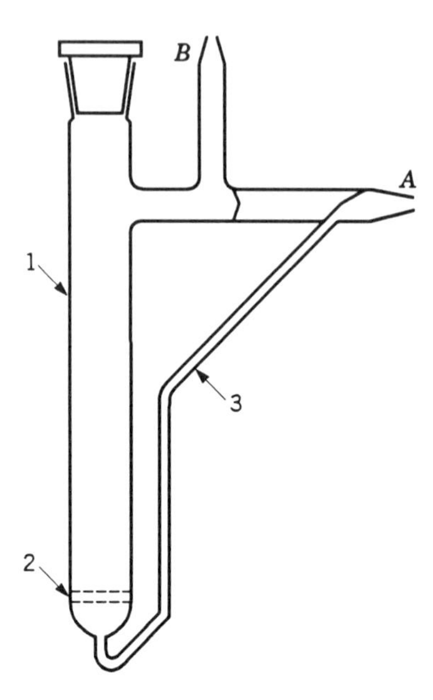

Figure 1. Apparatus for washing and drying silver oxides: (1) glass tube (height = 150 mm, diameter = 13 mm); (2) glass filter; (3) glass capillary diameter = 1.5 nm).

joint onto the glass filter of the apparatus, which is immersed in an icebath. To remove the distilled water used for washing, opening *A* is connected to a water-jet pump. For drying, opening *B* is connected via a cold trap to an oil pump, while opening *A* is joined via a throttle valve (may be a glass capillary) to a source of dry oxygen. When pumping, a fluid bed is formed, and drying is finished in $\frac{1}{2}$ h.

Scale-up of the procedures is possible by increasing the volume of the electrolyte, while maintaining *current density* and the time of electrolysis. Consequently, a larger volume of electrolyte will require an anode with a large surface.

A. Ag_2O_3

$$2Ag^+ + 3H_2O = Ag_2O_3 + 6H^+ + 4e^-$$

$$4Ag^+ + 4e^- = 4Ag$$

The electrolyte is prepared as follows. Sodium perchlorate monohydrate (25 g, 0.178 mol) is added to 10 mL of the 1.00*M* silver perchlorate stock solution and made up to 100 mL.

The electrolysis cell consists of a 100-mL beaker made of glass or polyethylene, a cylindrical net of platinum [e.g., 20 mesh, 0.2 mm diameter); cylinder: height = 70 mm, diameter = 50 mm] as cathode, and a platinum wire with 0.5 mm diameter and 15 mm length as anode, with the lower end wound to a spiral (Fig. 2). The cathode may also consist of silver foil with the negligible disadvantage of lower mechanical stability.

The silver perchlorate solution is transferred into the cell and cooled down to -2°. The temperature can be maintained by means of ice rock/salt mixtures; however, it is more convenient to use a thermobeaker connected to a cryostat.

A current-stabilized power supply (0–5 V; 0–100 mA) is connected to the electrodes. The current is raised from 10 to 30 mA in the course of a linear sweep or may be raised in five steps. The time of electrolysis is 120 min, corresponding to a total charge transfer of 40 mA · h. After electrolysis, the silver content of the electrolyte will have decreased by 20% and the removed Ag^+ cations will have been replaced by protons. Increasing the electrolysis time is not recommended for the deposited Ag_2O_3 will dissolve as the acid concentration becomes too high. The yield of Ag_2O_3 is 50 mg per charge.

Purity of the product is monitored by X-ray powder techniques. The strongest lines ($I > 20\%$, $d > 1.6$ Å) are d = 3.341, 2.743, 2.692, 2.623, 2.482, 2.179, 2.065, 1.7619 and 1.6761 Å.

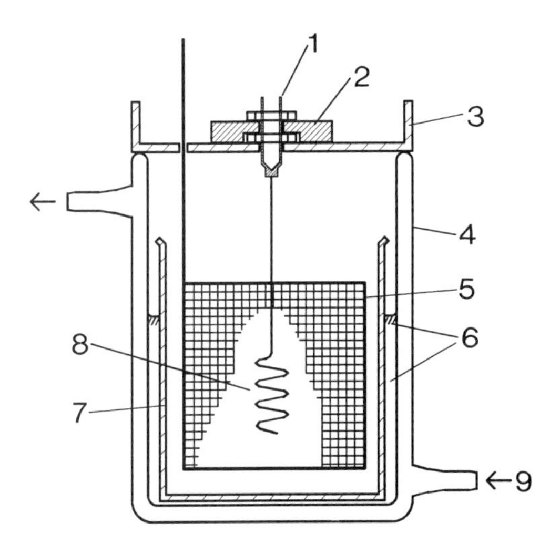

Figure 2. Thermostatted electrolysis cell. (1) steel connector; (2) anode support (PMMA); (3) glass cap; (4) thermobeaker; (5) platinum net cathode; (6) 2*M* sulfuric acid; (7) beaker (polyethylene); (8) platinum anode; (9) cooling liquid inlet.

B. Ag_3O_4

$$3Ag^+ + 4H_2O = Ag_3O_4 + 8H^+ + 5e^-$$

$$5Ag^+ + 5e^- = 5Ag$$

Potassium fluoride (12 g, 0.316 mol) is dissolved in 100 mL of doubly distilled water. Of this solution, 30 mL is combined with 10 mL of the 1.00*M* silver perchlorate stock solution, and potassium perchlorate precipitates. The precipitate is separated by filtration through a small sintered crucible and washed repeatedly with the rest of the KF solution until the total volume equals 100 mL. The same electrolysis cell as for the preparation of Ag_2O_3 (part A) is used. The cell is filled with the silver fluoride solution and subsequently cooled to $-2°$ (cf. part A). During electrolysis the current is raised from 10 to 40 mA within 120 min in a linear sweep or in five steps at minimum. Then the current is kept constant at 40 mA for another 120 min. The total charge transferred is 130 mA · h; thus, 80% of the Ag^+ cations will have been exchanged by protons. Typical yields are 380 mg per charge.

Purity of the product is monitored by X-ray powder techniques. The strongest lines ($I > 30\%$, $d > 1.6$ Å) are $d = 3.220$, 3.156, 2.675, 2.503, 2.414, 2.264, 2.198, and 1.5640 Å.

C. AgO

$$Ag^+ + H_2O = AgO + 2H^+ + 2e^-$$

$$2Ag^+ + 2e^- = 2Ag$$

Potassium fluoride (6 g, 0.158 mol) is dissolved in 100 mL of doubly distilled water. Of this solution, 30 mL is combined with 10 mL of the 1.00*M* silver perchlorate stock solution, and potassium perchlorate precipitates. The precipitate is separated by filtration through a small sintered crucible and washed repeatedly with the rest of the KF solution until the total volume equals 100 mL. Adding 200 mL doubly distilled water gives the electrolyte to be used for the preparation of AgO.

The electrolysis cell, a cylindrical box made of PTFE with a screw cap, uses a silver foil cathode and as anode a platinum wire (0.3 mm of diameter of 20 mm of length), soldered to a steel connector (see Fig. 3).

The cell is filled with 100 mL of the electrolyte, closed with the screw cap, and placed below a large dewar vessel. The cell is then heated by steam guided through a tube into the dewar. During electrolysis the current is swept from 5 to 10 mA within 90 min and kept constant at 10 mA for 120 min. The total charge transferred is 31.3 mA · h. The dewar is removed and the cell is cautiously opened.

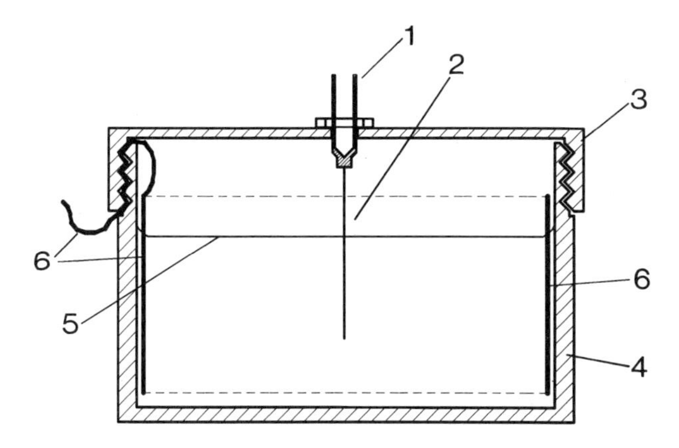

Figure 3. PTFE cell for synthesis of AgO: (1) steel connector; (2) platinum anode; (3) PTFE screw cap; (4) PTFE beaker; (5) electrolyte surface; (6) silver foil (0.1 mm) as cathode.

■ **Caution.** *The solution contains HF following the reaction. Rubber gloves should be worn, and splashing should be avoided.*

The yield is about 100 mg of metal-like crystals per charge. Purity of the product is monitored by X-ray powder technics. The strongest lines ($I > 20\%$, $d > 1.6$ Å) are $d = 2.794$, 2.771, 2.416, 2.286, 1.7016 and 1.6221 Å.

Properties

All silver oxides, AgO_x, $x \geq 1$, are very strong oxidants. They are soluble in cold nitric acid yielding dark brown solutions with the Ag^{2+} ion complexed by NO_3^-. These solutions will even oxidize saturated hydrocarbons. Both Ag_2O_3 and Ag_3O_4 decompose rapidly at temperatures of 50° and 60°, respectively. Even at room temperature, decomposition to AgO within a couple of hours is observed. Well-crystallized AgO, as obtained by electrocrystallization, is more stable and decomposes only above 160°.

References

1. B. Standke and M. Jansen, *Angew. Chem.*, **97**, 114 (1985); *Int. Ed. Engl.*, **24**, 118 (1985).
2. B. Standke and M. Jansen, *Z. Anorg. Allg. Chem.*, **535**, 39 (1986).
3. B. Standke and M. Jansen, *Angew. Chem.*, **98**, 78 (1986); *Int. Ed. Engl.*, **25**, 77 (1986).
4. B. Standke and M. Jansen, *J. Solid State Chem.*, **67**, 278 (1987).
5. M. Jansen and P. Fischer, *J. Less Common Metals*, **137**, 123 (1988).

12. TRISODIUM HEPTAPHOSPHIDE AND TRISODIUM UNDECAPHOSPHIDE

$$3Na(l) + \frac{7}{4} P_4(g) \rightarrow Na_3P_7(s)$$

$$3Na(l) + \frac{11}{4} P_4(g) \rightarrow Na_3P_{11}(s)$$

$$Na_3P_{11}(s) \rightleftarrows Na_3P_7(s) + P_4(g)$$

Submitted by WOLFGANG HÖNLE and HANS GEORG VON SCHNERING*
Checked by D. W. MURPHY†

The compounds Na_3P_7 and Na_3P_{11} are Zintl phases,[1] consisting of alkali-metal cations and polycyclic polyanions P_7^{3-} and P_{11}^{3-}. These anions are suitable synthons for the preparation of compounds such as P_7R_3 and $P_{11}R_3$ (R = $SiMe_3$ and similar). Besides Li_3P_7 the sodium compounds have been characterized to some extent.[2] Both compounds undergo a first-order phase transition from the crystalline low-temperature α type to the plastically crystalline high-temperature β type. Their synthesis can be performed in either solidex glass ampules or welded niobium containers heated in fused-silica jackets under vacuum.

Procedure

Standard vacuum line and inert-gas (argon) supply must be used for all manipulations, as adducts and products are sensitive to oxygen and moisture. A dry box is very helpful and expedites the synthesis. The inert gas must be free of moisture ($H_2O < 1$ ppm) and oxygen ($O_2 < 1$ ppm). This is achieved by P_2O_5 (e.g., Sicapent as final stage) and CuO (BTS–catalyst). The dry box normally operates at concentrations half of the values given above.

■ **Caution.** *All operations should be done in a hood or dry box. When working with glassware under vacuum, safety goggles must be worn. Fire-extinguishing dry sand and fire-resistant gloves should be on hand to deal with the possibility of a phosphorus fire. White phosphorus may ignite if milled mechanically under ambient conditions, and even the highly purified red phosphorous may contain small inclusions of white phosphorus. Therefore, all grinding of phosphorus should be done under inert atmosphere. All products synthesized here must be handled in inert atmosphere. Contact with wet air or*

* Max-Planck-Institut für Festkörperforschung, Heisenbergstr. 1, 70569 Stuttgart, Germany.
† AT&T Bell Laboratories, Murray Hill, NJ, 08812.

wet inert gas produces a phosphane odor, and the self-ignitable diphosphane P_2H_4 may be formed under some conditions.

Chemical Disposal. Small amounts of phosphides may be transferred into $CuSO_4$ solution in small portions under stirring. Alternatively, the material may be oxidized with Br_2/alcohol solution (**Caution:** *Splashes*!) An odor-free method of oxidation involves the use of a small autoclave, filled with HNO_3 (compare with the digestion of phosphides for ICP analysis[9]).

A. SYNTHESIS OF α-Na_3P_7 AND α-Na_3P_{11} IN SOLIDEX GLASS AMPULES

The α forms are the thermodynamically stable forms, which occur under the conditions given here. The β forms are metastable at normal conditions and can be obtained by heating.

Sodium is cut under paraffin (as supplied) into pea-size pieces, thus removing the oxide layers. After washing with absolute diethylether to remove the paraffin, the pieces are kept in a storage Schlenk container under inert gas (Fig. 1*a*).

Phosphorus is used in electronic-grade purity in the red, amorphous form. The pieces should be ground in an agate mortar under inert gas down to ~1 mm in size. It is also kept in a storage Schlenk container.

Borosilicate glass 3.3 ampules (Fig. 1*b*) can be used up to 550°. A glass ampule tube (15 mm in diameter, wall thickness 1–2 mm) is heated under vacuum (oil pump) with the yellow flame of a gas burner. After cooling, a suitable amount of sodium is transferred under a counter current inert-gas stream.[3]

For the preparation in glass, the double-tube arrangement shown in Fig. 1*d* is used, and an excess of ~5% sodium is used to counterbalance the reaction of sodium with the glass wall. After sealing this inner ampule under atmospheric inert-gas pressure, it is rigorously shaken to provide an intimate mixture of P and Na. The inner ampule serves as reaction ampule, thus preventing the corrosion of the outer ampule. The outer ampule is also sealed under atmospheric pressure, in order to counterbalance the vapor pressure of phosphorus (~10 atm at 500°). Typical batch sizes are a maximum of 4.5 g Na (19.6 mmol) and 13.45 g P_{red} (43.5 mmol) in an inner reaction ampule (15 mm diameter, 20 cm in length). To facilitate the kinetically hindered vaporization of phosphorus,[11] a trace of iodine or sulfur is added directly in the reaction ampule.

■ **Caution.** *As the pressure stability of the ampules depends not only on the type of glass itself but also on the quality of the seal, we recommend starting*

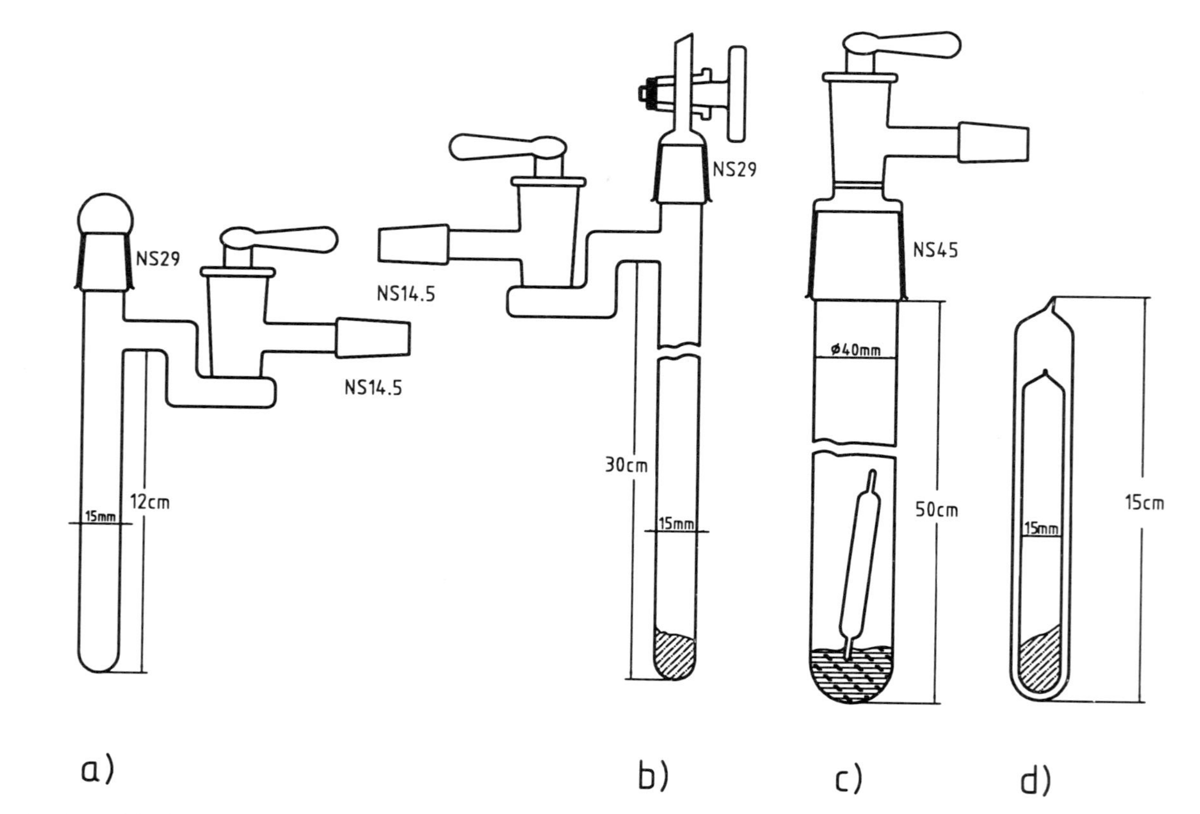

Figure 1. (*a*) Schlenk storage container; (*b*) arrangement for the sealing of Solidex ampules; (*c*) fused-silica jacket for heating Nb ampules (note the fused-silica wool as support); (*d*) solidex ampule with inner reaction ampule.

with smaller batch sizes (~25% of the values given above) and increasing the scale stepwise. Only ampules cooled to room temperatures should be shaken; otherwise the reaction between P and Na will start at the hot glass wall!

Heating. The heating is performed in vertically positioned tubular furnaces. The given temperature is valid for the central part of the ampule. Over 3 days the temperature is increased from room temperature to 500°. After keeping the ampule at 500° for 4 days, it is cooled over 4 days. The ampule is opened under inert (e.g., dry-box) conditions and the entire product is ground with a mortar and pestle. The second heating can be performed in a single glass ampule without a glass jacket. Excess P (occurring either when more than the excess sodium has reacted with the wall or with the decomposition of Na_3P_{11}) may be sublimed from the product by using a longer ampule (20 cm in length). Excess P sublimes to the cold end, mostly in the form of white P. A longer heating time transforms white P into a red form. (**Caution.** *See disposal instructions noted earlier.*)

Purification of α-Na_3P_7. α-Na_3P_7 can be sublimed at 550° and 5×10^{-6} torr (high-vacuum turbomolecular pump). It undergoes a congruent, dissociative sublimation.[4] Using water as cooling agent in the cold finger, α-Na_3P_7 is condensed as fine yellow crystalline needles. Applying liquid nitrogen in the cold finger, the phosphide condenses as a black amorphous powder.

This black powder has the analytical composition Na_3P_7 and transforms under heating into yellow crystalline α-Na_3P_7.

Preparation of α-Na_3P_{11}. α-Na_3P_{11} can either be synthesized by the reaction of α-Na_3P_7 and P_4 or directly from the elements. The synthesis from the elements follows the same route as indicated for α-Na_3P_7. However, because of the decomposition, α-Na_3P_{11} cannot be further purified by sublimation. Therefore, special care must be taken to exclude oxygen and moisture.

B. SYNTHESIS OF α-Na_3P_{11} FROM α-Na_3P_7 AND P_4

For this reaction white phosphorus is used. P_4 is distilled directly into the reaction ampule and then the necessary amount of α-Na_3P_7 is added. There is no need for excess sodium, as the reaction of the ampule with elemental sodium does not occur at temperatures up to 480°. The heating is done in 36 h up to 480°, keeping the ampule for 4 days at that temperature and then cooling within 1 day to room temperature.

C. PREPARATION IN NIOBIUM TUBES

The preparation in niobium tubes has the following advantages:

- Nearly no wall reaction of sodium in the initial reaction step
- Easy crimping and welding
- Simple opening after use with a spin cutter (same as used for cutting copper pipes)

The niobium tube is crimped and welded at one side[4] (see Fig. 2*a*, Ref. 5), then cleaned.[5] After filling with sodium and red amorphous phosphorus under inert conditions, it is crimped and welded at the other side.

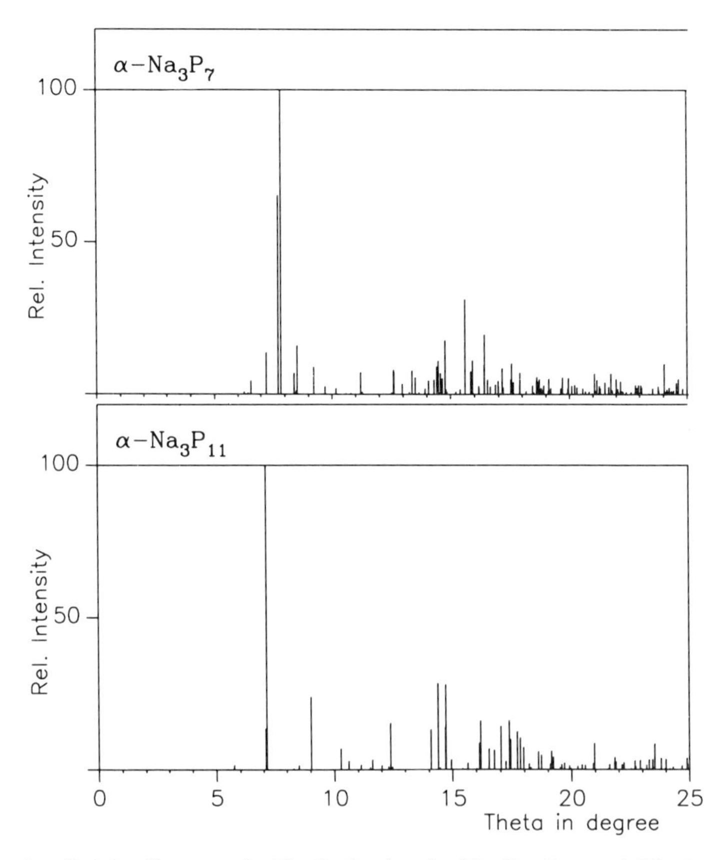

Figure 2. Guinier diagram of α-Na_3P_7 (top) and α-Na_3P_{11} (bottom). The intensities have been calculated[10] for the Guinier geometry and $\lambda(CuK_\alpha) = 154.056$ pm.

TABLE I. Values of *h*, *k*, *l*, and *d* (pm) and Relative Intensity for α-Na_3P_7 (left) and α-Na_3P_{11} (right) down to *d* > 250 pm and *d* > 200 pm, respectively

α-Na_3P_7					α-Na_3P_{11}				
h	*k*	*l*	*d*	*I*	*h*	*k*	*l*	*d*	*I*
0	0	2	673.7	4	0	2	0	623.3	13
2	0	1	613.0	14	1	1	1	620.0	100
2	1	0	574.4	71	0	0	2	491.4	24
0	1	2	565.8	100	2	1	1	431.5	7
2	1	1	528.4	7	1	1	2	418.6	3
0	2	0	521.1	16	0	3	1	382.7	3
2	0	2	481.5	9	1	3	1	359.2	15
1	2	1	458.3	2	3	1	1	316.5	13
2	2	1	397.0	7	2	2	2	310.0	28
2	1	3	353.8	10	1	3	2	303.5	14
2	2	2	353.6	7	3	2	0	303.2	9
3	2	0	344.4	3	1	1	3	303.1	19
3	2	1	333.7	5	1	4	0	298.6	3
4	0	1	333.5	4	1	4	1	285.7	2
1	2	3	330.3	5	2	0	3	277.3	9
4	1	1	317.6	5	3	1	2	276.4	16
1	1	4	312.2	5	2	1	3	270.6	7
2	3	0	310.1	10	2	4	0	267.4	6
0	3	2	308.8	12	0	4	2	263.2	14
3	2	2	306.7	9	4	0	0	260.2	3
2	2	3	305.0	6	3	2	2	258.0	4
2	0	4	302.6	19	2	4	1	258.0	12
4	2	0	287.2	34	0	3	3	257.3	10
0	2	4	282.9	8	2	2	3	253.3	12
2	3	2	281.7	12	4	0	1	251.6	10
1	2	4	2771.1	3	1	3	3	249.8	7
3	3	0	277.0	2	0	5	1	241.7	6
3	2	3	273.3	11	4	2	0	240.2	5
4	0	3	273.2	22	2	4	2	234.9	6
3	3	1	271.3	6	3	3	2	234.2	3
1	3	3	269.5	3	3	1	3	234.0	4
5	1	0	266.2	4	0	2	4	228.6	2
4	1	3	264.3	5	4	3	1	215.2	9
4	2	2	264.2	3	1	3	4	207.3	3
2	2	4	261.7	10	2	4	3	207.1	4
5	1	1	261.2	3	3	3	3	206.7	2
1	1	5	256.4	5					
3	3	2	256.2	11					
1	4	0	256.0	2					
2	3	3	255.2	5					
1	4	1	251.5	3					

The heating is performed in an evacuated fused-silica jacket (Fig. 1*c*). Typical heating temperatures are 50 h up to 600°, holding 2 days at that temperature, cooling within 1 day to room temperature.

Because of the less pronounced reaction of sodium with the niobium tube, no excess sodium is necessary. Grinding of the initial reaction product and a second reaction helps to achieve complete reaction.

Properties

α-Na_2P_7 occurs as a bright yellow [$E_{opt} = 3.4(1)$ eV] crystalline powder, α-Na_3P_{11} as orange-red powder. Both can be identified either by their characteristic X-ray powder pattern (Table I and Fig. 2). Further identification is possible via the Raman spectrum[6] of α-Na_3P_7 and the phase transition as monitored by DTA.[2,7] Raman spectra[8] of α-Na_3P_{11} have to be taken with care, because intensive light power leads to decomposition into Na_3P_7 and P_4.

The given experimental conditions have been tested during several years. Nevertheless, sometimes byproducts have been identified by X-ray crystallography. Typical byproducts are SiP_2, due to the reduction of SiO_2 by sodium and subsequent reaction of Si with P to SiP_2. Because oxygen is present, the formation of Na_3PO_4 is observed. NaP and NaP_7 occur sometimes as coexisting phases when the composition is not properly fixed, as, for example, if dirty sodium or phosphorus are used. If the temperature during the preparation in niobium tubes is raised too high, the formation of NbP can be observed.

References

1. H. G. von Schnering, *Angew. Chem.*, **93**, 44 (1981); *Angew. Chem. Int. Ed. Engl.*, **20**, 33 (1981).
2. W. Wichelhaus and H. G. von Schnering, *Naturwissenschaften*, **60**, 104 (1973); H. G. von Schnering, W., Hönle, V. Manriquez, Th. Meyer, Ch. Mensing, and W. Giering, *Proceedings of the Second European Conference on Solid State Chemistry.*, Veldhoven, The Netherlands, June 1982, *Studies in Inorganic Chemistry*, Vol. 3, Elsevier, Amsterdam (1983); V. Manriquez, W. Hönle, and H. G. von Schnering, *Z. Anorg. Allg. Chemie*, **539**, 95 (1986); W. Hönle, V. Manriquez, and H. G. von Schnering, *Z. Anorg. Allg. Chemie* (submitted).
3. For the transfer of alkali metals under inert conditions, see, e.g. (a) A. Simon, in *Handbuch der Präparativen Anorganischen Chemie*, Vol. 3, G. Brauer (ed.), Auflage, Ferdinand Enke Verlag, Stuttgart, 1978, Chapter 17; (b) A. Simon, W. Brämer, B. Hillenkötter, and H.-J. Kullman, *Z. Anorg. Allg. Chemie*, **419**, 253 (1976).
4. R. P. Santandrea, C. Mensing, and H. G. von Schnering, *Thermochim. Acta*, **98**, 301 (1986).
5. J. D. Corbett, *Inorg. Synth.*, **22**, 15 (1983).
6. W. Henkel, K. Strössner, H. D. Hochheimer, W. Hönle, and H. G. von Schnering, *Proceedings of the 8th International Conference of Raman Spectroscopy*, France, Sept. 1982, Wiley, New York, 1982 p. 459.

7. K. Tenschev, E. Gmelin, and W. Hönle, *Thermochim. Acta*, **85**, 151 (1985).
8. H. G. von Schnering, M. Somer, G. Kliche, W. Hönle, T. Meyer, J. Wolf, L. Ohse, and P. B. Kempa, *Z. Anorg. Allg. Chemie*, **601**, 13 (1991).
9. O. Buresch, W. Hönle, and H. G. von Schnering, *Fresenius Z. Anal. Chem.*, **325**, 607 (1986).
10. K. Yvon, W. Jeitschko, and E. Parthé, *J. Appl. Crystallogr.*, **10**, 73 (1977).
11. H. Schäfer and M. Trenkel, *Z. Anorg. Allg. Chemie*, **391**, 11 (1972).

Chapter Two

TERNARY COMPOUNDS

13. LEAD RUTHENIUM OXIDE, $Pb_2[Ru_{2-x}Pb_x^{4+}]O_{6.5}$

$$(2 + x)\,Pb(NO_3)_2 + (2 - x)\,Ru(NO_3)_3 + (3/4 + x/4)\,O_2 + \left(\frac{10 - x}{2}\right)H_2O$$

$$\xrightarrow[75^\circ]{pH\ 13} Pb_2[Ru_{2-x}Pb_x^{4+}]O_{6.5} + (10 - x)HNO_3$$

Submitted by H. S. HOROWITZ,* J. M. LONGO,* and J. T. LEWANDOWSKI*
Checked by J. MURPHY†

Reprinted from *Inorg. Synth.*, **22**, 69 (1983)

High-electronic-conductivity oxides with the pyrochlore structure have generated significant interest for both scientific and technological reasons. The existence of the pyrochlores $Pb_2Ru_2O_{7-y}$ and $Pb_2Ir_2O_{7-y}$ $(0 \leqslant y \leqslant 1)$ was first suggested by Randall and Ward.[1] These two compounds were later isolated and more fully characterized by Longo et al.[2] Sleight[3] has reported the synthesis of $Pb_2Ru_2O_{6.5}$ at elevated pressures. Bouchard and Gillson[4] first prepared the pyrochlores $Bi_2Ru_2O_7$ and $Bi_2Ir_2O_7$.

All the cited literature references to the above compounds have described solid-state syntheses at temperatures of 700–1200°. Such synthesis conditions

* Corporate Research, Exxon Research and Engineering Company, Linden, NJ 07036.
† Department of Chemistry, Oklahoma State University, Stillwater, OK 74078.

will always lead to pyrochlore structure compounds in which all the octahedrally coordinated sites are occupied by the noble metal cation, thus requiring the posttransition metal to noble-metal molar ratio always to be 1.0. This chapter focuses on solution medium syntheses at quite low temperatures ($\leqslant 75°$), thereby stabilizing a new class of pyrochlore compounds in which a variable fraction of the octahedrally coordinated sites are occupied by posttransition-element cations.[5,6] The specific example here involves the $Pb_2[Ru_{2-x}Pb_x^{4-}]O_{6.5}$ series. The synthesis conditions may be simply adapted, however, to accommodate preparation of a wider range of pyrochlores that can be described by the formula $A_2[B_{2-x}A_x]O_{7-y}$ where A is typically Pb or Bi, B is typically Ru or Ir and $0 \leqslant x \leqslant 1$, and $0 \leqslant y \leqslant 1$.

The synthesis method involves reacting the appropriate metal ions to yield a pyrochlore oxide by precipitation and subsequent crystallization of the precipitate in a liquid alkaline medium in the presence of oxygen.[7] The alkaline solution serves both as a precipitating agent and as a reaction medium for crystallizing the pyrochlore, thus eliminating the need for subsequent heat treatment.

Procedure

A pyrochlore of the approximate composition $Pb_2[Ru_{1.33}Pb_{0.67}^{4+}]O_{6.5}$ may be synthesized by first preparing an aqueous solution source of lead and ruthenium cations in a 2:1 lead-to-ruthenium ratio as follows: 5.0 g of commercially available* $Ru(NO_3)_3$ solution, 10% (wt.) ruthenium metal (0.005 mol ruthenium metal) is diluted with 25 mL of distilled water. Reagent-grade $Pb(NO_3)_2$ (3.277 g, 0.010 mol) is dissolved in 100 mL of distilled water and added to the ruthenium aqueous solution. This aqueous solution of lead and ruthenium is then stirred for approximately 10 min.

The aqueous solution of lead and ruthenium is then added with stirring to approximately 400 mL of 0.3*M* KOH contained in a polyolefin beaker (600-mL capacity). A black precipitate immediately forms. The pH and volume of this reaction medium are then adjusted to 13.0 and 600 mL, respectively, by adding appropriate amounts of KOH and distilled water. The reaction medium is kept agitated with a magnetic stirring bar for the duration of this synthesis. The beaker is then heated so that the reaction medium attains a temperature of 75°, and oxygen is bubbled though it at ~1.0 SCFH for the duration of the synthesis by means of a plastic gas bubbling tube beneath the surface of the liquid. The exact flow rate of oxygen is not considered critical; it is desirable merely to maintain a positive flow of oxygen through the system. The most important requirement of this synthesis

* Engelhard Industries, 429 Delancy Street, Newark, NJ.

procedure is that the reaction medium be kept at an oxidizing potential sufficient to stabilize the tetravalent lead that partially occupies the octahedrally coordinate site. More specifically, it is found that an oxidizing potential within the range of 1.0–1.1 V versus a reversible hydrogen electrode must be maintained, and this can easily be accomplished by keeping the above described reaction medium sparged with oxygen. The beaker is equipped with a cover and any large leaks are sealed with parafilm. A leaktight system is not necessary. The essential requirement is that the pH be maintained at a roughly constant level, and for this reason excessive evaporation losses should be minimized. If large evaporative losses are observed, the pH can still be maintained at an acceptably constant level by replacing the lost volume with water at any time necessary during the synthesis. The reaction medium is maintained in this condition for 4–5 days.

The precipitate is then separated from the still-hot reaction medium by vacuum filtration using a fritted glass filter funnel (350-mL Pyrex, fine frit). At the point when the precipitate is still submerged under a minimal volume of alkaline liquid, distilled water (at $\geqslant 75°$) is added to the filtration funnel and allowed to wash the precipitate, always taking care to keep the precipitate submerged under liquid until the filtration is complete. This procedure is carried out in the manner specified until all KOH is removed from the precipitate. (Phenolphthalein can be used to test the filtrate.) The precipitate is then dried at 100° for approximately 18 h. To ensure that any residual water or hydroxide is removed, it is recommended that the solid be fired at 2 h at 400° in air. X-ray diffraction confirms that the solid is a single-phase pyrochlore. The yield is approximately 2.9 g.

Properties

The resulting solid is black in color. Energy-dispersive X-ray fluorescence indicates a lead:ruthenium ratio of $2.05 \pm 0.05{:}1$. If a 2.05:1 lead:ruthenium ratio is assumed, thermogravimetric reduction in hydrogen (after *in situ* thermogravimetric drying at 400° in oxygen) gives an oxygen content consistent with the formula $Pb_2[Ru_{1.31}Pb^{4+}_{0.69}]O_{6.5 \pm 0.1}$. X-ray diffraction confirms that the material is a single-phase pyrochlore having a cubic unit cell parameter of 10.478 Å. The material is stable in oxygen to approximately 460°.

It should be noted that the exact cation stoichiometry of the product is highly sensitive to the exact metal concentration of the ruthenium source solution and temperature and pH of the reaction medium (inadvertent increases in both of these parameters lead to increased solubility of lead in the alkaline reaction medium and consequently yield solid products of lower lead:ruthenium ratios). While synthesis of a pure lead ruthenium oxide pyrochlore is relatively easy, the precise cation stoichiometry of the product is a property that is not always easy to control. A relatively quick check on the cation stoichiometry of the lead ruthenium oxide product can be obtained, however, by using the correlation between lattice parameter and composition

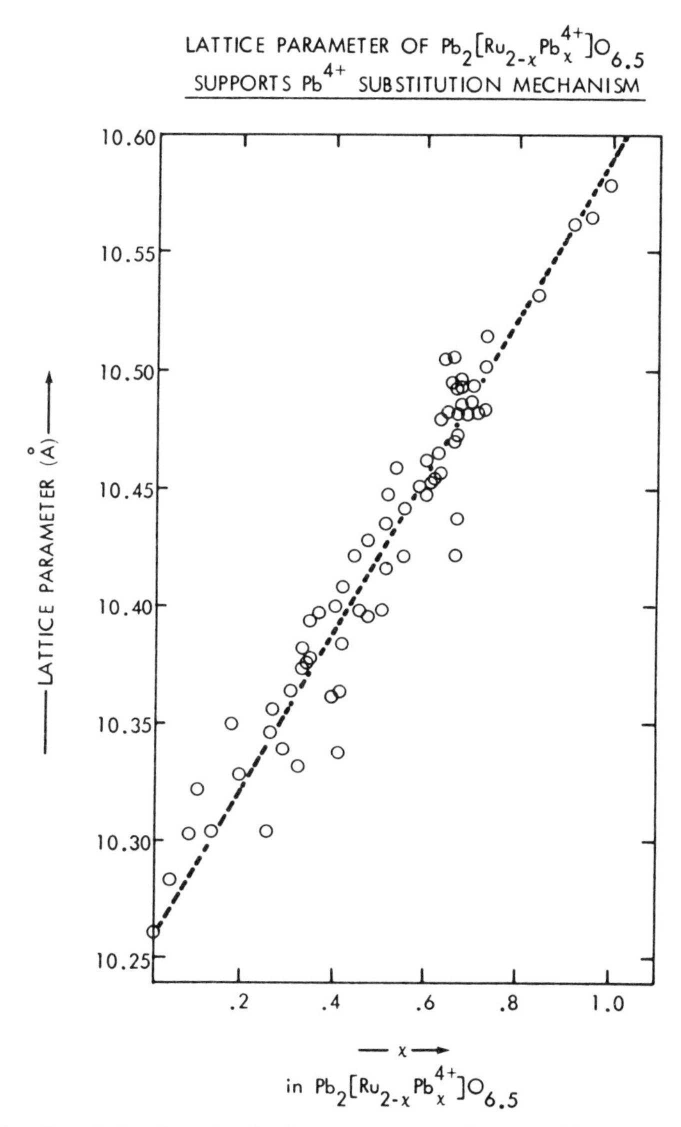

Figure 1. Correlation between lattice parameter and composition for the series, $Pb_2[Ru_{2-x}Pb_x^{4+}]O_{6.5}$.

that is displayed in Fig. 1. When lattice parameter and cation stoichiometry are independently determined, the relationship shown in Fig. 1 also provides an assessment of product purity since data points that show significant departures from the displayed linear correlation indicate the presence of impurity phases. The thermal stability of the lead ruthenium oxides decreases with increasing occupancy of tetravalent lead on the octahedrally coordinated site, but all of the ruthenium oxide pyrochlores described are stable to at least 350° in oxygen.

References

1. J. J. Randall and R. Ward, *J. Am. Chem. Soc.*, **81**, 2629 (1959).
2. J. M. Longo, P. M. Raccah, and J. B. Goodenough, *Mater. Res. Bull.*, **4**, 191 (1969).
3. A. W. Sleight, *Mater. Res. Bull.*, **6**, 775 (1971).
4. R. J. Bouchard and J. L. Gillson, *Mater. Res. Bull.*, **6**, 669 (1971).
5. H. S. Horowitz, J. M. Longo, and J. I. Haberman, U.S. Patent 4,124,539 to Exxon Research and Engineering Co., 1978.
6. H. S. Horowitz, J. M. Longo, and J. T. Lewandowski, U.S. Patent, 4,163,706, to Exxon Research and Engineering Co., 1979.
7. H. S. Horowitz, J. M. Longo, and J. T. Lewandowski, U.S. Patent 4,129,525 to Exxon Research and Engineering Co., 1978.

14. CALCIUM MANGANESE OXIDE, $Ca_2Mn_3O_8$

$$2Ca^{2+} + 3Mn^{2+} + 5CO_3^{2-} \rightarrow Ca_2Mn_3(CO_3)_5$$

$$Ca_2Mn_3(CO_3)_5 + \tfrac{3}{2}O_2 \rightarrow Ca_2Mn_3O_8 + 5CO_2$$

Submitted by HAROLD S. HOROWITZ* and JOHN M. LONGO*
Checked by CARLYE BOOTH† and CHRISTOPHER CASE‡

Reprinted from *Inorg. Synth.*, **22**, 73 (1983)

Historically, the research done on the Ca–Mn–O system has been performed at high ($\geqslant$1000°) temperatures.[1–9] The traditional ceramic synthesis approach to these complex oxides involves repeated high-temperature firing of the component oxides with frequent regrindings. These severe reaction conditions are necessary to obtain a single-phase product because of the diffusional limitations of solid-state reactions. Such high-temperature syn-

* Corporate Research, Exxon Research and Engineering Company, Linden, NJ 07036.
† Department of Chemistry, University of Georgia, Athens, GA 30602.
‡ Department of Engineering, Brown University, Providence, RI 02912.

theses naturally lead to crystalline, low-surface-area materials and often preclude the preparation of mixed metal oxides that are stable only at relatively low temperatures.

Achieving complete solid state reaction at low temperatures in a refractory oxide system, such as the one under investigation, is a formidable problem. Even fine reagent powders of approximately 10-μm particle size still represent diffusion distances on the order of approximately 10^4 unit-cell dimensions. The use of techniques such as freeze drying[10,11] or coprecipitation[12,13] improves reactivity of the component oxides or salts because these methods can give initial particles of about several hundred Å in diameter. But this still means that diffusion must occur across 10–50 unit cells. To achieve complete solid-state reaction in the shortest time and at the lowest possible temperatures, one would prefer to have homogeneous mixing of the component cations on an atomic scale. Compound precursors[14,15] do achieve this objective. Unfortunately, the stoichiometry of the precursors often does not coincide with the stoichiometry of the desired product. The use of solid-solution precursors[16–18] provides all the advantages of compound precursors but avoids the stoichiometry limitations.

Because both $CaCO_3$ and $MnCO_3$ have the calcite crystal structure, it is possible to prepare a complete series of Ca–Mn carbonate solid solutions. These solid solutions are then used as precursors for subsequent reaction to Ca–Mn oxides. In this way, precursors are obtained in which the reactant cations are on the order of 10 Å apart, regardless of the precursor particle size. Since solid solution in this system is complete, it is possible continuously to vary the cation composition in the precursor structure without restriction. While the specific example described here involves the synthesis of $Ca_2Mn_3O_8$, the compounds Ca_2MnO_4, $CaMnO_3$, $CaMn_3O_8$, and $CaMn_7O_{12}$ may also prepared in a similar manner.[17,18] Not only does this synthesis technique give a route to higher-surface-area complex oxides, but it also provides an approach to the preparation of several pure mixed metal oxides that are not stable at the high temperatures usually necessary for conventional solid-state reactions between small particles.

Procedure

A solid-solution precursor is first synthesized by preparing an aqueous solution containing a 2:3 molar ratio of calcium to manganese cations. Specifically 7.006 g (0.07 mol) $CaCO_3$ and 12.069 g (0.105 mol) $MnCO_3$ are dissolved in 100 mL of distilled water plus sufficient nitric acid to effect complete solution (pH of this solution should be 1–5). The $CaCO_3$ used is reagent grade and should be well dried. The $MnCO_3$ used is freshly precipitated from a manganese nitrate solution with a large excess of ammonium carbonate, dried at 100° in a vacuum oven and stored in a container sealed under inert gas until used. Commercially available reagent-grade $MnCO_3$ is unacceptable since it usually contains significant amounts of oxidized manganese products as evidenced by its brown color. The aqueous solution of calcium and manganese cations (described above) is then added, with stirring, to 102.936 g (1.072 mol) $(NH_4)_2CO_3$ dissolved in 500 mL of distilled water. The resulting precipitate is then separated from the aqueous phase by vacuum filtration. It is dried in a vacuum oven, an inert atmosphere drying oven, or for short periods of time in a microwave oven, taking care that the precipitate is not subjected to high temperatures in the presence of oxygen, which could cause divalent manganese to be oxidized. To further prevent premature oxidation, the precipitates are usually stored in an inert atmosphere. X-ray diffraction indicates that the precipitate is a single-phase solid-solution Ca–Mn carbonate with the calcite structure.

Approximately 3–5 g of the $Ca_2Mn_3(CO_3)_5$ solid structure is fired at 700° under flowing oxygen in a tube furnace for 1–48 h to yield the pure product, $Ca_2Mn_3O_8$.

Properties

$Ca_2Mn_3O_8$, as prepared above, is a dark brown powder. The structure is monoclinic ($C2/m$, $a = 10.02$ Å, $b = 5.848$ Å, $c = 4.942$ Å, $\beta = 109.80°$) and is isostructural with Mn_5O_8 and $Cd_2Mn_3O_8$.[19] The cation stoichiometry of $Ca_2Mn_3O_8$ is determined by reducing the compound for about 2 h in hydrogen at 1000°. The cubic lattice parameter of the resulting single-phase Ca–Mn rock salt is refined to ± 0.001 Å. Since the lattice parameter variation in the Ca–Mn–O system is linear,[20] it is possible to determine the atom % of calcium and manganese present, respectively, to $\pm 0.3\%$. The cation stoichiometry, experimentally determined in this way, is 40.66% Ca, 59.34% Mn as compared to a calculated stoichiometry of 40.0% Ca, 60.0% Mn. Thermogravimetric reduction in hydrogen gives a formula of $Ca_2Mn_3O_{8.0 \pm 0.1}$ on the basis of a 2:3 ratio of calcium to manganese. The average manganese valence determination, obtained by measuring the reducible cation content using the oxalate method,[21] is in agreement, giving a formula of $Ca_2Mn_3O_{8.02 \pm 0.01}$. The decomposition temperature at $P_{O_2} =$ 1.0 atm is 890°. The measured resistivity of pressed powder samples of $Ca_2Mn_3O_8$ is $10^5\,\Omega \cdot$cm, and the Mn(IV) moments are found to order antiferromagnetically near 60 K. Figure 1 shows the placement of $Ca_2Mn_3O_8$ in the subsolidus system along with the thermal stabilities of the other low-temperature Ca–Mn–O phases accessible by the solid-solution precursor route.[17]

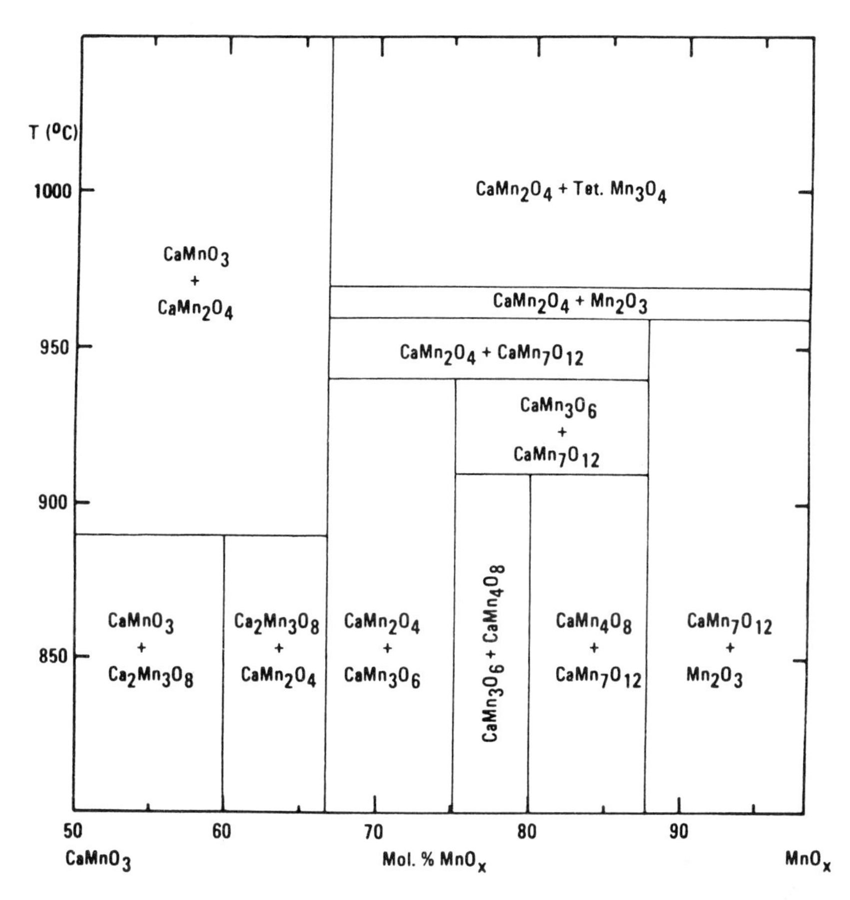

Figure 1. Isobaric (P_{O_2} = 1.0 atm) subsolidus phase relations in the manganese-rich portion of the Ca–Mn–O system.

References

1. P. V. Ribaud and A. Maun, *J. Am. Ceram. Soc.*, **46**, 33 (1963).
2. C. Brisi and M. Lucco-Borlera, *Atti Acad. Sci. Torino*, **96**, 805 (1962).
3. C. Brisi and M. Lucco-Borlera, *J. Inorg. Nucl. Chem.*, **27**, 2129 (1965).
4. G. H. Jonker and J. H. Van Santen, *Physica*, **16**, 337 (1950).
5. H. L. Yakel, *Acta Cryst.*, **8**, 394 (1955).
6. E. O. Wollan and Koehler, *Phys. Rev.*, **100**, 545 (1955).
7. S. N. Ruddlesden and P. Popper, *Acta Cryst.*, **10**, 538 (1957).
8. C. Brisi, *Ann. Chim.*, (Rome) **51**, 1399 (1961).
9. B. Bochu, J. Chenevas, J. C. Joubert, and M. Marezio, *J. Solid State Chem.*, **11**, 88 (1974).

10. F. J. Schnettler, F. R. Monforte, and W. W. Rhodes, in *Science of Ceramics*, Vol. 4, G. H. Stewart (ed.), The British Ceramic Soc., 1968, p. 79.
11. Y. S. Kim and F. R. Monforte, *Am. Ceram. Soc. Bull.*, **50**, 532 (1971).
12. A. L. Stuijts, in *Science of Ceramics*, Vol. 5, C. Brosset and E. Knapp (eds.), Swedish Institute of Silicate Research, 1970, p. 335.
13. T. Sato, C. Kuroda, and M. Saito, in *Ferrites, Proc. Int. Conf.*, G. Hoshino, S. Iida, and M. Sugimoto (eds.), Univ. Park Press, Baltimore, 1970, p. 72.
14. W. S. Calbaugh, E. M. Swiggard, and R. Gilchrist, *J. Res. Natl. Bur Standards*, **56**, 289 (1956).
15. P. K. Gallagher and D. W. Johnson, Jr., *Thermochim. Acta.* **4**, 283 (1972).
16. L. R. Clavenna, J. M. Longo, and H. S. Horowitz, U.S. Patent 4,060,500 to Exxon Research and Engineering Co., November 29, 1977.
17. H. S. Horowitz and J. M. Longo, *Mater. Res. Bull.*, **13**, 1359 (1978).
18. J. M. Longo, H. S. Horowitz, and L. R. Clavenna, in *Solid State Chemistry: A Contemporary Overview*, Adv. Chem. Series No. 186, S. L. Holt, J. B. Milstein, and M. Robbins (eds.), American Chemical Society, Washington, D.C., 1980.
19. G. B. Ansell, H. S. Horowitz, and J. M. Longo, *Acta Cryst.*, **34A**, S157 (abstract) (1978).
20. A. H. Jay and K. W. Andrews, *J. Iron Steel Inst.*, **152**, 15 (1946).
21. D. A. Pantony and A. Siddiqui, *Talanta*, **9**, 811 (1962).

15. TERNARY CHLORIDES AND BROMIDES OF THE RARE-EARTH ELEMENTS

Submitted by GERD MEYER*
Checked by S.-J. HWU† and J. D. CORBETT†

Reprinted from *Inorg. Synth.*, **22**, 1 (1983)

Compound formation in the alkali-metal halide/rare-earth metal trihalide systems seems to be limited to the four following formula types:[1-4]

$$A^{I}RE^{III}_{2}X_{7}, \qquad A^{I}_{3}RE^{III}_{2}X_{9}, \qquad A^{I}_{2}RE^{III}X_{5}, \qquad A^{I}_{3}RE^{III}X_{6}$$

Most of the existing ARE_2X_7 and A_3REX_6 phases (with A = K, Rb, Cs and X = Cl, Br) melt congruently, and those of the A_2REX_5 and $A_3RE_2X_9$ types incongruently. They may be prepared by heating together appropriate amounts of the anhydrous halides AX and REX_3.

The synthetic route described here, starting with aqueous solutions of the appropriate amounts of alkali-metal halides and rare-earth metal halides,

* Institut für Anorganische und Analytische Chemie I, Justus-Liebig-Universitaet Giessen, Heinrich-Buff-Ring 58, 6300 Giessen, Germany.

† Department of Chemistry and Ames Laboratory, Iowa State University, Ames, IA 50011. The Ames Laboratory is operated for the U.S. Department of Energy by Iowa State University under Contract No. W-7405-ENG-82.

makes use of a high concentration of hydrogen halide gas as a driving force in a flow system for the dehydration of a mixture that contains binary and ternary compounds (hydrates). It therefore avoids the complicated and time-intensive preparation of the anhydrous rare-earth metal trihalides that have to be sublimed at least twice to get rid of their contamination with oxyhalides. For example, $Cs_3Sc_2Cl_9$[5,6] has been used as a starting material for the preparation of $CsScCl_3$ with divalent scandium (reduction with Sc metal in sealed tantalum containers[7]).

A. $A^IRE_2^{III}X_7$: CESIUM PRASEODYMIUM CHLORIDE ($CsPr_2Cl_7$) AND POTASSIUM DYSPROSIUM CHLORIDE (KDy_2Cl_7)

$$CsCl + 2PrCl_3 \cdot 7H_2O \xrightarrow{HCl(aq)} (CsPr_2Cl_7)(hyd) \xrightarrow[-H_2O]{HCl(g)\ 500^\circ} CsPr_2Cl_7$$

$$KCl + Dy_2O_3 \xrightarrow{HCl(aq)} (KDy_2Cl_7(hyd) \xrightarrow[-H_2O]{HCl(g)\ 500^\circ} KDy_2Cl_7$$

Procedure

■ **Caution.** *Inhalation of HCl or HBr gas is harmful to the respiratory system; all operations should be carried out in a well-ventilated hood.*

1. For the preparation of $CsPr_2Cl_7$, 168.4 mg CsCl (1 mmol) and 746.7 mg $PrCl_3 \cdot 7H_2O$ (2 mmol, commercially available from various sources; "Pr_6O_{11}" may also be used in appropriate amounts, but the Pr content should be determined by analysis) are dissolved by adding a small amount of concentrated hydrochloric acid, which should be at least reagent grade and in the case of optical investigations, completely free of transition-metal ions, especially iron.
2. For the preparation of KDy_2Cl_7, 74.6 mg KCl (1 mmol) and 373.0 mg Dy_2O_3 (1 mmol) are used as starting materials. Here and in the following syntheses, the rare-earth sesquioxides should be freshly ignited before weighing, because they are suspected of slowly forming carbonates when exposed to air for a long period of time. Concentrated hydrochloric acid ($\approx$100 mL) is added to the reactants. A homogeneous solution is obtained after some minutes of boiling.

The clear solutions in 1 or 2 are slowly evaporated to dryness by using a heating plate or an infrared lamp. The residue is then transferred to a corundum crucible (size: 10 × 60 mm), and this is inserted into a fused-quartz tube in the center of a tubular furnace. (To prepare larger quantities, e.g.,

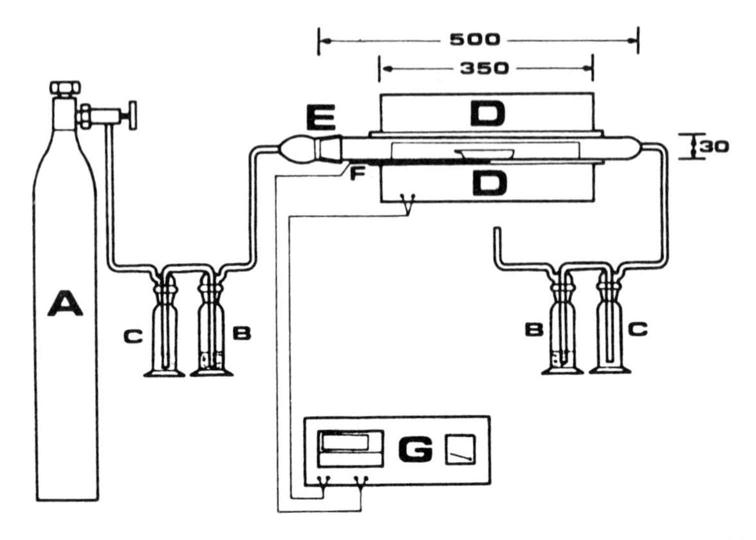

Figure 1. Apparatus for the synthesis of anhydrous ternary and quaternary chlorides and bromides. Sizes are given in millimeters; suitable for a mmolar scale. (*A*) Steel cylinder or lecture bottle with HCl or HBr gas; (*B*) washing bottles filled with concentrated sulfuric acid (for HCl only) or P_2O_5 and paraffin oil, respectively; (*C*) empty washing bottles to prevent suck-back; (*D*) tubular furnace; (*E*) quartz tube; (*F*) thermocouple, e.g., Chromel-alumel or Pt-Pt/Rh; (*G*) temperature control and measuring device (a programming unit may be attached for slow heating and cooling).

10 mmol, a 100 × 15 × 10-mm Alundum boat is more convenient.) The sizes of the tubular furnace and the fused quartz tube should be sufficient that the sample can be heated without a temperature gradient (see Fig. 1). A slow flow of HCl gas (available in steel cylinders or lecture bottles in high purity, e.g., "transistor grade"), one to two bubbles per second, dried with concentrated sulfuric acid or a trap cooled with dry ice/acetone, is then commenced and the sample heated to 500°. After 1–2 days, the HCl gas stream may be replaced by a dry inert-gas stream (N_2 or Ar) and the sample cooled by turning off the power to the furnace. Yields: approximately 100%; 662.9 mg $CsPr_2Cl_7$ and 612.3 mg KDy_2Cl_7, respectively.

Properties

The products $CsPr_2Cl_7$ (light green) and KDy_2Cl_7 (colorless) are obtained as slightly sintered, moisture-sensitive powders. They should be handled in a dry box (dry N_2 or Ar atmosphere) and sealed in Pyrex tubes for storage.

For characterization, common X-ray techniques (Debye–Scherrer or, better, Guinier patterns) are useful. The compound $CsPr_2Cl_7$ crystallizes in the orthorhombic space group $P222_1$, with $a = 1657.1$, $b = 963.7$, and $c = 1486.6$ pm and 8 formula units per unit cell. Some characteristic d values are (relative intensities on a 1–100 scale are given in parentheses): 4.840, 4.832(16); 4.262, 4.260(46); 4.055, 4.050, 4.032, 4.026(100); 3.649, 3.644(16); 3.394, 3.393(9) Å. The crystal structure contains $[PrCl_8]$ groups (partially disordered) linked together via common corners, edges, and faces.

The crystal structures of both KDy_2Cl_7 and $RbDy_2Cl_7$ have been determined from single-crystal X-ray data.[8] $RbDy_2Cl_7$ represents the orthorhombic aristotype: $a = 1288.1(3)$, $b = 693.5(2)$, $c = 1267.2(3)$ pm, space group $Pnma$, $Z = 4$. The compounds $RbRE_2Cl_7$, as well as the room-temperature modifications of $CsRE_2Cl_7$ that are indicated in Fig. 2 by filled circles, are isotypic to $RbDy_2Cl_7$. The other KRE_2Cl_7 compounds obtained so far

TYPE:	ARE_2X_7						$A_3RE_2X_9$						A_2REX_5						A_3REX_6					
RE \ A	K		Rb		Cs		K		Rb		Cs		K		Rb		Cs		K		Rb		Cs	
	Cl	Br	Cl	Br	Cl	Br	Cl	Br	Cl	Br	Cl	Br	Cl	Br	Cl	Br	Cl	Br	Cl	Br	Cl	Br	Cl	Br
La	○				●	○	–				–	–	○				–	–	–				○	○
Ce	○						○	–			–		○				–		○				○	
Pr	–	○	○		●	○	○	–			–	–	○	○			–	–	○	○			○	○
Nd	–	○	◑	○		○	○	–	–	–	–	–	○	○	○	○	–	–	○	○	○	○	○	○
Sm	◑	○	●	○	○	○	–	–	–	–	–	●	○	○	○	○	–	○	○	○	○	○	○	○
Eu	◑		●				–				–	●	○				–		○					
Gd	●	○	●	○		○	–	–		–	–	●	○	○		○	–	–	○	○		○		○
Tb	●		●				–				–	●	○				–		○					
Dy	●	–	●	○	●		–	–			–	●	–	○		○	●		○	○		○	○	
Y	●		●		●		–				●	●	–				●		◑		◑		◑	
Ho	●		●		●						●	●					●		◑					
Er	●	–	●	–	●	–	–	–		●	●	●	–	○	●	○	●	–	◑	○		○	○	○
Tm	●		●		●					●	●	●			●		●		◑					
Yb	–	–	●	–	●	–	–	–		●	●	●	–	○		–	●	–	◑	○		○	○	○
Lu	–						–			●	●	●	–		●		●		●					
Sc	–	–	–	–	–	–	–	○	●	●	●	●	○	–	–	–	–	–	●	○	◑	○	◑	○

Figure 2. Compound formation in the alkali halide/rare-earth trihalide systems (chlorides and bromides); ● crystal structure known; see Tables I and II [A_2REX_5 = type compounds with A = K, Rb; RE = La-Dy; X = Cl, Br are isotypic with K_2PrCl_5; see G. Meyer and E. Hüttl, *Z. Anorg. Allgem.* (1983)]; ○ observed in the phase diagram; ◑ phase investigated by X-ray diffraction, crystal structure not known so far; – not observed in the phase diagram; no entry means that the respective phase diagram was not investigated.

crystallize as the monoclinic hettotype with (for KDy_2Cl_7) $a = 1273.9(8)$, $b = 688.1(5)$, $c = 1262.1(6)$ pm, $\beta = 89.36(3)°$, space group $P112_1/a$ (an unconventional setting of $P2_1/c$), $Z = 4$. In both structures (Fig. 3), monocapped trigonal prisms $(DyCl_7]$ are connected via a triangular face and two edges with other prisms to form layers that are stacked in the [100] direction. These are held together by alkali metal cations (K^+, Rb^+), which are essentially 12-coordinated (bicapped pentagonal prism). Characteristic *d* values (in A) are: for $RbDy_2Cl_7$, 6.440(70), 6.084(27), 4.423(100), 3.474(64); for KDy_2Cl_7, 6.374(100), 6.063(52), 4.417(36), 4 370(40), 3.471(34). Numbers in parentheses are relative intensities on a 1–100 scale.

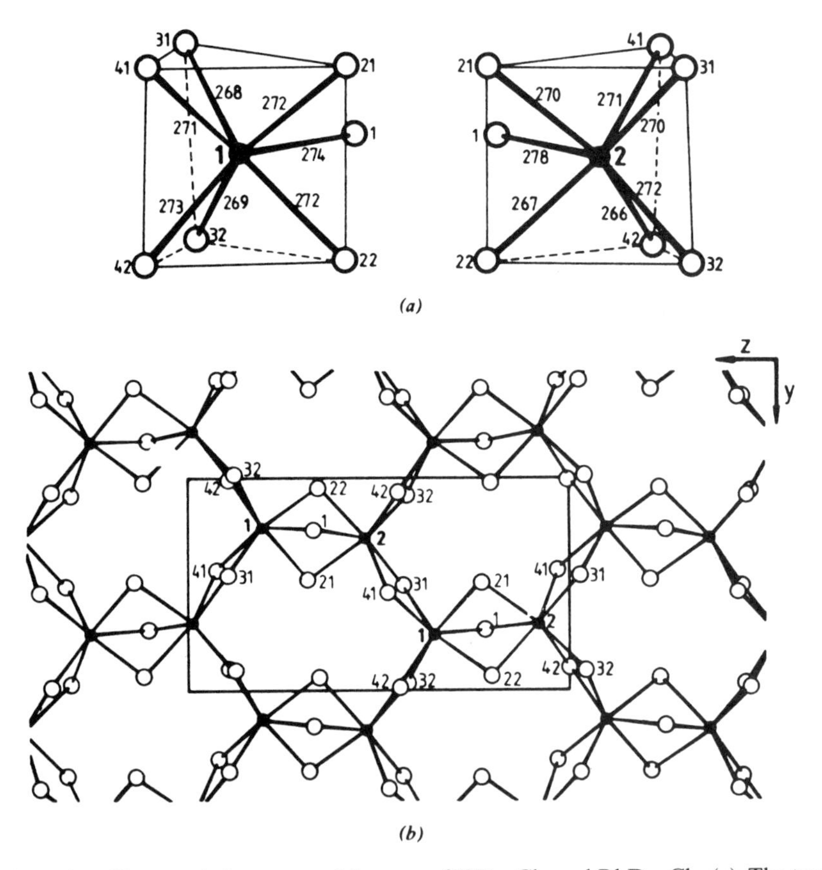

Figure 3. Characteristic structural features of KDy_2Cl_7 and $RbDy_2Cl_7$. (*a*) The two crystallographically independent monocapped trigonal prisms $[DyCl_7]$; (*b*) part of one ${}^2_\infty[Dy_2Cl_7]$ layer.

The compounds CsY_2Cl_7 may serve as an example for the cesium compounds with the heavier rare earths (those with La–Nd have the $CsPr_2Cl_7$ type structure). Lattice constants are: $a = 1335.6(3)$, $b = 696.6(2)$, $c = 1266.1(4)$ pm, *Pnma*, $Z = 4$. Characteristic d values are: 6.678(16), 4.594(16), 4.505(100), 3.610(20), 3.485(35) Å.

For the preparation of the analogous bromides, all the chloride-containing starting materials and HCl gas as well are replaced by bromides and HBr gas, respectively.

B. $A^I_3RE^{III}_2X_9$, $A^I_2RE^{III}X_5$, $A^I_3RE^{III}X_6$: THE CESIUM LUTETIUM CHLORIDES, $Cs_3Lu_2Cl_9$, Cs_2LuCl_5, Cs_3LuCl_6

$$4CsCl + Lu_2O_3 \xrightarrow{HCl(aq)} 2Cs_2LuCl_5 \cdot H_2O \xrightarrow[-H_2O]{HCl(g)\ 500^\circ} 2Cs_2LuCl_5$$

$$3CsCl + Lu_2O_3 \xrightarrow{HCl(aq)} 1.5Cs_2LuCl_5 \cdot H_2O(+0.5LuCl_3 \cdot hyd)$$

$$\xrightarrow[-H_2O]{HCl(g)\ 500^\circ} Cs_3Lu_2Cl_9$$

$$6CsCl + Lu_2O_3 \xrightarrow{HCl(aq)} 2Cs_2LuCl_5 \cdot H_2O + 2CsCl$$

$$\xrightarrow[-H_2O]{HCl(g)\ 500^\circ} 2Cs_3LuCl_6$$

Procedure

Appropriate molar amounts of CsCl and Lu_2O_3 (CsCl should be reagent-grade or ultrapure, Lu_2O_3 at least 99.9%; both are available from various sources) are used as starting materials:

1. for Cs_2LuCl_5: 673.4 mg CsCl (4 mmol) and 397.9 mg Lu_2O_3 (1 mmol)
2. for $Cs_3Lu_2Cl_9$: 505.1 m CsCl (3 mmol) and 397.9 mg Lu_2O_3 (1 mmol)
3. for Cs_3LuCl_6: 1010.1 mg (6 mmol) and 397.9 mg Lu_2O_3 (1 mmol)

Approximately 100 mL hydrochloric acid (reagent-grade) is added to the reactants (1, 2, or 3). A homogeneous solution is obtained after some minutes of boiling. After that, a heating plate or an infrared lamp is used for slow evaporation to dryness. For large (e.g., 10 mmol) quantities it might be necessary to grind the product from the initial evaporation and put it in a 100° oven overnight to reduce the degree of hydration and considerably speed the final step. The procedure following this is identical to that described in Section A. The yields approach the theoretical values: 1.236 g (2 mmol) Cs_2LuCl_5, 1.068 g (1 mmol) $Cs_3Lu_2Cl_9$, 1.573 g (2 mmol) Cs_3LuCl_6.

Properties

The compounds Cs_2LuCl_5, $Cs_3Lu_2Cl_9$, and Cs_3LuCl_6 are obtained as slightly sintered, moisture-sensitive, colorless powders. They should be stored and handled in a dry box and for longer periods of time sealed under inert gas (N_2 or Ar) in Pyrex tubes to prevent hydration. Principal impurities in the samples obtained are most likely oxyhalides such as LuOCl. These are usually not detectable at a sensitivity level of a few percent with the Guinier method.

X-ray powder diffraction techniques are suitable for the characterization of Cs_2LuCl_5 and $Cs_3Lu_2Cl_9$:Cs_2Lu_5[9] is othorhombic, displaying the Cs_2DyCl_5[10] structure type (Table I). The first six strongest lines (in Å) are: d = 4.178(100), 4.021(44), 3.995(32), 3.858(31), 3.700(54), 3.531(34); further strongest lines are observed with d = 2.862(34), 2.720(57), 2.524(50); intensities are give in parentheses on a 1–100 scale.

The compound $Cs_3Lu_2Cl_9$[6] belongs to the trigonal $Cs_3Ti_2Cl_9$[11–13] structure type. The strongest X-ray lines have d values equal to 4.443(45), 3.734(88), 2.856(100), 2.758(36), 2.219(48), 1.867(40) Å.

Little is known about Cs_3LuCl_6 and other A_3REX_6-type compounds. Preliminary investigations show a reversible but slow phase transition above 300° for Cs_3LuCl_6. The compound $Cs_2LuCl_5 \cdot H_2O$,[9] detected as an intermediate in all three cases, crystallizes with the erythrosiderite type

TABLE I. Lattice Constants of A_2RECl_5-Type Compounds[9] (Orthorhombic, *Pbnm*, $Z = 4$)

	a/pm	b/pm	c/pm
Cs_2DyCl_5	1523.3(3)	954.9(3)	749.7(3)
Cs_2YCl_5	1522.6(2)	953.3(1)	746.9(1)
Cs_2HoCl_5	1520.2(2)	951.6(2)	745.4(1)
Cs_2ErCl_5	1519.1(2)	949.9(2)	744.2(1)
Cs_2TmCl_5	1517.7(2)	948.1(2)	741.8(1)
Cs_2YbCl_5	1514.7(5)	945.6(3)	740.8(2)
Cs_2LuCl_5	1514.2(2)	944.8(2)	738.5(1)
Rb_2ErCl_5	1466.6(3)	951.3(2)	727.4(2)
Rb_2TmCl_5	1462.1(3)	946.5(4)	727.1(3)
Rb_2LuCl_5	1460.9(2)	939.8(2)	724.6(1)

($K_2FeCl_5 \cdot H_2O$-type[14], orthorhombic, space group *Pnma*) structure with a = 1456.4(4), b = 1046.6(3), c = 752.0(3) pm. The d values of the strongest X-ray lines are: 6.099(45), 5.966(44), 3.698(100), 3.638(40), 2.989(35), 2.614(86) Å.

Investigations of the alkali-metal halide/rare-earth metal halide phase diagrams are still fragmentary, as may be seen from Fig. 2.

Compounds of the A_2REX_5 type, indicated in Fig. 2 by filled circles, together with their lattice constants as determined from Guinier powder patterns are listed in Table I, and those of $A_3RE_2X_9$ type in Table II.

Like the elpasolites $A^I_2B^IRE^{III}1X_6$ (see Section 2) A_3REX_6-type compounds contain "isolated" [REX_6] "octahedra," presumably of low symmetry in a low-symmetry crystal structure. The compounds K_3ScCl_6 and K_3LuCl_6[15] crystallize with the K_3MoCl_6 type[16] (monoclinic, $P2_1/a$, $Z = 4$). Lattice constants, characteristic d values in Å and relative intensities on a 1–100 scale are: for K_3ScCl_6, a = 1226.5(6), b = 758.0(1), c = 1282.1(5) pm, β = 109.03(2)°; d(I), 6.085(50), 5.250(34), 3.723(27), 2.968(39), 2.745(55),

TABLE II. Lattice Constants of $A_3RE_2X_9$-type Compounds[a] (Mostly $Cs_3Tl_2Cl_9$ structure type, trigonal, R$\bar{3}$c, Z < 6)

	Chlorides, X = C1		Bromides, X = Br	
	a/pm	c/pm	a/pm	c/pm
$Cs_3Sm_2X_9$			1379.7	1937
$Cs_3Gd_2X_9$			1372.1	1939
$Cs_3Tb_2X_9$			1367.2	1938
$Cs_3Dy_2X_9$			1363.8	1935
$Cs_3Y_2X_9$	1306.9	1831		
$Cs_3Ho_2X_9$	1305.3	1832	1360.3	1936
$Cs_3Er_2X_9$	1301.5	1829	1357.7	1937
$Cs_3Tm_2X_9$	1299.3	1828	1355.1	1934
$Cs_3Yb_2X_9$	1296.9	1829	1352.0	1932
$Cs_3Lu_2X_9$	1293.9	1829	1350.0	1931
$Cs_3Sc_2X_9$	1270.4	1810.9	768.0[b]	1926
Rb_3Er_2Xg			1340.2(4)	1916 (1)
$Rb_3Tm_2X_9$			1338.3(2)	1913.4(3)
$Rb_3Yb_2X_9$			1334.3(2)	1914.1(3)
$Rb_3Lu_2X_9$			1331.8(1)	1916.3(2)
$Rb_3Sc_2X_9$	1239.8(1)	1789.3(2)	756.8(1)[b]	1914.9(8)

[a] Data on Cs compounds were taken from Reference 6, Rb compounds from Reference 15.

[b] $Cs_3Cr_2Cl_9$ structure type, hexagonal $P6_3/mmc$, $Z = 2$.

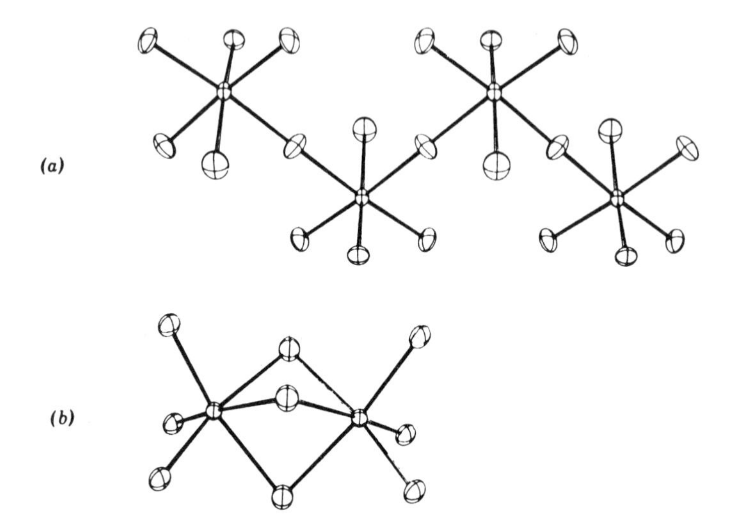

Figure 4. Increasing octahedral "condensation" with decreasing alkali-metal halide content. (*a*) A section of the infinite-*cis*-connected zigzag chain of octahedra in Cs_2RECl_5; (*b*) a confacial bioctahedron as the characteristic structural feature of $A^{I}_3RE^{III}_2X_9$-type compounds in both the $Cs_3Tl_2Cl_9$ and $Cs_3Cr_2Cl_9$ structure types.

2.625(100) Å; for K_3LuCl_6, $a = 1251.5(4)$, $b = 767.6(2)$, $c = 1301.1(5)$ pm, $\beta = 109.79(3)°$; *d*(I), 6.183(100), 6.121(38), 5.888(30), 5.304(70), 3.776(40), 3.665(50), 2.652(56). The characteristic feature of Cs_2RECl_5-type compounds is an extended, infinite zweier–single zigzag chain of octahedra, *cis*-connected via common corners, while $A_3RE_3X_9$-type compounds contain "isolated" confacial bioctahedra $[RE_2X_9]$ (Fig. 4). Two structure types are found: $Cs_3Tl_2Cl_9$[11–13] (stacking sequence of $CsCl_3$ layers: ABABAB . . .) and $Cs_3Cr_2Cl_9$[17] (ABACBC . . .).

References

1. *Gmelins Handbuch aer Anorganischen Chemie*, 8th ed., system no. 39, part C5, Springer Verlag, New York, 1977.
2. J. Kutscher and A. Schneider, *Z. Anorg. Allgm. Chem.*, **408**, 135 (1974).
3. R. Blachnik and D. Selle, *Z. Anorg. Allgm. Chem.*, **454**, 90 (1979).
4. R. Blachnik and A. Jaeger-Kasper, *Z. Anorg. Allgm. Chem.*, **461**, 74 (1980).
5. K. R. Poeppelmeier, J. D. Corbett, T. P. McMullen, D. R. Torgeson, and R. G. Barnes, *Inorg. Chem.*, **19**, 129 (1980).
6. G. Meyer and A. Schoenemund, *Mater. Res. Bull.*, **15**, 89 (1980).
7. K. R. Poeppelmeier and J. D. Corbett, *Inorg. Synth.*, this volume, Chapter 2, Section 16.
8. G. Meyer, *Z. Anorg. Allg. Chem.*, **491**, 217 (1982).
9. G. Meyer, J. Soose, and A. Moritz, unpublished work (1980/81).

10. G. Meyer, *Z. Anorg. Allg. Chem.*, **469**, 49 (1980).
11. J. L. Hoard and L. Goldstein, *J. Chem. Phys.*, **3**, 199 (1935).
12. H. M. Powell and A. F. Wells, *J. Chem. Soc.*, 1008 (1935).
13. G. Meyer, *Z. Anorg. Allg. Chem.*, **445**, 140 (1978).
14. A. Bellanca, *Period. Mineral.*, **17**, 59 (1948); see also C. O'Connor, B. S. Deaver, Jr., and E. Sinn, *J. Chem. Phys.*, **70**, 5161 (1979).
15. G. Meyer and U. Strack, unpublished results (1981).
16. Z. Amilius, B. van Laar, and H. M. Rietveld, *Acta Cryst.*, **B25**, 400 (1969).
17. G. J. Wessel and D. J. W. Ijdo, *Acta Cryst.*, **10**, 466 (1957).

16. CESIUM SCANDIUM(II) TRICHLORIDE

$$3CsCl + 2ScCl_3 + Sc \rightarrow 3CsScCl_3$$

$$Cs_3Sc_2Cl_9 + Sc \rightarrow 3CsScCl_3$$

Submitted by K. R. POEPPELMEIER* and J. D. CORBETT*
Checked by GERD MEYER†

Reprinted from *Inorg. Synth.*, **22**, 23 (1983)

The synthesis of the title compound[1] is a good example of a reaction that is made possible through the proper choice of the container material. In this preparation, the tantalum reaction vessel serves to contain both the potentially reactive and volatile scandium(III) chloride melt and the strongly reducing scandium metal under the conditions necessary for complete reaction. This choice of components is required since the simple binary scandium(II) compound $ScCl_2$ has not been prepared; thus, the direct reaction of $ScCl_2$ with alkali-metal halide cannot be used to obtain $CsScCl_3$. The reduction starts instead with either the separate components $CsCl + ScCl_3$ (method A), or the compound $CsSc_2Cl_9$ obtained by indirect means[2] (method B). The ternary halide $CsScCl_3$ is of value since it exhibits an unusual nonstoichiometry on the metal-poor side and contains the electronically interesting d^1 ion of scandium(II).

* Ames Laboratory and Department of Chemistry, Iowa State University, Ames, Iowa 50011. The Ames Laboratory is operated by the U.S. Department of Energy by Iowa State University under contract No. W-7405-ENG-82.

† Institut für Anorganische und Analytische Chemie I, Justus-Liebig-Universität, 6300 Giessen, Germany.

Procedure

Standard vacuum line and dry-box techniques must be used for manipulation and storage of the starting materials and the product; both moisture and oxygen must be scrupulously excluded to avoid contamination by ScOCl.

A. SYNTHESIS WITH CsCl

The CsCl ($\geqslant$99.9% purity) is dried at 200° under high vacuum and then melted in a silica container. The latter step reduces the possibility of recontamination by surface moisture and makes weighing and complete sample transfer easier in the dry box.

Sublimed $ScCl_3$ (750°, 10^{-5} torr in a tantalum jacket), prepared by treating the metal with HCl, that has received minimum exposure to the dry-box atmosphere will give the best product[3]. [Although anhydrous scandium(III) chloride is available from a few suppliers, those samples have not been sufficiently pure. Sublimation and a check for other metals, especially Fe, are advisable.] The product after sublimation will give typical recoveries for both chloride and metal of 100.0 ± 0.2 wt. % with $Cl:Sc = 3.00 \pm 0.001$. Scandium is titrated with EDTA using Xylenol Orange as the indicator[4] and chloride is determined gravimetrically as AgCl. However, since the analysis is not nearly sensitive enough to detect deleterious amounts of moisture or hydrolysis products and the freedom of the product from ScOCl, the $ScCl_3$ melting point ($967 \pm 3°$) is probably a better measure of purity.

The next step in the preparation is to combine in the dry box the stoichiometric amounts (molar ratio 3:2:1) of CsCl, $ScCl_3$, and Sc* (powder, if available; otherwise small chunks or foil) respectively, the total quantities conveniently varying from one to several grams depending on the size of the tantalum tube and the precision of weighing possible.† The tantalum container is crimped and sealed by arc welding[5] and the welded tube next sealed under vacuum in a fused-silica tube. The sample is now heated to 700° for 2–3

* High-purity metals are available from a variety of suppliers. These include Atomergic-Chemical Corp., 100 Fairchild Avenue, Plainview, NY 11803; Leico Industries, Inc., 250 West 57th Street, New York, NY 10019; Lunex Company, P.O. Box 493, Pleasant Valley, IA 52767; Molycorp, Inc., 6 Corporate Park Drive, White Plains, NY 10604; Research Chemicals Division, Nuclear Corporation of America, P.O. Box 14588, Phoenix, AZ 85031; Ronson Metals Corporation, 45–65 Manufacturers Place, Newark, NJ 07105; Rare Earth Products, Ltd., Waterloo Road, Widnes, Lancashire, England (United States representative of rare-earth products is United Mineral & Chemical Corp., 129 Hudson St., New York, NY 10013; and Kolon Trading Co., Inc., 540 Madison Avenue New York, NY 10022.

† A useful electronic weighing sensor with remote output is available from Scientech, Inc., Boulder, CO.

days and then cooled to 650° and annealed for at least 24 h. The latter step is necessary since the fully reduced phase appears to melt somewhat incongruently. The reaction product should be examined closely to determine whether pieces of unreacted scandium metal can be detected. If so, the sample should be ground, resealed, and the procedure repeated. (Lines at 6.99, 4.70, 4.19, and 3.78 Å distinguish $Cs_3Sc_2Cl_9$ from the similar lattice of $CsScCl_3$,[1] but these are weak to very weak even with the pure phase and so are not very useful in checking for complete reduction.) The reaction can also be run with excess metal if it is in a form (foil, etc.) that can easily be recovered.

Either procedure when properly applied will give an essentially quantitative yield. The compound ScOCl, which has a light pink tinge in reducing systems, is the most likely impurity but can be avoided by careful work. The stronger lines in the ScOCl powder pattern (with intensities in parentheses) are about: 8.2(vs), 3.56(vs), 2.63(vs), 2.023(ms), 1.878(s), 1.585(ms) Å. It is estimated that ScOCl can be detected at $\leqslant 1\%$ visually and $\leqslant 2\%$ by careful Guinier work.

B. SYNTHESIS WITH $Cs_3Sc_2Cl_9$

Alternatively, $Cs_3Sc_2Cl_9$ can be used as the starting material for reduction. This is prepared by dehydration of a mixture of the hydrated chloride and $Cs_2ScCl_5 . H_2O$ under flowing HCl according to the procedure of Meyer (this volume, Chapter 2, Section 15.B). The driving force of the formation of the anhydrous ternary compound $Cs_3Sc_2Cl_9$ avoids the production of ScOCl, which would otherwise result when $ScCl_3$ preparation is attempted by this method in the absence of CsCl.

Properties

The compound $CsScCl_3$ has the $2H$ (hexagonal perovskite) structure of $CsNiCl_3$, space group $P6_3/mmc$, $a = 7.350(2)$ Å, $c = 6.045(3)$ Å, and represents the first example of a scandium(II) compound, although many more reduced metal–metal bonded phases are known.[6] The material is black in color in bulk and blue when ground. The material is relatively stable to moisture for a reduced scandium compound but still readily reacts with water with the evolution of hydrogen gas.

The stoichiometry of $CsScCl_3$ can evidently be varied in single phase between the $6R$-type ($Cs_3Tl_2Cl_9$) structure of the scandium(III) salt $Cs_3Sc_2Cl_9$ and $Cs_3Sc_3Cl_9$($=CsScCl_3$) according to the reaction

$$Cs_3Sc_2Cl_9 + x\text{Sc} \rightarrow Cs_3Sc_{2+x}Cl_9, \qquad 0 \leqslant x \leqslant 1.0$$

without the formation of coherent superstructures according to powder X-ray diffraction (Guinier technique). The material is a semiconductor, as there is evidently only a weak interaction between the Sc(II) ions even though they form an infinite chain (surrounded by confacial chloride trigonal antiprisms) with a separation of only 3.02 Å.

Analogous phases exist with $CsScBr_3$, $CsScI_3$, $RbScCl_3$, and $RbScBr_3$ but have not been obtained with potassium. But the stoichiometry range noted above exists only with $Rb_3Sc_2Cl_3$ because the other $M^I_3R_2X_9$ phases have a different and less favorable structure ($Cs_3Cr_2Cl_9$ type).

References

1. K. R. Poeppelmeier, J. D. Corbett, T. P. McMullen, R. G. Barnes, and D. R. Torgeson, *Inorg. Chem.*, **19**, 129 (1980).
2. G. Meyer and A. Schönemund, *Mater. Res. Bull.*, **15**, 89 (1980).
3. J. D. Corbett, *Inorg. Synth.*, **22**, 39 (1983).
4. J. Korbl and R. Pribil, *Chemist-Analyst*, **45**, 102 (1956).
5. J. D. Corbett, *Inorg. Synth.*, **22**, 15 (1983).
6. K. R. Poeppelmeier and J. D. Corbett, *J. Am. Chem. Soc.*, **100**, 5039 (1978).

17. A ONE-DIMENSIONAL POLYCHALCOGENIDE, $K_4Ti_3S_{14}$[1]

$$2K_2S + Ti + 6S \rightarrow K_4Ti_3S_{14}$$

Submitted by STEVEN A. SUNSHINE,† DORIS KANG,* and JAMES A. IBERS*
Checked by EDWARD G. GILLAN‡

Molten salts of the type A_2Q/Q (A = alkali metal, Q = S, Se, or Te) are useful and may be necessary in the synthesis of one-dimensional solid-state polychalcogenides.[1] Sodium polysulfide melts have previously been utilized by Scheel[2] as fluxes to recrystallize metal sulfides. However, synthesis of $K_4Ti_3S_{14}$ employs the molten salt not only as a "flux" for the reaction but also as a reactive reagent. There has since been an increased interest in the use of this preparatory method in the synthesis of new ternary polychalcogenides.[3-6] This approach is of general utility and is illustrated here for the synthesis of the one-dimensional semiconductor $K_4Ti_3S_{14}$.

* Department of Chemistry, Northwestern University, Evanston, IL 60208-3113.
† Raychem Corporation, Menlo Park, CA.
‡ Department of Chemistry, University of California at Los Angeles, Los Angeles, CA 90024.

Procedure

In an Ar atmosphere, Ti powder (0.023 g, 0.480 mmol) (AESAR, 99.9%) with K_2S (0.107 g, 0.971 mmol) (Alfa, approximately 44% K_2S, remainder other potassium sulfides), and S powder (0.0932 g, 2.912 mmol) (Aldrich Gold Label, 99.999%) are ground together to a fine powder and subsequently loaded into a fused-silica tube (i.d. 10 mm, o.d. 12 mm, length 12 cm). The tube is then evacuated to $< 10^{-4}$ torr and sealed. The reaction mixture is then heated in a furnace at 375° for 50 hr, or up to 890° for 5 hr, and then cooled to 375° for extended heating. The latter procedure produces better-quality crystals. This compound forms as hexagonal shaped needles within the K_2S/S melt. The excess melt is dissolved in distilled water. The resulting crystals are black to reflected light and dark burgundy to transmitted light. Crystals up to 1 mm in length are formed. These crystals are not soluble in organic solvents or in mineral acids. They are marginally suitable for X-ray diffraction analysis.

Properties

$K_4Ti_3S_{14}$ crystallizes with four formula units in the space group C_{2h}^6-$C2/c$ of the monoclinic system in a cell of dimensions $a = 20.91(2)$, $b = 7.916(8)$, $c = 12.84(1)$ Å, $\beta = 112.20(3)°$, $V = 1967$ Å^3 at $-160°$. $K_4Ti_3S_{14}$ is a semiconductor with room-temperature conductivity $\approx 1.7\ 10^{-5}\ \Omega^{-1}\text{cm}^{-1}$. The temperature dependence of its conductivity could not be measured owing to the fact that the magnitude of the conductivity is at the experimental limit of our apparatus. This structure is composed of one-dimensional chains of Ti/S that are separated by K^+ cations. These chains contain not only the S^{2-} ligand but also the S_2^{2-} ligand. The compound may be formulated as $K_4[Ti_3(S_2)_6(S)_2]$.

References

1. S. A. Sunshine, D. Kang, and J. A. Ibers, *J. Am. Chem. Soc.*, **109**, 6202 (1987).
2. H. J. Scheel, *J. Crys. Growth*, **24**, 669 (1974).
3. S. Schreiner, L. E. Aleandri, D. Kang, and J. A. Ibers, *Inorg. Chem.*, **28**, 392 (1989).
4. D. Kang and J. A. Ibers, *Inorg. Chem.*, **27**, 549 (1988).
5. M. G. Kanatzidis and Y. Park, *J. Am. Chem. Soc.*, **111**, 3767 (1989).
6. Y. Park and M. G. Kanatzidis, *Angew. Chem., Int. Ed. Engl.*, **29**, 914 (1990).

18. SYNTHESIS OF $K_4M_3Te_{17}$ (M=Zr, Hf)[1] AND $K_3CuNb_2Se_{12}$[2]

$$\text{“}K_2Te_4\text{”} + \tfrac{1}{2}M + 2Te \rightarrow K_4M_3Te_{17}$$

$$3K_2Se_5 + 4Nb + 2Cu + 9Se \rightarrow 2K_3CuNb_2Se_{12}$$

Submitted by PATRICIA M. KEANE,* YING-JIE LU,* and JAMES A. IBERS*
Checked by JOHN B. WILEY†

Since the use of reactive molten or fluxes of the type A_2Q/Q (A = alkali metal, Q = S or Se) were described in the synthesis of ternary polychalcogenides,[3] there have been a number of sulfides and selenides reported that employ this synthetic technique.[4-7] The compounds typically exhibit low dimensionality, novel structure types, and interesting chalcogen–chalcogen bonding modes. Unlike the sulfides and selenides, no tellurides had been synthesized with this molten-salt method. The formation of intercalation compounds of the type $A_xM_yQ_z$ (M = group IV, V, or VI; Q = S, Se, or Te)[8-10] and the stable binary metal tellurides MTe_3[11] and MTe_5[12] (M = Zr, Hf) hinder the formation of new ternary polytellurides. Through the use of a K_2Te/Te flux, the one-dimensional polytelluride $K_4M_3Te_{17}$ has been synthesized.

Molten alkali-metal polyselenide fluxes also prove to be useful in the syntheses of unusual quaternaries, here illustrated by the synthesis of the one-dimensional polyselenide $K_3CuNb_2Se_{12}$.

Both of these polychalcogenides are synthesized at 870° or above. The use of low temperatures (<450°) is *not* a necessary condition for the synthesis of metal polychalcogenides, as has been implied.[13]

Procedure

Elemental K (Alfa, 99%) and elemental Te (AESAR, 99.5%) in the stoichiometric ratio of 1:2 are prereacted at 650° for 3 days in an evacuated fused-silica tube. The resultant black material is air- and moisture-sensitive. In an argon atmosphere, 0.157 g of this mixture, 0.025 g (0.14 mmol) of Hf powder (AESAR, 99.6%), and 0.068 g (0.53 mmol) of Te powder are ground together and then loaded into a fused-silica tube (i.d. 4 mm, o.d. 6 mm and approximately 18 mm long). The tube is subsequently evacuated ($\sim 10^{-4}$ torr) and sealed. The reaction mixture is heated at 650° for 6 days, then ramped to 900° to heat for 4 days. The tube is then cooled at the rate of 3°/h to 450° and

* Department of Chemistry, Northwestern University, Evanston, IL 60208-3113.
† Department of Chemistry, University of California at Los Angeles, Los Angeles, CA 90024.

finally to room temperature at 90°/h. Dull black, needle-shaped crystals of $K_4Hf_3Te_{17}$ are prevalent at the surface and within the melt. Single crystals may be manually extracted from the melt. These crystals are air-stable.

$K_4Zr_3Te_{17}$ is prepared by the same route with 0.165 g of the K/Te mixture, 0.0135 (0.15 mmol) of Zr powder (AESAR, 99%), and 0.071 g (0.56 mmol) Te. The product contains the dull black needles formed in the melt.

K_2Se_5 can be prepared by reaction of stoichiometric quantities of elemental potassium (fresh) with selenium in liquid ammonia.[14] $K_3CuNb_2Se_{12}$ can be synthesized from a reaction in an argon atmosphere of K_2Se_5 with elemental Nb, Cu, and Se in the ratio of K_2Se_5:Nb:Cu:Se = 3:4:2:9 (Nb powder 99.8%. AESAR; Cu powder 99%, ALFA; Se powder 99.5% Aldrich). The starting materials are loaded into a fused-silica tube that is then evacuated to 10^{-4} torr. The tube is sealed and placed in a furnace that is heated from room temperature to 870° in 12 h, then kept at 870° for 4 days, before it is slowly cooled to room temperature at a rate of 4°/h. The yield of crystalline material is close to 100%.

Properties

The compounds $K_4M_3Te_{17}$ crystallize in space group C_{2h}^5-$P2_1/c$ of the monoclinic system with four formula units in cells: $a = 10.148(6)$, $b = 29.889(17)$, $c = 11.626(7)$ Å, $\beta = 115.21(2)°$ ($T = 107$ K) for $K_4Hf_3Te_{17}$, $a = 10.146(2)$, $b1 \sim 29.98(1)$, $c = 11.669(4)$ Å, $\beta = 115.01(3)°$ ($T = 153$ K) for $K_4Zr_3Te_{17}$. The $K_4Hf_3Te_{17}$ structure comprises infinite, one-dimensional chains of Hf-centered polyhedra that are separated from each other by K^+ ions. Each Hf is eight-coordinate. There are six Te_2^{2-} ligands, one μ_2-η-Te_2^{2-} unit, and a μ-Te_3^{2-} unit. The composition of the infinite chain is $^1_\infty[Hf_3(Te_3)(Te_2)_7^{4-}]$, with the Hf atoms present in the +4 oxidation state.

$K_3CuNb_2Se_{12}$ is obtained as black needles. The compound contains an infinite one-dimensional mixed-metal chain. The material is monoclinic, space group C_{2h}^5-P_1/n with cell dimensions $a = 9.510(6)$, $b = 13.390(9)$, $c = 15.334(10)$ Å, $\beta = 96.09(4)°$.

References

1. P. M. Keane and J. A. Ibers, *Inorg. Chem.*, **30**, 1327 (1991).
2. Y.-J. Lu and J. A. Ibers, *Inorg. Chem.*, **30**, 3317 (1991).
3. S. A. Sunshine, D. Kang, and J. A. Ibers, *J. Am. Chem. Soc.*, **109**, 6202 (1987).
4. S. Schreiner, L. E. Aleandri, D. Kang, and J. A. Ibers, *Inorg. Chem.*, **28**, 392 (1989).
5. D. Kang and J. A. Ibers, *Inorg. Chem.*, **27**, 549 (1988).
6. M. G. Kanatzidis and Y. Park, *J. Am. Chem. Soc.*, **111**, 3767 (1989).
7. Y. Park and M. G. Kanatzidis, *Angew. Chem., Int. Ed. Engl.*, **29**, 914 (1990).
8. G. Huan and M. Greenblatt, *Mater. Res. Bull.*, **22**, 505 (1987).

9. G. Huan and M. Greenblatt, *Mater. Res. Bull.*, **22**, 943 (1987).
10. W. Bronger, in *Crystallography and Crystal Chemistry of Materials with Layered Structures*, F. Lévy (ed.), Reidel, Dordrecht-Netherlands, 1976, pp. 93–127.
11. L. Brattås and A. Kjekshus, *Acta Chem. Scand.*, **26**, 3441 (1972).
12. S. Furuseth, L. Brattås, and A. Kjekshus, *Acta Chem. Scand.*, **27**, 2367 (1973).
13. A. Stein, S. W. Keller, and T. E. Mallouk, *Science*, **259**, 1558 (1993).
14. D. Nicholls, *Inorganic Chemistry in Liquid Ammonia*, Elsevier, Amsterdam, 1979, Chapter 7.

19. SYNTHESIS OF TERNARY CHALCOGENIDES IN MOLTEN POLYCHALCOGENIDE SALTS: α-$KCuQ_4$, $KAuS_5$, $NaBiS_2$, $KFeQ_2$ (Q = S, Se)

Submitted by YOUNBONG PARK,* TIMOTHY J. McCARTHY*, ANTHONY C. SUTORIK,* and MERCOURI G. KANATZIDIS*
Checked by EDWARD G. GILLAN†

Solid-state chalcogenide compounds constitute a broad class of materials with diverse properties such as semiconducting, metallic, superconducting, nonlinear optical, charge storage, and catalytic.[1] Synthetic procedures for these compounds vary depending on the compound and the form in which it is needed. Synthesis from a flux such as alkali-metal polychalcogenide is a known synthetic technique and offers an alternative way of producing ternary metal chalcogenides with alkali ions at relatively lower temperatures.[2] The compounds α-$KCuQ_4$ (Q = S, Se), $KAuS_5$, represent some of the few examples known to contain long polychalcogenide fragments in the solid sate,[3] while $KFEQ_2$[4] (Q = S, Se), and $NaBiS_2$[5] possess one- and two-dimensional 1D, 2D structures with potentially interesting semiconducting and magnetic properties.[6] Here we give detailed syntheses for these compounds using the alkali-metal polychalcogenide flux technique.

A. α-POTASSIUM TETRASULFIDOCUPRATE(I)

$$2K_2S + Cu + 8S \rightarrow \alpha\text{-}KCuS_4 + K_2S_x$$

Procedure

0.064 g (1.00 mmol) of Cu powder (– 200 mesh), 0.256 g (8.00 mmol) of S powder, and 0.221 g (2.00 mmol) of K_2S are weighed and mixed thoroughly

* Department of Chemistry and Center for Fundamental Materials Research, Michigan State University, East Lansing, MI 48824.
† Department of Chemistry and Biochemistry and Solid State Science Center, University of California, Los Angeles, CA 90024.

and then transferred to a Pyrex tube (6 mL in volume) inside a N_2 = filled glovebox. The Pyrex tube is then flame-sealed under vacuum at the pressure of $\sim 10^{-3}$ torr. The reaction tube is placed in a programmable furnace, and the reaction temperature is programmed as follows:

	T_1 (°C)	T_2 (°C)	Time (h)
Step 1	50	215	12
Step 2	215	215	96
Step 3	215	50	100

To isolate the product, the tube is opened with a glass cutter and the contents are placed into a 250-mL flask under N_2. Excess K_2S_x is removed by adding degassed water (150 mL) with occasional stirring, resulting in a yellow solution dissolving the potassium polysulfide flux. After the remaining solid settles completely, the solution is slowly decanted. The washings with degassed water are repeated until the solution remains colorless, indicating total removal of the potassium polysulfide flux. Then, the product is washed with ethanol and ether. Orange-yellow needle-like crystals of α-$KCuS_4$ are obtained. The yield is 0.166 g (72% based on Cu).

Properties

Observed X-ray powder pattern spacings (Å, Cu K_α): 7.05(vs), 6.35(m), 5.07(s), 3.99(m), 3.29(m), 3.15(vs), 3.06(m), 2.94(m), 2.91(m), 2.73(m), 2.70(m), 2.58(s), 2.55(m), 2.47(m), 2.42(m), 2.32(m), 2.15(w), 2.09(s), 1.94(m), 1.87(w), 1.76(m), 1.74(w), 1.66(m), 1.62(w), 1.58(w). Fourier transform–infrared (FT-IR) spectra (cm^{-1}, CsI pellet): 481(m), 435(m), 412(m), 286(m), 259(m).

B. α-POTASSIUM TETRASELENIDOCUPRATE(I)

$$2K_2Se + Cu + 8Se \rightarrow \alpha\text{-}KCuSe_4 + [K_2Se_x]$$

Procedure

First, 0.032 g (0.50 mmol) of Cu powder (−200 mesh), 0.316 g (4.00 mmol) of Se powder, and 0.157 g (1.00 mmol) of K_2Se are weighed and mixed thoroughly and then transferred to a Pyrex tube (6 mL in volume) inside an N_2-filled glovebox. The Pyrex tube is then flame-sealed under vacuum at the pressure of $\sim 10^{-3}$ torr. The reaction tube is placed in a programmable furnace and the reaction temperature is programmed as follows:

	T_1 (°C)	T_2 (°C)	Time (h)
Step 1	50	250	12
Step 2	250	250	114
Step 3	250	50	100

To isolate the product, the tube is opened with a glass cutter and the contents are placed into a 250-mL flask under N_2. Excess K_2Se_x is removed by adding degassed water (150 mL) with occasional stirring, resulting in a red solution on dissolving the potassium polyselenide flux. After the remaining solid settles completely, the solution is slowly decanted. The washings with degassed water are repeated until the solution remains colorless, indicating the total removal of the potassium polyselenide flux. The product is finally washed with ethanol and ether. Dark red needle-like crystals of α-$KCuSe_4$* are obtained. The yield is 0.159 g (76% based on Cu).

Properties

Observed X-ray powder pattern spacings (Å, Cu K_α): 7.26(m), 6.49(m), 5.21(vs), 3.87(w), 3.80(w), 3.43(w), 3.31(m), 3.24(s), 3.17(w), 3.06(m), 3.03(s), 2.83(m), 2.79(m), 2.68(m), 2.65(m), 2.59(m), 2.52(w), 2.44(w), 2.30(w), 2.25(m), 2.17(m), 2.15(w), 2.04(w), 2.01(m), 1.98(w), 1.93(w), 1.89(w), 1.82(m), 1.74(m), 1.70(m), 1.68 ~ 1.67 (m, diffuse). FT–IR spectra (cm^{-1}, CsI pellet): 249(m), 241(m), 181(m), 145(m).

C. POTASSIUM PENTASULFIDOAURATE(I)

$$1.8K_2S + Au + 8S \rightarrow KAuS_5 + [K_2S_x]$$

Procedure

First, 0.196 g (1.00 mmol) of Au powder (−325 mesh), 0.256 g (8.00 mmol) of S powder, and 0.198 g (1.80 mmol) of K_2S are weighed and mixed thoroughly and then transferred to a Pyrex tube (6 mL in volume) inside an N_2-filled glovebox. The Pyrex tube is then flame-sealed under vacuum at the pressure

* α-$KCuSe_4$ can also be prepared using a 1–4:1:8 ratio of K_2Se:Cu:Se in the temperature region 250–390°.

of ~10^{-3} torr. The reaction tube is placed in a programmable furnace, and the reaction temperature is programmed as follows:

	T_1 (°C)	T_2 (°C)	Time (h)
Step 1	50	250	12
Step 2	250	250	99
Step 3	250	50	100

To isolate the product, the tube is opened with a glass cutter and the contents are placed into a 250-mL flask under N_2. Excess K_2S_x is removed by adding degassed water (150 mL) with occasional stirring, resulting in a yellow solution on dissolving the potassium polysulfide flux. After the remaining solid settles completely, the solution is slowly decanted. The washings with degassed water are repeated until the solution remains colorless, indicating total removal of the potassium polysulfide flux. The product is finally washed with ethanol and ether. Golden yellow needle-like crystals of $KAuS_5$ are obtained. The yield is 0.310 g (78% based on Au).

Properties

Observed X-ray powder pattern spacings (Å, Cu K_α): 7.93(m), 6.73(s), 5.48(m), 5.08(m), 4.22(m), 3.91(w), 3.81(w), 3.70(m), 3.33(m), 3.06(vs), 2.86(s), 2.71(s), 2.56–2.52(m, diffuse), 2.28(w), 2.22–2.21(m, diffuse), 2.18(m), 2.12(m), 2.09(m), 2.04(w), 2.02(w), 1.96(m), 1.91(w), 1.87(w), 1.84(w), 1.80(w), 1.71(w), 1.68(w), 1.65(w), 1.62(w).

D. SODIUM DISULFIDOBISMUTHATE(III)

$$4Na_2S + Bi + 8S \rightarrow NaBiS_2 + [Na_2S_x]$$

Procedure

First, 0.064 g (2.00 mmol) of S powder, 0.052 g (0.25 mmol) of Bi, and 0.078 g (1.00 mmol) of Na_2S are weighed and mixed thoroughly and then transferred to a Pyrex tube (6 mL in volume) inside a N_2-filled glovebox. Thc tube is then flame-sealed under vacuum at the pressure of ~10^{-3} torr. The reaction tube is placed in a programmable furnace, and the reaction temperature is programmed as follows:

	T_1 (°C)	T_2 (°C)	Time (h)
Step 1	50	290	12
Step 2	290	290	96
Step 3	290	80	105
Step 4	80	50	2

To isolate the product, the tube is opened with a glass cutter and the contents are placed in a 250 mL flask under N_2. Excess Na_2S_x is removed by adding degassed water (150 mL) with occasional stirring, resulting in a yellow solution on dissolving the sodium polysulfide flux. After the remaining solid settles completely, the water is slowly decanted. The washings with degassed water are repeated until the solution remains colorless, indicating total removal of sodium polysulfide flux. The product then is washed with isopropanol (20 mL) and finally ether (20 mL). Gray microcrystals of $NaBiS_2$* are obtained. The yield is 0.062 g (84% based on Bi).

Properties

Observed X-ray powder pattern spacings (Å, Cu K_α): 3.36(s), 2.90(s), 2.05(s), 1.75(s), 1.67(s).

E. POTASSIUM DISULFIDOFERRATE(III)

$$4K_2S + 1.1Fe + 8S \rightarrow KFeS_2 + [K_2S_x]$$

Procedure

First, 0.015 (0.26 mmol) of Fe powder (325 mesh), 0.064 g (2.00 mmol) of S powder, and 0.110 g (1 mmol) of K_2S are weighed and mixed thoroughly and then transferred to a Pyrex tube (6 mL in volume) inside a N_2-filled glovebox. The tube is then flame-sealed under vacuum at the pressure of $\sim 10^{-3}$ torr. The reaction tube is placed in a programmable furnace and the reaction

* $NaBiS_2$ can also be prepared using 1–6:1:8 ratio of Na_2S:Bi:S at 250°.

temperature is programmed as follows:

	T_1 (°C)	T_2 (°C)	Time (h)
Step 1	50	350	12
Step 2	350	350	192
Step 3	350	80	135
Step 4	80	50	1

To isolate the product, the tubes are opened in a N_2-filled glovebox and the contents are loosened with a portion of degassed water and transferred to a fritted funnel. The potassium polysulfide flux is removed by adding successive portions of degassed water to the sample (each removed by suction filtration), which forms a yellow solution on dissolving the potassium polysulfide flux. This is continued until the added water remains colorless, indicating the potassium polysulfide flux has been totally removed. The product is finally washed with ethanol and ether. Brown microneedle crystals of $KFeS_2$ are obtained. The yield is 0.026 g (57% based on Fe).

Properties

Observed X-ray powder pattern spacings (Å Cu K_α): 5.66(vs), 3.26(s), 3.22*(m), 2.92(s), 2.83*(w), 2.51(m), 2.48*(w), 2.28(w), 2.20(m), 2.14(m), 2.06(w), 2.03*(w), 1.88(m), 1.83(w), 1.74(m), 1.70(w), 1.68*(w), 1.65(w), 1.63*(m), 1.59(w), 1.57*(s).

Those peaks marked with (*) are not present in the reported powder pattern of this compound[4] but were confirmed as belonging to $KFeS_2$ by comparison to a powder pattern calculated based on the published coordinates.[4]

F. POTASSIUM DISELENIDOFERRATE(III)

$$6K_2Se + Fe + 8Se \rightarrow KFeSe_2 + [K_2Se_x]$$

Procedure

First, 0.021 g (0.038 mmol) of Fe powder (325 mesh), 0.237 g (3.00 mmol) of Se powder, and 0.356 g (0.75 mmol) of K_2Se are weighed and mixed thoroughly and then transferred to a Pyrex tube (6 mL in volume) inside a N_2-filled glovebox. The tube is then flame-sealed under vacuum at the pressure of $\sim 10^{-3}$ torr. The reaction tube is placed in a programmable furnace, and the reaction temperature is programmed as follows:

	T_1 (°C)	T_2 (°C)	Time (h)
Step 1	50	400	12
Step 2	400	400	192
Step 3	400	120	140
Step 4	120	50	1

To isolate the product, the tubes are opened in a N_2-filled glovebox and the contents are loosened with a portion of degassed water and transferred to a fitted funnel. The potassium polyselenide flux is removed by adding successive portions of degassed water to the sample (each removed by suction filtration), which forms a dark red solution on dissolving the potassium polyselenide flux. This is continued until the added water remains colorless, indicating the potassium polyselenide flux has been totally removed. The product is finally washed with ethanol and ether. The resulting crystals are black or dark purple chunks, the largest of which are 1 mm^3. If present, the black color indicates the presence of small amounts of elemental Se, but this is easily removed by washing the crystals with portions of ethylenediamine (ethanol/ether rinse), leaving the entire sample dark purple. On gently crushing the chunks, it can be seen that they are in fact composed of bundles of needles, the expected form of $KFeSe_2$ crystals. The yield is 0.024 g (24% based on Fe).

Properties

Observed X-ray powder pattern spacings (Å, Cu K_α): 5.93(vs), 3.41(m), 3.35(m), 3.04(m), 2.95(s), 2.61(w), 2.29(w), 2.22(m), 1.95(m), 1.80(w), 1.69(w), 1.62(w).

General Comments

All compounds are insoluble in all organic solvents. They are air stable for a limited time (days). However, as with most chalcogenide compounds, they should be kept under inert environment for long-term storage.

Acknowledgment

Financial support from the Center for Fundamental Materials Research (CFMR) Michigan State University and the Donors of the Petroleum

Research Fund, administered by the American Chemical Society is gratefully acknowledged.

References

1. (a) J. Rouxel, in *Crystal Chemistry and Properties of Materials with Quasi One-Dimensional Structures*, J. Rouxel (ed.), Reidel, 1986, and references cited therein; (b) R. R. Chianelli, T. A. Pecoraro, T. R. Halbert, W.-H. Pan, and E.I. Stiefel, *J. Catal.*, **86**, 226 (1984); (c) N. Yamada, N. Ohno, N. Akahira, K. Nishiuchi, K. Nagata, and M. Takao, *Proc. Int. Symp. Optical Memory, 1987, Jpn. J. Appl. Phys.,* **26**, Suppl. 26-4, 61 (1987).
2. (a) H. J. Scheel, *J. Cryst. Growth*, **24/25**, 669 (1974). (b) S. A. Sunshine, D. Kang, and J. A. Ibers, *J. Am. Chem. Soc.*, **109**, 6202 (1987).
3. (a) M. G. Kanatzidis and Y. Park, *J. Am. Chem. Soc.*, **111**, 3767 (1989). (b) Y. Park and M. G. Kanatzidis, *Angew. Chem., Int. Ed. Engl.*, **29**, 914 (1990).
4. (a) K. Preis, *J. Prakt. Chem.*, **107**, 12 (1969). (b) R. Schneider, *J. Prakt. Chem.*, **108**, 16 (1869); (c) W. Bronger, *Naturwissenschaften*, **53**, 525 (1966).
5. (a) J. W. Boon, *Rec. Trav. Chim. Pays-Bas*, **63**, 32 (1944); (b) *X-ray Powder Data File*, American Society for Testing Materials, **8**, 406 (1955); (c) V. B. Lazarev, A. V. Salov, and S. I. Berul, *Zh. Neorg. Khim.*, **24**, 563 (1979); (d) E. Y. Peresh, M. I. Golovei, and S. I. Berul, *Inorg. Mater.*, **7**, 29 (1971).
6. (a) W. Bronger, *Z. Anorg. Allg. Chem.*, **359**, 225 (1968); (b) W. Bronger, and P. Muller, *J. Less-Common Met.*, **70**, 253 (1980); (c) H. P. Nissen and K. Nagorny, *Z. Phys. Chem. NF*, **95**, 301 (1975); (d) H. P. Nissen and K. Nagorny, *Z. Phys. Chem. NF*, **99**, 14 (1976); (e) W. V. Sweeney and R. E. Coffman, *Biochim. Biophys. Acta*, **286**, 26 (1972); (f) D. C. Johnston, S. C. Mraw, and A. J. Jacobson, *Solid State Commun.*, **44**, 225 (1982); (g) *X-ray Powder Data File*, American Society for Testing Materials, **3**, 363 (1960); (h) W. Bronger, A Kyas, and P. Muller, *J. Solid. State Chem.*, **70**, 262 (1987).

Chapter Three

CRYSTAL GROWTH

20. SINGLE CRYSTALS OF TRANSITION-METAL DIOXIDES

Submitted by D. B. ROGERS,* S. R. BUTLER,*† and R. D. SHANNON*
Checked by A. WOLD‡ and R. KERSHAW‡

Reprinted from *Inorg. Synth.*, **13**, 135 (1972)

Most of the dioxides of the transition metals crystallize in structural types that are closely related to that of the rutile form of titanium dioxide. The series is notable for the wide variety of physical properties and modifications in structure that occur as functions of *d*-electron number.[1] Electrical transport properties range from insulating to metallic; magnetic properties from Pauli paramagnetic to ferromagnetic. Titanium dioxide has a high index of refraction and is useful as a white pigment and as a dielectric; chromium dioxide is ferromagnetic and has found application in magnetic recording tapes; vanadium dioxide exhibits electrical switching at 68° as conductivity-type changes reversibly from semiconducting to metallic. Many of the heavier dioxides are remarkably good electrical conductors. De Haas–Van Alphen oscillations recently observed[2] in crystals of the dioxides, RuO_2, IrO_2, and

* Central Research Department, Experimental Station, E. I. du Pont de Nemours & Company, Wilmington, DE 19898.

† Present address: Department of Metallurgy and Materials Science, Lehigh University, Bethlehem, PA 18015.

‡ Brown University, Providence, RI 02912.

OsO_2 were the first such observations on oxidic compounds and provide strong evidence for the formation in these oxides of a primarily *d*-character conduction band that possesses a discrete Fermi surface.

Several high-temperature procedures have been described in the literature for the preparation of the transition-metal dioxides. Direct oxidation of the metals, lower oxides, chlorides, or nitrate precursors provides a convenient route to the dioxides of several metals: Ti, Mn, Ru, Rh, Os, Ir, and Pt.[1,3–5] (Syntheses of the rutile forms of rhodium and platinum dioxides by direct oxidation requires application of high pressures.[5]) Reduction of higher oxides is the most common method of synthesis for these dioxides: VO_2, NbO_2, MoO, and β-ReO_2.[4,6–8] Stoichiometry in these reactions is most readily controlled by use of the respective metal or a lower oxide as reductant. Chromium dioxide is normally synthesized by hydrothermal reduction of the trioxide.[9]

Except for platinum and rhodium, which have low thermal stabilities, and for technetium, single-crystal growth has been achieved for all the transition-metal dioxides with rutile-related structures. Techniques include flame fusion,[10] electrolytic or thermal reduction from fused salts,[11,12] chemical transport,[13] and extremely high-pressure or hydrothermal recrystallization.[1,14,15] Of these, chemical transport and hydrothermal procedures have the most general applicability and appear to lead to products of highest purity. Chemical transport is preferred because it is convenient and utilizes relatively simple laboratory equipment. This technique has been found to be applicable for the crystal growth of titanium dioxide using hydrogen chloride[13,16] or titanium tetrachloride[13] as transporting agents of ruthenium, osmium, and iridium using oxygen,[13] and of niobium, tungsten, and rhenium dioxides using iodine.[1,17] Tungsten dioxide has also been grown by the chemical transport of tungsten trioxide in a reducing hydrogen–water stream.[18] It seems likely that this rather general technique would also be useful in the growth of crystals of molybdenum and technetium dioxides, perhaps using iodine as the transporting agent, in view of the similar chemistries of these entities to those of tungsten and rhenium, respectively.

Procedures are given here for the crystal growth of the following dioxides: RuO_2, IrO_2, OsO_2, and WO_2, using chemical transport techniques. The procedures described for the dioxides of ruthenium, iridium, and osmium are elaborations of those previously given by Schäfer et al.[13,19,20]

A. RUTHENIUM AND IRIDIUM DIOXIDES

$$RuO_2(s) + \tfrac{1}{2}O_2(g) \rightleftarrows RuO_3(g)$$

$$RuO_2(s) + O_2(g) \rightleftarrows RuO_4(g)$$

$$IrO_2(s) + \tfrac{1}{2}O_2(g) \rightleftarrows IrO_3(g)$$

Procedure

■ **Caution.** *The volatile higher oxides of ruthenium and iridium are highly toxic. The gas trains used in this procedure should be vented into an efficient fume hood.*

Ruthenium and iridium dioxides can be grown by essentially identical procedures. These procedures take advantage of chemical transport from a hotter temperature (T_2) to a cooler one (T_1) via volatile higher oxides. Starting reagents are the polycrystalline dioxides prepared by direct oxidation of ruthenium* and iridium† metal sponges. A silica boat containing about 5 g of the respective metal‡ is placed in a silica combustion tube that is fitted via lubricated standard-taper joints to an inlet tube for dry oxygen and an exit tube leading to a water bubbler at the opposite end. The tube is then placed in a horizontal tube furnace and heated in a slow stream of dry oxygen at 1000° for 24 h. During this process, partial volatilization of the metal oxides will occur as evidenced by the formation of small ($\sim \frac{1}{2}$–1-mm) crystals on the downstream end of the silica boat and inner wall of the combustion tube. However, maximum crystal size and quality require more careful control of transport conditions. The yield of polycrystalline dioxide remaining in the boat is about 90–95%. In the case of iridium, oxygen diffusion in the solid is slow, and the process does not give the dioxide quantitatively. This fact is not important for the subsequent conversion of the non-stoichiometric, polycrystalline product to stoichiometric crystals.

Single-Crystal Growth

The apparatus used for crystal growth is shown in Fig. 1 and essentially consists of an oxygen flow system that permits monitoring of flow rate and a gradient furnace that permits careful control of temperature gradients in the region of growth. The growth portion of the flow system is fabricated of fused silica and consists of an outer combustion tube (130 cm in length and 18 mm i.d.) and an inner tube (60 cm in length and 15 mm i.d.) that is split into two semicylindrical halves along its length. The function of the split inner tube is to facilitate removal of the product crystals, which grow on the inner wall of this tube. The outer combustion tube is equipped at both ends with standard-taper joints that connect on the inlet end to a monitored source of dry oxygen and on the exit end to a water bubbler and final ventilation in a fume hood.

* Engelhard Industries, Newark, NJ 07114.

† United Mineral and Chemical Company, New York, NY 10013.

‡ The checkers used 2 g of the relevant metal.

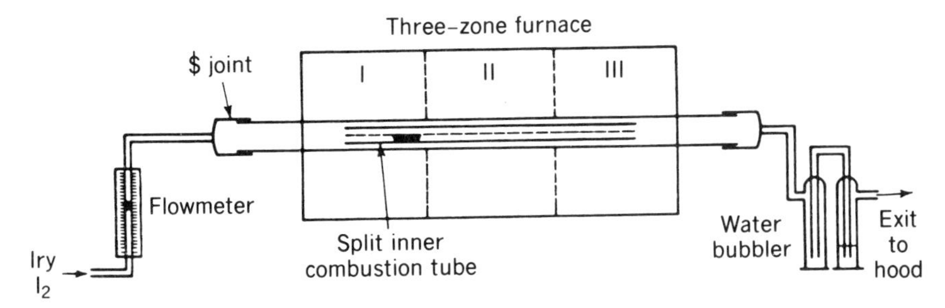

Figure 1. Apparatus for crystal growth by chemical transport.

The gradient furnace can be of variable design, but should provide a source region at a temperature of 1190° and a growth region at 1090°, with a gradient decreasing from about 8°/cm in the source region to about 1°/cm in the growth region. In the apparatus of Fig. 1, these conditions are achieved using a three-zone furnace. Zone I is the source region of highest temperature (1190°). Crystals of maximum quality grow in zone II, where a shallow gradient (~1°/cm) can be maintained over several centimeters at 1090° using a buffer zone (III) set at a slightly higher temperature (~1110°).

The combustion tubes are thoroughly washed in distilled water and dried before use. About 2 g* of ruthenium or iridium dioxide is placed in a $2\frac{3}{4}$-in. boat of recrystallized alumina, and the sample is positioned in zone I of the apparatus, as shown in Fig. 1. Transport is then accomplished with a source-region temperature (T_2) of 1190° and a growth-region temperature (T_1) of 1090° under a stream of oxygen flowing at a rate of about 15–20 cm^3/min. Transport is complete in about 15 days, and the operation is terminated by decreasing the flow rate of oxygen to about 2 cm^3/min. and turning off all furnace power. When the furnace has cooled to room temperature (usually overnight), the split, inner combustion tube is removed and the semi-cylindrical halves are separated. Crystals will be found all along the down-stream length of the tube; however, those of maximum size and quality occur in that portion of the tube exposed to the shallow gradient at 1090°. Generally, these crystals can be removed by gentle tapping of the tube walls.

Properties

Ruthenium dioxide is blue-black, and crystals formed in the growth region of the apparatus described above are tabular and about 3–4 mm in length. Iridium dioxide is somewhat darker (almost black), the normal crystal habit

* The checkers used about 1 g of the dioxide and carried out the transport for 4 days. The resulting crystals were smaller than those reported by the authors, but had the same properties.

is needle-like, and the crystals are smaller than those of ruthenium dioxide. X-ray powder diffraction patterns taken on ground samples of the crystals can be indexed on the basis of tetragonal unit cells for both dioxides with $a_0 = 4.4906$ Å, $c_0 = 3.1064$ Å. For RuO_2, and $a_0 = 4.4990$ Å, $c_0 = 3.1546$ Å for IrO_2. Both dioxides are Pauli paramagnetic and exhibit metallic conductivity ($\rho_{300\text{ K}} \approx 4 \times 10^{-5}\ \Omega \cdot \text{cm}$).

Anal. Weight percent oxygen calcd. for RuO_2: 24.05. Found: 24.24. Calcd. for IrO_2: 14.27. Found: 14.40.

B. OSMIUM DIOXIDE

$$3\text{Os} + 2\text{NaClO}_3 \rightarrow 3\text{OsO}_2 + 2\text{NaCl}$$

$$\text{OsO}_2(s) + \text{O}_2(g) \rightleftarrows \text{OsO}_4(g)$$

Procedure

■ **Caution.** *The tetraoxide of osmium, which is involved in the transport process, is more stable, volatile, and toxic than that of ruthenium. It is recommended that the operations described here, which involve the handling of the metal and the oxide at elevated temperatures, be carried out in an efficient fume hood.*

Both the preparation of polycrystalline osmium dioxide reagent and its subsequent growth into single-crystal form are conveniently carried out in a single reaction ampule. The ampule is fabricated from a 25-cm length of 13-mm i.d. silica tubing that has been closed at one end. After the tube has been thoroughly washed in distilled water and dried, 0.15 g of osmium metal powder* and 0.065 g of sodium chlorate (about 10% excess over the amount needed for complete conversion of the osmium to the dioxide) are added through a long-stemmed funnel and the tube is connected via rubber tubing to any common vacuum system. When the pressure in the system has been reduced to about 10^{-3} mm, the transport tube is sealed at a length of 15 cm. with an oxyhydrogen flame.

■ **Caution.** *The end of the tube containing sodium chlorate and osmium must not be heated during the sealing procedure. This can be prevented by immersing the lower 5 cm of the tube in a beaker of water or by wrapping such a length in wet asbestos.*

* Electronic Space Products, Inc., Los Angeles, CA 90035.

The sealed ampule is then slowly heated (at a rate of about 50°/h) in a muffle furnace to 300° and left overnight.

■ **Caution.** *Rapid heating at this point must be avoided.*

The temperature is then raised to 650° for an additional 3 h. This treatment results in complete decomposition of the chlorate and formation of golden osmium dioxide powder in one end of the tube. An oxygen pressure of about 0.2 atm results from the excess of sodium chlorate, and about 0.036 g of sodium chloride is present as the byproduct of chlorate decomposition. The oxygen is useful as the transporting agent during subsequent crystal growth; sodium chloride serves no useful function in the reaction, but is not detrimental.

Single-Crystal Growth

Chemical transport of osmium dioxide is carried out in a transport furnace wired to provide two independently controllable zones. The transport tube containing oxide, oxygen, and byproduct sodium chloride is centered in the two-zone furnace with half of the tube in each of the separate zones. Thermocouples for temperature control are placed at the ends of the tube. It is important for optimum quality and size of product crystals that the growth (empty) zone of the transport tube be free of nucleation sites. To ensure that no microscopic seeds of osmium dioxide are in the growth zone, reverse transport conditions are imposed by heating the growth zone to 960°, while holding the charge at a lower temperature (900°). After several hours of back-transport, the temperatures of the zones are reversed and growth is allowed to proceed with the charge maintained at 960° (T_2) and the growth zone at 900° (T_1). After 2 days of growth, the furnace is turned off and allowed to cool to room temperature (overnight), and the transport tube is removed and opened.

Properties

Osmium dioxide is golden and crystals formed in the growth zone have an equidimensional habit and are about 2 mm across a polyhedral face. X-ray powder diffraction patterns taken on powdered crystals can be indexed on the basis of a tetragonal unit cell with $a_0 = 4.4968$ Å and $c_0 = 3.1820$ Å. The oxide exhibits metallic conductivity ($\rho_{300\,K} \approx 6 \times 10^{-5}\ \Omega \cdot cm$) and is Pauli paramagnetic. Resistivity ratios ($\rho_{300\,K}/\rho_{4.2\,K}$) on typical crystals are about 200–300.

Anal. Weight percent oxygen calcd. for OsO_2: 14.40. Found: 14.49.

C. TUNGSTEN DIOXIDE AND β-RHENIUM DIOXIDE

$$W + 2WO_3 \rightarrow 3WO_2$$

$$Re + 2ReO_3 \rightarrow 3ReO_2$$

Procedure

The dioxides of tungsten and rhenium are conveniently prepared in powder form for subsequent conversion to single crystals by direct reactions between their respective metal powders and trioxides in a sealed, evacuated system. both of the metal powders should be freshly reduced in a stream of hydrogen for 3 h at 1000°.

■ **Caution.** *Hydrogen forms explosive mixtures with air. The combustion system used for reduction should be thoroughly flushed with nitrogen before admitting hydrogen, and provision for venting the exit gas must be made.*

Tungsten trioxide is predried at 550° for about 2 h prior to use; rhenium trioxide*† can be used without pretreatment. For the preparation of rhenium dioxide, 2.0 g of rhenium trioxide and 0.7942 g of rhenium‡ are added by means of a long-stemmed funnel to a precleaned ampule fabricated by closing one end of a 25-cm length of 13-mm-i.d. silica tubing. The ampule is then attached *via* rubber tubing to a common vacuum system, the pressure is reduced to about 10^{-3} mm, and finally, the tube is sealed at a length of about 10 cm using an oxyhydrogen flame. Reaction inside the sealed ampule to form the dioxide is then accomplished by heating§ the ampule in a muffle furnace at 500° for about 24 h. The general procedure for the preparation of tungsten dioxide is the same as for rhenium dioxide. However, in this case, 2.0 g of trioxide is reacted with 0.7932 g of tungsten at 1100° for 24 h. Reaction rate is improved by adding 1 or 2 mg of iodine as a mineralizer to the tungsten reagents prior to evacuation and sealing of the reaction ampule. The products of these reactions are a gray-black powder in the case of rhenium and a golden-brown powder in the case of tungsten.

Single-Crystal Growth

Single crystals of tungsten and rhenium dioxides are grown by a procedure that is analogous to that previously described for osmium dioxide, except

* Alfa Inorganics, Inc., Beverly, MA 01915.
† The checker redried the ReO_3 for 12 h at 110°.
‡ Electronic Space Products, Inc., Los Angeles, CA 90035.
§ The tube was slowly heated to 500° at the rate of 15°/h in order to prevent explosions.

that iodine is used as the transport agent. The reaction is presumed to involve an oxyiodide and to be of the type

$$MO_2(s) + I_2(g) \rightleftarrows MO_2I_2(g) \qquad (M = W \text{ or } Re)$$

However, the vapor species involved have not been identified. About 0.5 g of powdered dioxide and 0.003 g of iodine are added by means of a long-stemmed funnel to a precleaned, silica ampule fabricated by closing one end of a 25-cm length of 13-mm-i.d. tubing. The ampule is evacuated to a pressure of about 10^{-3} mm, sealed at a length of 15 cm, and centered in a two-zone furnace as described in the procedure for growth of OsO_2 crystals. In the case of rhenium dioxide, back-transport to remove stray nuclei is accomplished by heating the growth end of the ampule to 850° while maintaining the charge at 825°; for tungsten dioxide, this is done at temperatures of 1000 and 960° for growth and charge ends, respectively. After back-transport for several hours, the temperatures of the zones are reversed and growth is allowed to proceed. After 3 days of growth, the furnace is turned off, allowed to cool to room temperature, and the transport ampule is removed and opened. Rhenium dioxide crystals should be black; however, the crystals as recovered occasionally have a red-black mottled appearance. This is due to slight surface oxidation with the formation of red trioxide during the cooling procedure. This surface impurity is readily removed by etching the crystals in cold, dilute nitric acid.

Properties

Crystals of tungsten dioxide are golden, and when grown by the procedure described above, they are equidimensional with approximately 2-mm polyhedral faces. The crystallographic symmetry is monoclinic with unit cell parameters $a_0 = 5.5607$ Å, $b_0 = 4.9006$ Å, $c_0 = 5.6631$ Å, and $\beta = 120.44°$. Tungsten dioxide exhibits metallic conductivity ($\rho_{300\,K} \approx 3 \times 10^{-3}\ \Omega \cdot cm$). The resistivity ratio ($\rho_{300\,K}/\rho_{4.2\,K}$) measured for a typical crystal is about 20.

Anal. Weight percent oxygen calcd. for WO_2: 14.81. Found: 14.89.

Two crystallographic forms of rhenium dioxide are known. When synthesized below 300°, the structural modification (α) is isostructural with tungsten dioxide. Above $\sim$300°, this modification transforms irreversibly to an orthorhombic form (β), and when initial synthesis is carried out at temperatures greater than 300°, the β-modification is invariably recovered. Therefore, the crystals grown by the process described above are β-rhenium dioxide and have an orthorhombic unit cell with $a_0 = 4.809$ Å, $b_0 = 5.643$ Å, and $c_0 = 4.601$ Å. The crystals are black, possess a columnar habit, are about

2–3 mm in length, and usually are twinned. They exhibit metallic conductivity with $\rho_{300\,K} \approx 10^{-4}\,\Omega\cdot$cm.; however, crystals grown by this process have a relatively low resistivity ratio ($\lesssim 10$).

Anal. Weight percent oxygen calcd. for ReO_2: 14.67. Found: 14.73.

References

1. D. B. Rogers, R. D. Shannon, A. W. Sleight, and J. L. Gillson, *Inorg. Chem.*, **8**, 841 (1969).
2. S. M. Marcus and S. R. Butler, *Phys. Lett.*, **26A**, 518 (1968); S. M. Marcus, paper presented at Am. Phys. Soc. Meeting, Philadelphia, March, 1969.
3. H. Remy and M. Kohn, *Z. Anorg. Allg. Chem.*, **137**, 381 (1924).
4. G. Brauer, *Handbook of Preparative Inorganic Chemistry*, Vol. 2, Academic Press, New York, 1965.
5. R. D. Shannon, *Solid State Commun.*, **6**, 139 (1968).
6. C. Friedheim and M. K. Hoffman, *Ber. Deut. Chem. Ges.*, **35**, 792 (1902).
7. O. Glemser and H. Sauer, *Z. Anorg. Allg. Chem.*, **252**, 151 (1943).
8. P. Gibart, *Compt. Rend.*, **261**, 1525 (1965).
9. P. Arthur, Jr., U.S. Patent 2,956,955 (1960).
10. J. B. MacChesney and H. J. Guggenheim, *J. Phys. Chem. Solids*, **30**, 225 (1969).
11. D. S. Perloff and A. Wold, in *Crystal Growth*, H. S. Peiser (ed.), Pergamon Press, London, 1967, p. 361.
12. A. M. Vernoux, J. Giordano, and M. Foex, ibid., p. 67.
13. H. Schäfer, *Chemical Transport Reactions*, Academic Press, New York, 1964.
14. B. L. Chamberland, *Mater. Res. Bull.*, **2**, 827 (1967).
15. M. L. Harvill and R. Roy, in *Crystal Growth*, H. S. Peiser (ed.), Pergamon Press, London, 1967, p. 563.
16. H. Sainte-Claire Deville, *Ann. Chem.*, **120**, 176 (1861).
17. H. Schäfer and M. Hüesker, *Z. Anorg. Allg. Chem.*, **317**, 321 (1962).
18. T. Millner and J. Neugebauer, *Nature*, **163**, 601 (1949).
19. H. Schäfer and H. J. Heitland, *Z. Anorg. Allg. Chem.*, **304**, 249 (1960).
20. H. Schäfer, G. Schneidereit, and W. Gerhardt, ibid., **319**, 327 (1963).

21. MOLYBDENUM(IV) OXIDE AND TUNGSTEN(IV) OXIDE SINGLE CRYSTALS

$$2MO_3 + M \rightarrow 3MO_2$$

$$MO_2 + I_2 \underset{T_1}{\overset{T_2}{\rightleftharpoons}} MO_2I_2 \qquad (M = Mo, W)$$

Submitted by LAWRENCE E. CONROY* and LINA BEN-DOR†
Checked by R. KERSHAW‡ and A. WOLD‡

Reprinted from *Inorg. Synth.*, **14**, 149 (1973)

The dioxides of molybdenum and tungsten crystallize in distorted forms of the rutile structure. The usual method of preparation is the reduction of the trioxide with the metal at 900–1000 °C *in vacuo* or under an inert atmosphere. This procedure yields only microcrystalline products. Extended periods of reaction are necessary to avoid contamination with oxides intermediate between the trioxide and dioxide. Hägg and Magneli[1, 2] have shown that a series of such oxides can be identified in the products of partial reduction of the trioxides. Single crystals of molybdenum(IV) oxide may be produced by electrolytic reduction of Na_2NoO_4–MoO_3 melts.[3] The synthesis described below makes use of the chemical-transport technique[4] to convert the microcrystalline dioxide to single crystals sufficiently large for electrical and crystallographic investigations. The principles of this method are described in earlier volumes of this series.[5–7]

Procedure

The pure molybdenum(VI) or tungsten(VI) oxide is dried by heating in air at 500° for one hour. Approximately 10–15 g of the 2:1 molar mixture of the trioxide and corresponding metal is mixed thoroughly and ground together in a mortar. The mixture is transferred to an alumina or silica combustion boat and heated under purified argon gas. Alternatively, the sample may be sealed *in vacuo* in a silica ampule. The $2MoO_3$–Mo mixture is converted to molybdenum(IV) oxide in 70 h at 800°. The $2WO_3$–W mixture is converted

* University of Minnesota, Minneapolis, MN 55455.
† The Hebrew University, Jerusalem, Israel.
‡ Brown University, Providence, RI 02912.

§ These extended heating periods are necessary for complete conversion to the dioxides. During the early stages of heating, the primary products are the phases Mo_4O_{11} and $W_{18}O_{49}$,[1] which transport more readily than MoO_2 and WO_2 and thus interfere with the growth of good single crystals of the last two compounds.

to tungsten(IV) oxide in 40 h§ at 900°. Molybdenum(IV) oxide is produced as a brown-violet powder, and tungsten(IV) oxide as a bronze powder.

■ **Caution.** *The danger of implosion is always present when sealed glass vessels are heated, however carefully. Use protective eye and face covering or a protective shield for the apparatus.*

Silica or Vycor ampules are satisfactory for chemical transport. Optimum dimensions are 15 cm long and 2.5 cm o.d., holding a sealed volume of approximately 75 mL. Of the metal dioxide, 4 g is added to the ampule through a long-stemmed funnel, along with the required quantity (see Table I) of purified iodine. Several techniques for addition of iodine to a transport ampule are described in Volume 12 (p. 161) of *Inorganic Syntheses*. The ampule is then sealed.* To promote the growth of large crystals, it is desirable to transfer the charge of dioxide to the seal-off end (hot zone) of the ampule and to allow crystals to grow at the hemispherical end (growth zone). The ampule is then placed in a cold two-zone transport furnace.† The growth zone is heated to 900° for at least 12 h, while maintaining the charge zone at 500–600°.

This procedure minimizes seed sites to promote the growth of a small number of large crystals. The temperature of the charge zone is then increased to the appropriate temperature listed in Table I, and the growth zone is cooled at the rate of 10°/h for the stipulated time period. For the growth of

TABLE I. Optimum Growth Conditions for MoO_2 and WO_2

Ampule Dimensions	MoO_2 15 cm Long × 2.5 cm o.d.	WO_2 15 cm Long × 2.5 cm o.d.
Ampule volume	~75 mL	~75 mL
Quantity of MO_2 powder	4 g	4 g
Iodine concentration	4 mg/mL	1 mg/mL
Transport temperatures	900 → 700°	900°800°
Transport time	6–7 days	3 days
Cooling rate, growth zone	10°/h	10°/h
Temperature programming	Necessary	Not necessary

* Vycor tubing that is constricted to < 10-mm o.d. may be sealed with an oxygen-gas flame. Silica tubing or larger-diameter Vycor tubing may require an oxyhydrogen flame.

† Such furnaces are described in *Inorg. Synth.*, **11**, 5 (1968); **12**, 161 (1970).

larger crystals of molybdenum(IV) oxide, it is essential that the cooling be very uniform. Therefore, some type of programmed cooling procedure is necessary. Many commercial temperature controllers permit such programmed procedures. The growth of tungsten(IV) oxide crystals is much less sensitive than that of molybdenum(IV) oxide crystals to the rate of cooling. Under the transport conditions specified in Table I, the transport of the dioxide powder is essentially quantitative. The dimensions of the largest crystal obtained by this procedure are roughly $1 \times 3 \times 3$ mm^3.

Properties

The molybdenum(IV) oxide crystals produced by this method are thick, oblong needles or thick platelets having a brown-violet metallic luster. The cell edges are close to those reported for the polycrystalline powders,[7] monoclinic crystals with $a = 5.60$ Å, $b = 4.86$ Å, $c = 5.63$ Å, $\beta = 120.95°$. The crystals are excellent metallic conductors.

The tungsten(IV) oxide crystals obtained by this procedure are polyhedra having bronze metallic luster. The crystallographic data agree well with those reported for the polycrystalline compounds.[8] Monoclinic crystals with $a = 5.56$ Å, $b = 4.89$ Å, $c = 5.66$ Å, $\beta = 120.5°$. Tungsten(IV) oxide is a good metallic conductor.

Chemical analyses of the oxygen content of these crystals, carried out by (1) careful oxidation to molybdenum(VI) oxide in a stream of oxygen and (2) carbon reduction of the metal, yielded the following data.

Anal. Calcd. for MoO_2: O, 25.01. Found, 25.1. Mo:O ratio, 1:2.01. Calc. for WO_2:O, 14.83. Found, 14.9. W:O ratio, 1:2.01.

References

1. G. Hägg and A. Magneli, *Arkiv. Kemi*, **19A**(2), 1 (1944).
2. A. Magneli, *Nova Acta Regiae Soc. Sci. Upsal.*, **14**(8), 13 (1949).
3. A. Wold, W. Kunnmann, R. J. Arnott, and A. Ferretti, *Inorg. Chem.*, **3**, 545 (1964).
4. H. Schäfer, *Chemical Transport Reactions*, Academic Press, New York, 1964.
5. A. G. Karpedes and A. V. Cafiero, *Inorg. Synth.*, **11**, 5 (1968).
6. L. E. Conroy, ibid., **12**, 158 (1970).
7. D. B. Rogers, S. R. Butler, and R. D. Shannon, ibid., **13**, 135 (1972).
8. A. Magneli, G. Andersson, B. Blomberg, and L. Kihlborg, *Anal. Chem.*, **2**, 1998 (1952).

22. NIOBIUM MONOXIDE

$$3Nb + Nb_2O_5 \rightarrow 5NbO$$

Submitted by T. B. REED* and E. R. POLLARD*†
Checked by L. E. LONNEY,‡ R. E. LOEHMAN, ‡ and J. M. HONIG‡

Reprinted from *Inorg. Synth.*, **14**, 131 (1973)

The electric arc has been used since the time of Henri Moissan (1890) in the preparation of refractory inorganic materials. The more recent (1940) development of inert-gas, cold-hearth melting techniques[1] has made arc melting one of the simplest and most versatile methods for the synthesis of those inorganic compounds that are stable up to their melting points. Synthesis by this technique removes volatile impurities and produces dense, homogeneous samples, from which crystals of several millimeters on a side can often be isolated. Crystals thus obtained are suitable for the measurement of chemical and physical properties. The synthesis of NbO has been described below, but the same procedure with minor modifications has been used to prepare other refractory oxides of intermediate valence, such as TiO, Ti_2O_3, Ti_3O_5, VO, and V_2O_3.

Procedure

A laboratory-size arc furnace§ suitable for preparing NbO buttons of up to 30 g is shown schematically in Fig. 1. The power source is a dc (direct-current) welding power supply, capable of supplying at least 300 A, which has the drooping-voltage characteristic used in Heliarc welding. A thoriated tungsten electrode, $\frac{1}{8}$ or $\frac{3}{32}$ in. in diameter, serves as the cathode, and the water-cooled copper or graphite hearth as the anode. The charge is placed in the cylindrical cavity with a movable graphite piston at the bottom. Water cooling is sufficient to keep the hearth cool while the high-temperature reaction occurs, and to form a thin protective layer around the melt, which will protect it from contamination by the hearth material.

* Lincoln Laboratory, MIT, Lexington, MA 02173. Work supported by the U.S. Air Force.
† Present address, Raytheon Corporation, Burlington, MA.
‡ Department of Chemistry, Purdue University, W. Lafayette, IN 47907. Work supported by NSF grant, GP 8302.
§ Centorr Company, Suncook, NH 03275.

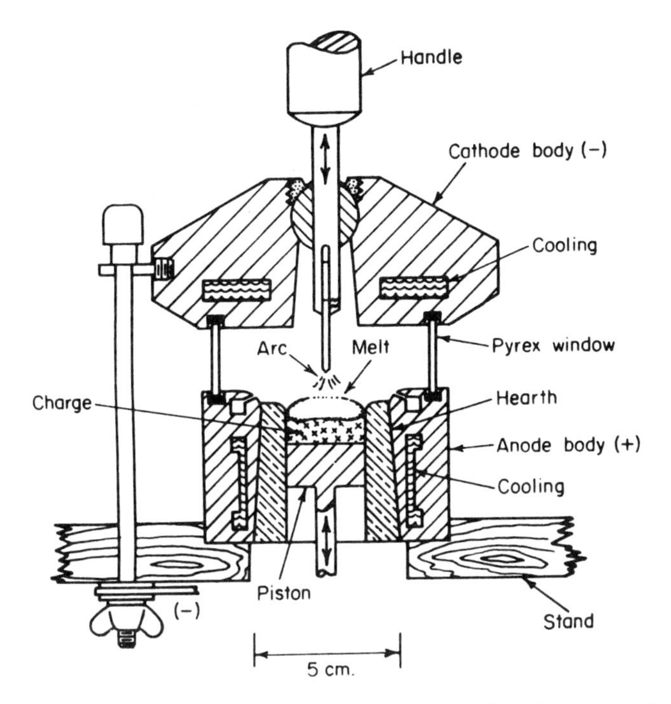

Figure 1. Schematic diagram of cold-hearth arc-melting furnace with movable piston.

The hearth cavity is filled with high-purity Nb_2O_5 powder and Nb metal rod* or powder in appropriate proportions. The entire assembly is flushed with argon gettered over Ti chips or foil at 900° to remove any O_2, N_2, or hydrocarbons. A continuous argon flow of approximately 2 L/min is maintained during operation to carry the volatile impurities out of the chamber. The arc is struck by touching the electrode momentarily to the hearth. The electrode is then raised and swiveled until the arc plasma bathes the mixed powders. As the metal melts and reacts with the oxide powder, the material sinks into the crucible and the piston is continuously raised to keep the melt in view. If a simple cup-shaped hearth is employed in place of the piston hearth, the electrode is lowered in the course of the reaction.

After the reaction is complete, the power is raised to melt the entire button and then gradually reduced to ensure slow cooling. After the arc is extinguished, it is desirable to flip the button over with the electrode and then to

* Materials Research Corporation, Orangeburg, NY 10962.

strike a new arc and remelt the button; this procedure ensures greater homogeneity.

The resulting button exhibits a bright silvery metallic sheen. As long as the oxygen/metal ratio x (in NbO_x) is within the homogeneity range of $0.98 \leqslant x \leqslant 1.02$, single-crystal grains up to 4 mm on an edge are formed.

The stoichiometry of the sample is determined readily by heating an aliquot portion in air for 16 h at 800° to form Nb_2O_5. The oxygen/metal ratio is calculated from

$$x = 2.500 - 8.3070\frac{g_2 - g_1}{g_2}$$

where g_1 and g_2 are the initial and final weights. If $g_1 = 0.5$ g, and g_1 and g_2 are measured to 0.1 mg, then x is known to within ± 0.002. Adjustments in x may be made by adding Nb or Nb_2O_5 to the button and remelting.

A typical analysis using the spark-source mass spectrograph shows the following impurities: N = 400, C = 300, Ta = 200, Fe = 90, Mo = 40, V = 10 atomic ppm. All other impurities are below the 10-ppm limit; it is interesting that only 1 ppm each of Cu and W have been detected, even though these are the major construction materials of the furnace used.

Properties

The congruently melting composition is $NbO_{1.006}$, which melts at 1940°. The lattice parameter of $NbO_{1.000}$ is 4.2111 ± 0.0002 Å at room temperature. The resistivity of NbO_x at 300° and 4.2 K is approximately 2×10^{-5} and $7 \times 10^{-7}\ \Omega\cdot$cm, respectively. The Seebeck coefficient at room temperature varies between $+1.0$ and $-1.0\ \mu$V/°, depending on x.

References

1. T. B. Reed, *Mater. Res. Bull.*, **2**, 349 (1967).
2. E. R. Pollard, Electronic Properties of Niobium Monoxide, Ph.D. thesis, Dept. Material Sciences, MIT (1968).

23. BARIUM TITANATE, $BaTiO_3$

$$BaCO_3 + TiO_2 \rightarrow BaTiO_3 + CO_2$$

Submitted by FRANCIS S. GALASSO*
Checked by MICHAEL KESTIGAN†

Reprinted from *Inorg. Synth.*, **14**, 142 (1973)

Barium titanate, $BaTiO_3$ [barium titanium(IV) oxide], is probably the most widely studied ferroelectric oxide. Extensive studies were conducted on this compound during World War II in the United States, England, Russia, and Japan, but the results were not revealed until after the war. Barium titanium(IV) oxide was found to be a ferroelectric up to a temperature of 120°, which is its Curie point. Above 120°, barium titanium(IV) oxide has the cubic perovskite structure, and below this temperature the oxygen and titanium ions are shifted and result in a tetragonal structure with the *c* axis approximately 1% longer than the *a* axis. Below 0°, the symmetry of barium titanate becomes orthorhombic, and below −90° it becomes trigonal.

Because barium titanate has interesting properties, many methods have been used to grow single crystals of this compound. One of the most popular techniques, using a potassium fluoride flux, was first employed by Remeika.[1]

Procedure

Crystals of barium titanate can be grown with a mixture of 30% barium titanate by weight and anhydrous potassium fluoride. In a specific procedure, 28 g of barium titanate [or 23.6 g (0.1 mol) of barium carbonate and 9.6 g (0.12 mol) of titanium(IV) oxide] is placed in a 50-mL platinum crucible and covered with 66 g of anhydrous potassium fluoride. The cover is placed on the crucible, and the crucible and contents are heated to 1160°. When barium carbonate and titanium(IV) oxide are used with potassium fluoride, the contents of the crucible are first slowly melted before the cover is placed on tightly and the mixture heated to 1160°. After being held for 12 h at 1160°, the crucible is cooled 25°/h at 900°, and the flux is poured off. The crystals then are annealed by slowly cooling them to room temperature, and they are removed by soaking the crucible contents in hot water.

* United Aircraft Research Laboratories, East Hartford, CT 06108.
† Sperry Rand Research Center, Sudbury, MA 01776.

Properties

Crystals grown by this technique are in a form called butterfly twins. The twins consist of two isosceles right triangles with a common hypotenuse.[2] Normally the composition plane is (1 1 1), the crystal faces are {1 0 0} and the angle between the wings is approximately 38°.

The dielectric constant of barium titanate, along [0 0 1] is about 200 and along [1 0 0] it is 4000 at room temperature.[3] The spontaneous polarization at room temperature is 26×10^{-6} C/cm^2, and the value of the coercive field has been found to vary from 500 to 2000 V/cm. The crystal structure of barium titanate at room temperature can be represented by a tetragonal unit cell with size of $a_0 = 3.992$ Å, and $c_0 = 4.036$ Å, but the symmetry becomes cubic above 120°, at which temperature the crystals no longer exhibit ferroelectrical properties.

References

1. J. P. Remeika, *J. Am. Chem. Soc.*, **76,** 940 (1954).
2. R. C. DeVries, *J. Amer. Cer. Soc.*, **42**, 547 (1959).
3. W. J. Merz, *Phys. Rev.*, **76**, 1221 (1949).

24. BISMUTH TITANATE, $Bi_4Ti_3O_{12}$

$$2Bi_2O_3 + 3TiO_2 \rightarrow Bi_4Ti_3O_{12}$$

Submitted by FRANCIS S. GALASSO*
Checked by MICHAEL KESTIGAN†

Reprinted from *Inorg. Synth.*, **14**, 144 (1973)

Bismuth titanate, $Bi_4Ti_3O_{12}$ [bismuth titanium oxide], is one of a class of ferroelectrics with the general formula $(Bi_2O_2)^{2+}(A_{x-1}B_xO_{3x+1})^{2-}$, where A is a monovalent or divalent element, B is Ti^{4+}, Nb^{5+}, or Ta^{5+}, and x can have values of 2, 3, 4, and so on.[1-3] The crystal structure of these compounds consists of Bi_2O_2 layers and stacks of perovskite with the layers perpendicular to the [0 0 1] axis. There are three "perovskite" layers between two bismuth oxygen layers in the structure of $Bi_4Ti_3O_{11}$, two perovskite layers in $ABi_2B_2O_9$ compounds, and four in $ABi_4Ti_4O_{15}$ compounds. Since the

* United Aircraft Research Laboratories, East Hartford, CT 06108.
† Sperry Rand Research Center, Sudbury, MA 01776.

perovskite layers consists of TiO_6, TaO_6, or NbO_6 octahedra, spontaneous polarization can take place in the plane of these layers.

Procedure

To form a $\frac{3}{8}$-in.-diameter pellet of $Bi_4Ti_3O_{12}$, a sample is first made by mixing 0.47 g (1 mmol) of bismuth oxide and 0.12 g (1.5 mmol) of titanium(IV) oxide. The mixture is pressed into pellets with a calorimeter pellet press. The pellets are placed in Alundum on zircon boats and heated at 700° for 4 h. Then the pellets are ground, pressed again, and reheated at a temperature of 920° for one hour.

These pellets can be ground for powder X-ray studies, or electrodes can be evaporated on the flat surfaces of the pellets to produce samples for ferroelectric measurements. However, for more meaningful measurements single crystals of bismuth titanate are required.

Single crystals of Bi_4Ti_{312} can be grown by the following technique.[4] A charge of 210 g (0.45 mol) of bismuth oxide and 11.0 g (0.14 mol) of titanium(IV) oxide in a 50-mL platinum crucible is heated at 1200° for one hour. The charge is placed in the crucible by filling the crucible, melting, adding more charge, and remelting. The furnace is slowly cooled at a rate of 40°/h. The crystals can be removed by dissolving the bismuth oxide in hydrochloric acid.

If a shallow platinum container ($6 \times 4 \times \frac{1}{2}$ in.) is used instead of a typical platinum crucible, larger, higher-optical-quality, single crystals are obtained.

Properties

The crystals grown in this manner are in the form of clear sheets. The symmetry of the crystals is pseudotetragonal with a cell size of $a_0 = 3.841$ Å and $c_0 = 32.83$ Å. Electrodes can be evaporated, or indium amalgam can be applied to the flat surfaces of the crystals, to produce samples for measurements. The dc resistance of the crystals is about $10^{12}\ \Omega \cdot cm$. They exhibit ferroelectric hysteresis loops up to the Curie temperature of 643°.

References

1. B. Aurivillius, *Arkiv. Kemi*, **1**, 463 (1949).
2. B. Aurivillius, ibid., 499 (1949).
3. B. Aurivillius, ibid., 519 (1959).
4. L. G. Van Uitert and L. Egerton, *J. Appl. Phys.*, **32**, 959 (1961).

25. SODIUM TUNGSTEN BRONZES

$$Na_2WO_4 + WO_3 \rightarrow 2Na_xWO_3 + O_2\uparrow$$

Submitted by C. T. HAUCK,* A. WOLD,* and E. BANKS†
Checked by R. SEIVER‡ and H. A. EICK‡

Reprinted from *Inorg. Synth.*, **12**, 153 (1970)

The sodium tungsten bronzes were first reported by F. W. Wohler in 1824.[1] The synthesis of these compounds has been reported by many workers, including Brown and Banks.[2] The various syntheses include reduction of a sodium tungstate melt by hydrogen,[1] as well as by tin, zinc, iron, and phosphorus.[3,4] Bronzes can also be formed by reduction of a melt of sodium tungstate and tungsten(VI) oxide with tungsten metal,[5] and by the electrolytic reduction of tungstate melts.[6] The bronzes, Na_xWO_3, have metallic properties and form solid solutions of varying sodium content. The products range from a blue tetragonal bronze, in which $x = 0.3$ to 0.4, to a yellow bronze of cubic structure, for which x can be as high as 0.9. In the method discussed in this synthesis, the bronzes are formed at elevated temperatures (780–800°) by cathodic reduction of a sodium tungstate-tungsten(VI) oxide melt.§

Procedure

Preparation of Reactants. About 300 g** of Na_2WO_4 is placed in a ceramic crucible and fused at 800° overnight. The dehydrated tungstate is then ground to a fine powder with a porcelain mortar and pestle and stored in a desiccator. Reagent-grade tungsten(VI) oxide was used in these preparations. The components were intimately ground together in the following proportions:

* Brown University, Providence, RI 02912.
† Polytechnic Institute of Brooklyn, NY 11201.
‡ Department of Chemistry, Michigan State University, East Lansing, MI 48823.

§ Simpler electrolytic procedures will result in the formation of bronzes, but the crystals will not be as large, homogeneous, or well formed.

** This weight is sufficient for several preparations.

Na_2WO_4 (g)	WO_3 (g)	Mole Ratio ($Na_2WO_4:WO_3$)
83.06	65.56	10:10
117.14	76.84	12:10
153.57	46.43	26:10

Preparation of the Electrolytic Cell. The cell (Fig. 1) consists of a 50-mL alumina crucible fitted with a platinum cap. The anode* is a rectangular platinum strip $\frac{1}{2} \times 2\frac{3}{4} \times 0.01$ in. suspended in the middle of the melt, approximately 1 in. from the bottom of the crucible. The cathode is a 1-in.-diameter platinum disk, 0.01 in. thick, placed on the bottom of the cell. Its platinum lead (0.016-in. Pt wire) is isolated from the melt by means of an alumina tube placed snugly against the cell wall. It is essential that the cell be packed very tightly to the top with the finely ground mixture of Na_2WO_4–WO_3. Both platinum electrodes protrude a few inches out of the cell.

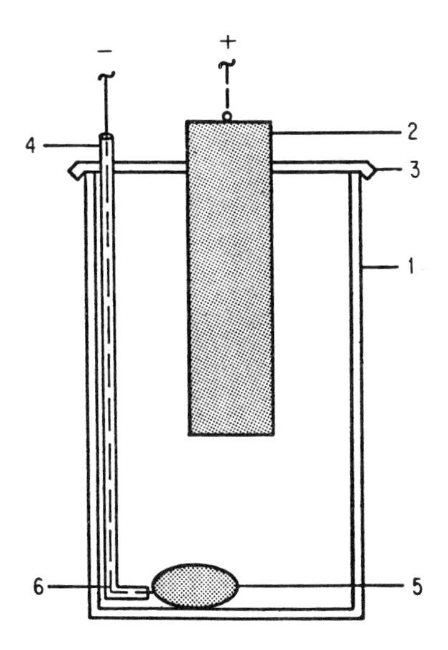

Figure 1. Electrolytic cell. (1) Alumina crucible; (2) anode (Pt); (3) platinum cap; (4) alumina tube; (5) cathode (Pt); (6) Pt lead wire.

* The Na_2WO_4–WO_3 melt is corrosive, and both electrodes as well as the crucible are subject to considerable attack.

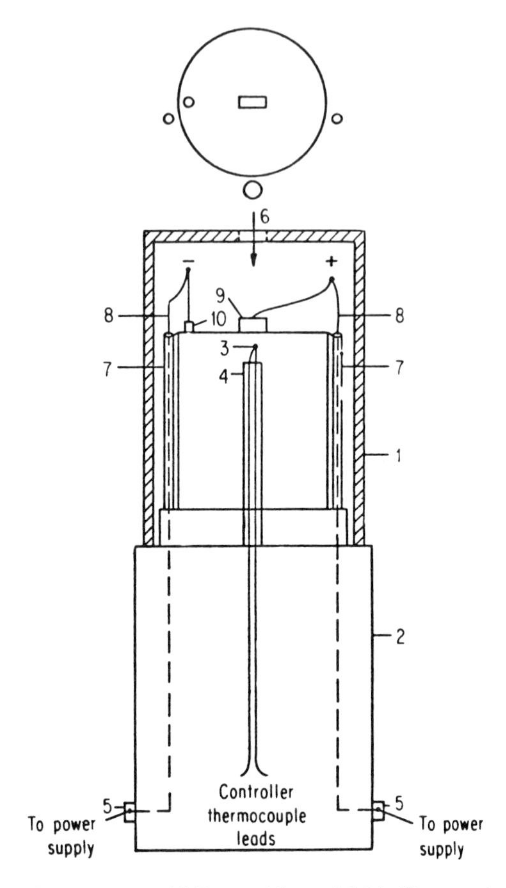

Figure 2. Electrolysis apparatus. (1) Inconel heat shield; (2) ceramic base; (3) monitor thermocouple; (4) alumina tube (two-hole); (5) terminal posts; (6) insert for calibration thermocouple; (7) alumina tubes; (8) Pt leads; (9) anode (Pt); (10) alumina tube for cathode lead wire.

The electrolytic cell is placed on a stand (Fig. 2), which is made of Alsimag 222 ceramic.* The anode and cathode are connected to platinum wires that run down the inside of the stand and are attached to terminal posts near the base of the stand.

Electrolytic Procedure. The cell is placed inside a vertical column, Hevi-Duty, multiple furnace† and is surrounded by an Inconel heat shield (Fig. 2).

* American Lava Corp., 219 Kruesi Building, Chattanooga, Tenn. 37405.
† Type MK-2012, Fisher Scientific Co., 711 Forbes Ave., Pittsburgh, PA 15219.

The temperature-control thermocouple (Pt vs. 90% Pt – 10% Rh) is located between the crucible and the heat shield and is protected by an outer ceramic tube. A calibration run is initially made without a charge in the crucible. A second thermocouple is inserted through the heat shield and centered on the cover of the electrolytic cell. This thermocouple is read directly by means of a potentiometer; the temperature of this thermocouple corresponds closely to that of the melt. The temperature controlled is adjusted until the second thermocouple indicates the desired temperature for the electrolysis.*

An electrolysis can be carried out by removing the second thermocouple, filling the crucible with a charge, and connecting the leads from the power supply to the cell. An Inconel plug is used to close the thermocouple hole in the heat shield. The dc constant-current power supply is turned on to warm up without any current fed to the cell. The temperature is then raised to the operating temperature of 870° at a rate of 200°/h. When the temperature has equilibrated, the voltage is applied, and the current is adjusted to 45 mA. The electrolysis is carried out for 5 days.

On completion of the run, the furnace is shut down, but the current from the power supply to the cell is left on. The resistance of the cell increases when the melt begins to solidify. This is indicated by a fluctuation of the current, and the power is then turned off. The reaction mixture is allowed to cool to 100° in the furnace, and the cell is disconnected. The crucible walls above the melt are broken off with a hammer, and the product is extracted with boiling water. This process is repeated until the flux has dissolved; the clean crystals are recovered and allowed to dry in air.

Preparation for X-ray Analysis. Lattice constants are calculated from patterns obtained on powder samples with a Norelco diffractometer using monochromatic radiation (AMR-202 Focusing Monochromator) from a high-intensity copper source. The crystals are powdered with a diamond† mortar and pestle, and the powder passed through a 74-μm sieve. Accurate lattice constants are calculated from the X-ray data.

* This calibration must be done since a thermocouple cannot be placed directly into the melt, and the heat shield prevents the control thermocouple from giving an accurate temperature reading.

† Plattner's diamond mortar and pestle of hard tool steel.

Properties

The following bronzes were obtained:

Na_xWO_3, x	Lattice Constant, a_0 (A.)	Mole Ratio of Reactants $Na_2WO_4:WO_3$	Color
0.58	3.832 ± 0.005	10:10	Magenta
0.59	3.833 ± 0.005	12:10	Red
0.79	3.849 ± 0.005	26:10	Yellow

The value for x are obtained from a lattice constant vs. nominal composition curve (Fig. 3) reported by Brown and Banks.[2] The linear portion of the curve can be represented by

$$a_0(\text{Å}) = 0.0819x + 3.7846.$$

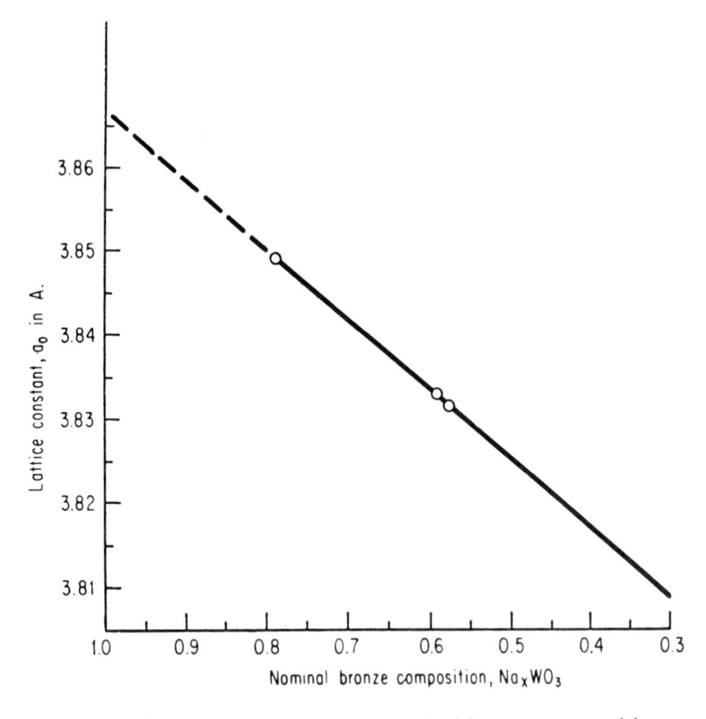

Figure 3. Lattice constant versus nominal bronze composition.

References

1. F. Wohler, *Ann. Chim. Phys* [2], **29**, 43 (1823).
2. B. W. Brown and E. Banks, *J. Am. Chem. Soc.*, **76**, 963 (1954).
3. H. Wright, *Liebigs Ann. Chem.*, **79**, 221 (1851).
4. J. Philip, *Berlk.*, **15**, 499 (1992); *Jahrb. Ber. Forte-Chem.*, **28**, 444 (1882).
5. M. E. Straumanis, *J. Am. Chem. Soc.*, **71**, 679 (1949).
6. L. D. Ellerbeck, H. R. Shanks, P. H. Sidles, and G. C. Danielson, *J. Chem. Phys.*, **35** (1), 298 (1961).

26. POTASSIUM MOLYBDENUM OXIDE "BLUE BRONZE": $K_{0.30}MoO_3$

$$0.15K_2MoO_4 + 0.85MoO_3 \rightarrow K_{0.30}MoO_3 + 0.0075O_2$$

Submitted by L. F. SCHNEEMEYER*
Checked by DAVID DICARLO†

The "blue bronze", $K_{0.30}MoO_3$, and its isomorph, $Rb_{0.30}MoO_3$, undergo charge–density–wave (CDW)-driven metal-to-semiconductor phase transitions at 181 K.[1,2]. The CDW instability is a consequence of the structure that leads to quasi-one-dimensional conductivity along infinite chains of corner-shared MoO_6 octahedra. Below the CDW onset temperature, a variety of phenomena associated with enhanced conductivity via a moving ("sliding") CDW are exhibited. A CDW, a periodic distortion of both the conduction electron density and the underlying lattice, is a charged object that when caused to move in an applied electric field, constitutes a new mechanism for charge transport. Among the phenomena observed in a CDW conductor is nonlinear dc conductivity. The conductivity of these materials has two components, the Ohmic ($I \propto V$) response of the normal conduction electrons, and, above a threshold field for depinning the CDW (the point at which it breaks free of impurities), an additional contribution from the moving CDW. Other phenomena which are observed include an enhanced ac conductivity, an ac response to a dc bias, and hysteresis and memory effects. An extensive review of the properties of molybdenum oxide bronzes centered around CDW instabilities has recently been published.[3]

Study of CDW instabilities in "blue bronze" is facilitated by the ease preparation of large, high-quality crystals by electrolysis of K_2MoO_4/MoO_3

* AT&T Bell Laboratories, Murray Hill, NJ 07974.
† Department of Physics, Cornell University, Ithaca, NY 14853.

melts[4] as described here. Crystals of $K_{0.30}MoO_3$ are obtained by the electrolysis of melts containing about 75 mol % MoO_3 as indicated in Fig. 1.[5] Crystals have also been obtained by a temperature gradient flux growth technique.[6] There are no reports of the successful growth of alkali molybdenum bronze crystals by vapor transport methods. The Rb analog, $Rb_{0.30}MoO_3$,[9,10] and a "red bronze", $K_{0.33}MoO_3$, can also be grown electrochemically.

Procedure

Commercial high-purity MoO_3 (typically 99.999%) (63.81 g, 0.4433 mol) and K_2MoO_4 (typically 99.9%) (31.51 g, 0.1323 mol) (3.35 MoO_3:1 K_2 MoO_4)

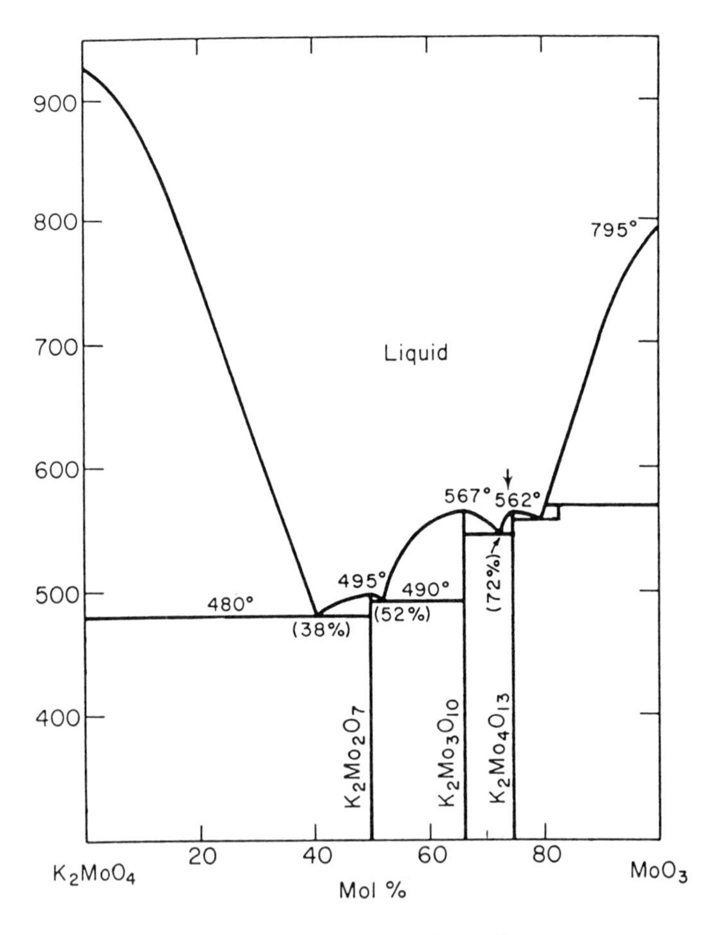

Figure 1. K_2MoO_4/MoO_3 phase diagram.

are ground together and packed into a 50-mL high-purity, high-density Al_2O_3 crucible (Coors). (A variety of crucible materials can be employed in potassium molybdenum bronze growth. As is typical of a flux growth-type technique,[7] the crucible can act as a source of contamination of product crystals. Platinum has traditionally been used in oxide bronze growth although Pt contamination at low levels has been documented in the case of tungsten bronze crystals. We have generally used high-purity, high-density alumina crucibles because they are inexpensive and easily obtained.) The crucible is then placed into a pot furnace configured as shown in Fig. 2 to allow atmosphere control and introduction of electrodes. The crucibles is supported in a machined graphite block which provides mechanical and thermal stability. A thermocouple is placed at the side of the crucible to monitor temperature. The system is flushed with nitrogen and a nitrogen overpressure is maintained. The sample is heated to $\approx 585°$ at approximately 200 °/h to melt the mixture. The temperature is then lowered to 570 °, approximately 5 ° above the melting temperature. The melt is monitored visually at look for freezing of the melt surface indicating the temperature is too low. Adequate time must be allowed for the system to come to thermal equilibrium.

To carry out the electrolysis, platinum electrodes are introduced into the melt. A large (several cm^2) Pt foil electrode serves simultaneously as the

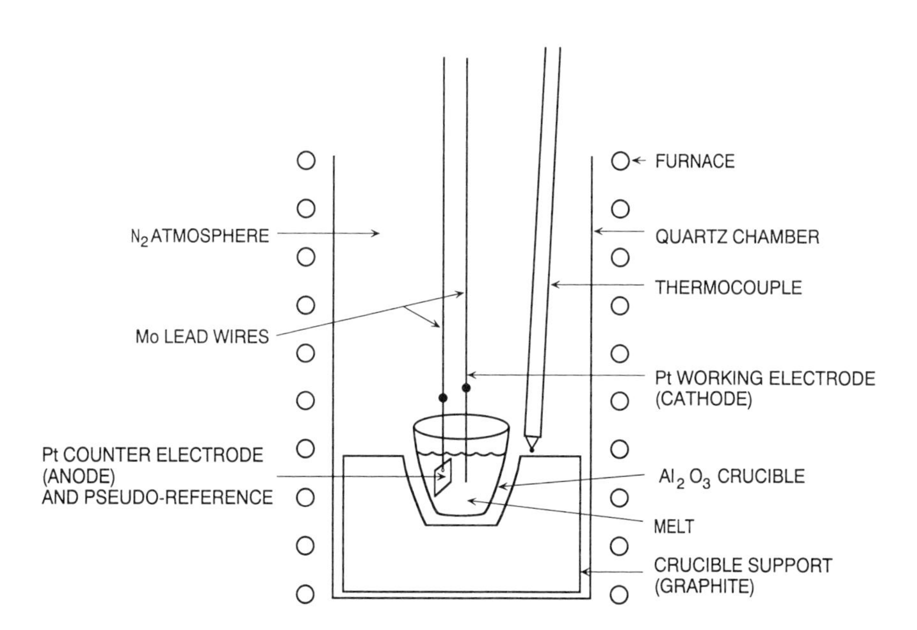

Figure 2. Experimental setup for electrochemical crystal growth.

counterelectrode and the pseudo reference electrode. Platinum wire, 1 mm diameter, is used as the working electrode and is introduced into the melt to a known depth, typically 0.5 cm. Care must be taken that the two electrodes do not touch in this one compartment cell.

Electrocrystallization is carried out under galvanostatic control as potentials are poorly defined in this system. High current densities are used to drive the reduction at competitive rates relative to back-reactions such as reoxidation or dissolution. An initial current density of 50 mA/cm^2 is usual. As the experiment proceeds, crystals grow on the cathode causing the current density to drop. Therefore, the current is stepped up over the course of approximately 2 h to a maximum of 200 mA/cm^2. The reaction is allowed to proceed for about an hour more. To end the experiment, the current is turned off and the electrode immediately raised above the surface of the melt. The electrode with attached crystals is carefully removed from the furnace.

The metallic blue crystals are detached from the electrode mechanically. Traces of attached flux (gray) are removed by boiling the crystals in 12*M* HCl (in a hood).

Properties

Crystals of $K_{0.30}MoO_3$ are shiny metallic blue and grow with a elongated tabular habit. Crystals are monoclinic, space group $C2/M$, with lattice constants, $a = 18.249(10)$ Å, $b = 7.560(5)$ Å, $c = 9.855(6)$ Å, and

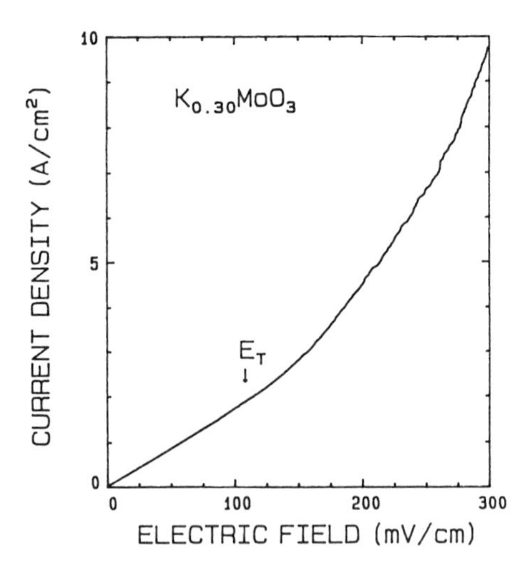

Figure 3. Current-voltage curve for $K_{0.30}MoO_3$ at 77 K. The arrow indicates the threshold field.

$\beta = 117.53(8)^\circ$ at 300 K[11,12]. The unit cell contains two $K_3\ Mo_{10}O_{30}$ units. The X-ray powder pattern is given in Table I.

To measure current–voltage characteristics, ultrasonically soldered indium contacts are applied in a four probe configuration. Figure 3 shows the current–voltage behavior for a $K_{0.30}MoO_3$ crystal at 77 K. Below the threshold field for depinning the CDW, E_T, indicated by the arrow in the figure, current is proportional to voltage, indicating Ohmic behavior. Above E_T, a *small* electric field of the order of tens of mV/cm, the CDW breaks free of the impurities (is "depinned") and provides an additional contribution to

TABLE I. Powder Pattern for $K_{0.30}$ MoO_3

h	*k*	*l*	2*θ*	*d* (Å)	*I*
0	0	1	10.084	8.772	w
2	0	−1	10.868	8.140	vs
1	1	0	12.872	6.877	m
1	1	−1	14.708	6.023	m
1	1	1	17.928	4.948	w
3	1	−1	18.840	4.710	vw
3	1	0	20.207	4.394	vw
4	0	0	21.921	4.054	s
0	2	0	23.527	3.781	s
2	2	0	25.988	3.428	s
1	1	2	26.228	3.398	vw
2	0	2	27.288	3.268	vs
4	0	1	28.248	3.159	m
2	2	1	29.788	2.999	vs
5	1	−3	32.406	2.763	vw
3	1	2	33.987	2.638	vw
2	2	2	36.347	2.472	s
4	2	1	37.067	2.425	vw
6	0	1	39.187	2.299	w
3	3	0	39.447	2.284	s
3	3	−2	40.807	2.211	vw
6	2	0	41.047	2.199	vw
6	0	2	47.088	1.930	w
2	0	4	47.828	1.902	w
0	4	0	48.167	1.889	m
5	3	−4	52.468	1.744	m
6	2	2	53.280	1.719	vw
2	4	2	56.238	1.636	w
4	4	1	56.752	1.622	w

the electrical conductivity.[8] The current is no longer proportional to the voltage, thus the behavior is non-Ohmic. The threshold field provides some measure of sample purity since impurities provide the pinning potential. Typically, "blue bronze" has a threshold field near 100 mV/cm although we have measured crystals where E_T was as low as 35 mV/cm.

References

1. J. Dumas, C. Schlenker, J. Marcus, and R. Buder, *Phys. Rev. Lett.*, **50**, 757 (1983).
2. J. P. Pouget, S. Kagoshima, C. Schlenker, and J. Marcus, *J. Phys.* (Paris) *Lett.*, **44**, L113 (1983).
3. C. Schlenker (ed.), *Low Dimensional electronic Properties of Molybdenum Bronzes and Oxides*, Kluwer Academic, MA, Boston 1989.
4. A. Wold, W. Kunnmann, R. J. Arnott, and A. Ferretti, *Inorg. Chem.*, **3**, 545 (1964).
5. P. Caillet, *Bull. Soc. Chim. Fr.*, **12**, 4753 (1967).
6. K. V. Ramanujachary, M. Greenblatt, and W. H. McCarroll, *J. Crystal Growth*, **70**, 476 (1984).
7. D. Elwell and H. J. Scheel, *Crystal Growth from High Temperature Solutions*, Academic Press, New York, 1975.
8. P. Strobel and M. Greenblatt, *J. Solid State Chem.*, **36**, 331 (1981).
9. J.-M. Reau, C. Fouassier, and P. Hagenuller, *Bull. Soc. Chim. France*, 2883 (1971).
10. L. F. Schneemeyer, S. E. Spengler, F. J. DiSalvo, and J. V. Waszczak, *Phys Rev. B*, **30**, 4297 (1984).
11. J. Graham and A. D. Wadsley, *Acta Cryst.*, **20**, 93 (1966).
12. R. M. Fleming, L. F. Schneemeyer, and D. E. Moncton, *Phys. Rev. B*, **31**, 899 (1985).

27. GROWTH AND CHARACTERIZATION OF SINGLE CRYSTALS OF ZINC FERRITES, $Fe_{3-x}Zn_xO_4$

$$\frac{3-x}{2} Fe_2O_3 + xZnO \rightarrow Fe_{3-x}Zn_xO_4 + 0.25\,(1-x)O_2$$

Submitted by M. WITTENAUER,* P. WANG,‡ P. METCALF,* Z. KĄKOL,*† and J. M. HONIG*
Checked by BRUCE F. COLLIER§ and J. E. GREEDAN§

Zinc ferrite, $FeZnO_4$, is a basic material that, when properly doped with other components, finds many industrial applications. It is therefore of

* Department of Chemistry, Purdue University, West Lafayette, IN 47907.

‡ Present address: Shanghai Institute of Ceramics, 1295 Dingxi Road, Shanghai 800050, People's Republic of China.

† Present address: Department of Physics, AGH Technical University, PL-30-059 Kraków, Poland.

§ Institute for Materials Science, McMaster University, Hamilton, Ontario, Canada L85 4M1.

interest to determine the physical properties of the zinc ferrite series $Fe_{3-x}Zn_xO_4$, especially since relatively little is known about the changes in characteristics with Zn content. As always, it is highly desirable to carry out such studies on well-characterized single crystals of uniform composition and oxygen content. The skull-melting procedure is well suited for single-crystal growth since it is a cold-crucible technique and thus avoids the problem of melt contamination. The naturally occurring temperature gradients in the skull melter give rise to convective stirring that ensures thorough mixing of the Fe and Zn components. However, it is not possible to control the oxygen content adequately during crystal growth, so that it is necessary to carry out a subsequent subsolidus annealing process to ensure uniformity of the oxygen composition and to achieve the ideal 4:3 oxygen:metal ratio.

We describe below the skull-melter technique and the subsequent annealing experiments which have been successfully employed to prepare single crystals of $(Fe_{3-x}Zn_x)O_{4+\delta}$ in the range $0 \leq x < 0.4$ and with $\delta \approx 0$. We also

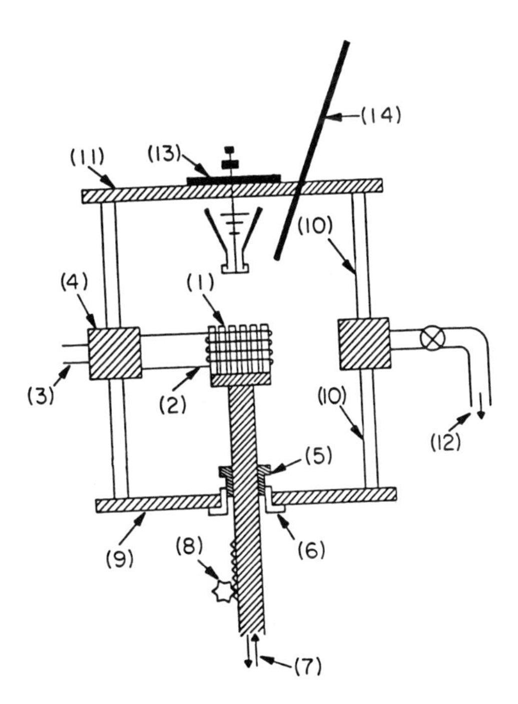

Figure 1. A schematic diagram of the skull-melting apparatus: (1) water-cooled crucible; (2) work coil; (3) to RG generator; (4) 12-port vacuum collar; (5) vacuum quick-connect coupling; (6) teflon insulating flange; (7) water (inlet and outlet); (8) crucible lowering mechanism; (9) baseplate and frame; (10) 18-in. diameter × 12-in-high Pyrex vacuum cylinder; (11) top plate; (12) to vacuum pump; (13) powder hopper; (14) Al_2O_3 poker.

report subsequent tests on the quality of the crystals so obtained. The procedures represent an adaptation of the method described in the literature for skull-melter growth of single crystals of magnetite of controlled oxygen content[1,2] as described in an earlier article in the present series.[3] Some details of the current work have been published elsewhere.[4] A schematic diagram of the skull-melter unit is displayed in Fig. 1; the reader is referred to this diagram in conjunction with the description offered below.

Procedure

The material to be reacted and grown as crystals is added in powdered form to a copper crucible displayed in Fig. 2. The unit consists of a split Cu base and a cylindrical array of vertically aligned Cu fingers, all of which are individually water-cooled. The fingers are sufficiently separated to permit penetration of the radiofrequency electromagnetic field and to minimize eddy currents in the various component sections of the crucible. On the other hand, the spacing is sufficiently close to keep the molten material confined to the interior. Electrical power is supplied by a commercial unit, at either 200 kHz or 3 MHz and at externally controlled power levels of up to 50 kW, to a work coil placed around the crucible. The assembly is enclosed by a vacuum chamber of 125-L capacity through which gas can be continually circulated by pumping, so as to maintain a carefully regulated oxygen partial

Figure 2. Photograph of skull crucible.

pressure. Gas controllers for admission of CO and CO_2, or other gas mixtures, are used to achieve appropriate gas mixing at the inlet. A hopper is available for refilling the crucible with additional powder during the run. The crucible can be moved out of the region of the stationary work coil by a rack and pinion mechanism which is operated at the conclusion of the process to achieve gradual cooling of the melt. A poker is also available to break "bridges" of encrustations that tend to form during the growth process.

Single crystals from a prior run are used to initiate the heating process; these couple to the applied field, thereby becoming hot and heating the surrounding powder to the point where it also couples to the field. An avalanching process then ensues in which all the remaining powder is heated to the melting point of the compound. Only powder close to the water-cooled copper walls remains at room temperature and thereby forms a thin skull of the same composition as the melt. This thin shell is interposed between the liquid material and the crucible. If single crystals are not initially available or do not couple to the field at room temperature a graphite ring may be used as a susceptor to start the thermal avalanching process. The ring later burns off completely in the slightly oxidizing atmosphere of the chamber.

While it is generally more difficult to initiate a run at 200 kHz than at 3 MHz, since the skin depth for field penetration is larger at the lower frequency, power is dissipated in a larger volume, which results in the formation of deeper melts with higher final yields. Initially, Fe_2O_3 and ZnO powders are loaded into the Cu vessel and CO_2 or O_2 is circulated through the chamber at 10 L/min for growth of $Fe_{3-x}Zn_xO_{4+\delta}$ for which x falls in the ranges $0 \leq x < 0.4$ or $x \geq 0.4$, respectively. Throughout the runs, considerable difficulty was experienced with ZnO evaporation, especially when working with compositions in the high-x region. To counteract this problem it was necessary to minimize the amount of molten liquid at the surface during crystal growth. This was done by reducing the power input, such that only small patches of liquid remained on the upper exposed surface. Vigorous stirring by convection occurred throughout the interior melt.

After maintaining the melt for at least 30 min to ensure proper mixing of the constituents, the content was cooled very slowly during the growth phase by automatic power reduction or by slow retraction of the crucible from the center of the work coil. One thereby obtained a frozen boule that was gradually cooled to room temperature through power reduction. Ordinarily, after the boule was opened by mechanical fracture, a large number of sizable single crystals were found in the interior. These were harvested and used in subsequent studies.

A typical boule obtained under the growth conditions described above is shown in Fig. 3. The bottom and sides consisted of a porous solid that was generally richer in ZnO than the center where the single-crystal specimens

Figure 3. Photograph of a typical boule of zinc ferrite $(Fe_3O_4)_{1-x}\cdot(Fe_2ZnO_4)_x$. Crystals are roughly 2 cm on an edge.

were encountered. Their size varied typically from $1 \times 1 \times 2\ cm^3$ for $x \approx 0.1$ to $\frac{1}{2} \times \frac{1}{2} \times 1\ cm^3$ for $x \approx 0.35$. A two-phase region containing hematite covered the top of the boule.

The single crystals obtained by the preceding method were oriented via the usual Laue back-scattering procedures and then cut to appropriate dimensions with a diamond saw. The crystals at that stage were inhomogeneous in oxygen content. It was therefore necessary to anneal these properly in order to achieve a uniform sample composition. The annealing requirements, based on a detailed thermodynamic analysis,[5] are summarized in Fig. 4. The center curve shows the oxygen partial pressure required to achieve the ideal oxygen stoichiometry as a function of the zinc content x in $Fe_{3-x}Zn_xO_4$ at 1100°. From the appropriate thermodynamic analysis,[5] one obtains the following expression for the equilibrium partial pressure P_{O_2} required to generate stoichiometric zinc ferrites:

$$\log(P_{O_2}, \text{atm}) = 10.88 - 25{,}879/T - 4\log(1 - x)$$

where T is in kelvins. The upper curve in Fig. 4 represents the oxidation boundary, whereas the lower curve, which is less well established, represents the reduction boundary. In order to minimize zinc oxide losses, an additional precaution was taken to carry out the crystal growth close to the oxidation boundary for which the oxygen fugacity is given by

$$\log(P_{O_2}, \text{atm}) = 13.96 - 24{,}634/T - 4\log(1 - x)$$

One must be very careful not to cross this boundary line either during crystal growth or on subsequent cooling. Calculations based on the cation vacancy diffusion rates in magnetite[2] have shown that annealing times of the order of 24–250 h suffice to homogenize the oxygen content at 1100–1400° in samples with thicknesses in the millimeter range.

The specimen was suspended from a thin Pt wire in a vertical tube furnace equipped with a ZrO_2/Y_2O_3 oxygen transfer cell,[6] which was placed next to the sample. The emf (electromotive force) of the transfer cell was used to monitor to oxygen partial pressure achieved with a CO_2/CO gas mixture circulated over the sample. The unit was calibrated against the oxygen partial pressure in equilibrium with the Ni/NiO phase boundary. The appropriate CO_2/CO content was then selected by reference to Fig. 4. The composition of

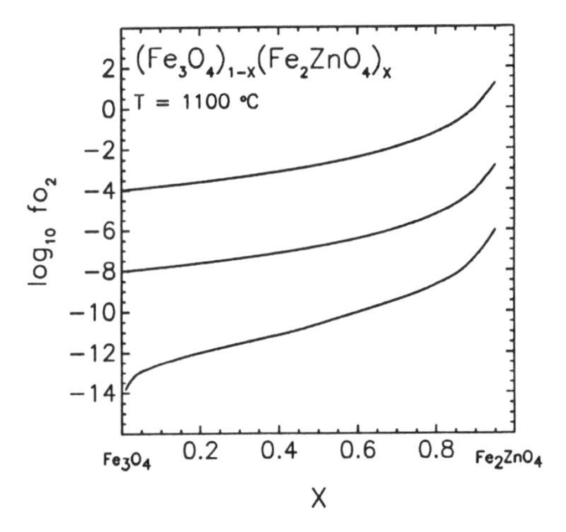

Figure 4. Plot of equilibrium partial pressure of oxygen, log f_{O_2}, versus x in $(Fe_3O_4)_{1-x}\cdot(Fe_2ZnO_4)_x$ at 1100°. This delineates the width of the phase field between reduction boundary (zinc ferrite, zinc wüstite) and oxidation boundary (zinc ferrite, zinc hematite). The center curve relates to the equilibrium fugacity for zinc ferrites with the ideal 4:3 oxygen:metal ratio. Pressures in atm.

the gaseous mixture was adjusted by use of two gas-flow controllers. The annealing temperature was selected from the range 1100–1400°, so as to achieve a reasonable rate of oxygen equilibration while minimizing zinc losses. After annealing under the conditions described above for a sufficient length of time a current pulse was used to sever the wire suspension, and the sample was quenched during its free fall onto a copper block. The exterior portions of the sample were then removed by abrasion to obtain a homogeneous core of material suitable for further testing.

Properties

Samples obtained as described above were subjected to a number of evaluations described below.

1. No impurities were encountered in the single crystals at the minimal detection level of the EDAX (energy-dispersive analysis by X rays) technique.
2. The uniformity of the zinc content in a given sample was checked by EDAX and by microprobe techniques. Single crystals were cut, mounted in epoxy resin, polished, and finally coated with a thin layer of carbon. A fully characterized $Zn_2 \cdot Y$ hexagonal ferrite crystal was used as a standard for the microprobe analysis; an internal standard was used for the EDAX analysis. Measurements were taken at several locations on the surface, avoiding surface imperfections due to polishing. The uniformity of the zinc distribution within the boule was checked by carrying out individual compositional analyses on several crystals obtained from various regions of the boule.

It was found that for $x < 0.4$ single crystals of $Fe_{3-x}Zn_xO_4$ were quite homogeneous in their Zn content; the observed fluctuations remained below the estimated experimental error of ± 0.004 in x. Compositional variation among different crystals of the same boule were larger, of the order $\pm$ 10–20%.

Results obtained by the EDAX and microprobe techniques were in substantial agreement. The actual Zn content of specimens was approximately 71% of the nominal value. A plot of actual–nominal compositions is shown in Figs. 5 and 6. These findings reflect the loss of ZnO by evaporation as well as the accumulation of excess ZnO at the outer layers of the boule. Samples with $x > 0.4$ exhibited large compositional fluctuations, thus rendering these specimens unsuitable for physical measurements. The utility of the skull-melting technique in growing single crystal zinc ferrites is thereby limited to the low end of the $Fe_{3-x}Zn_xO_4$ composition range.

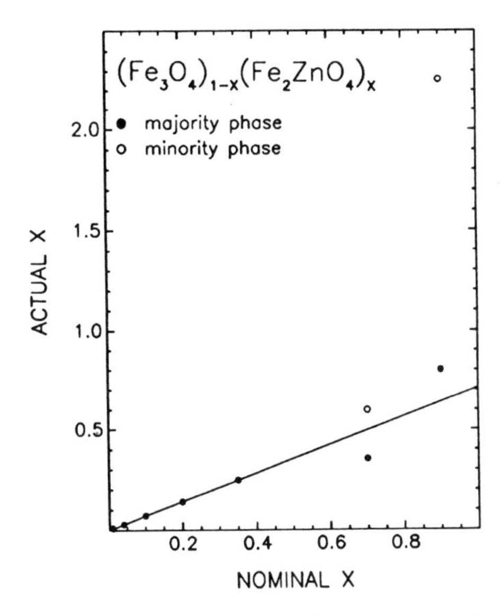

Figure 5. Plot of actual versus nominal zinc concentrations in $(Fe_3O_4)_{1-x}\cdot(Fe_2ZnO_4)_x$ over the entire x range, as obtained from microprobe analysis.

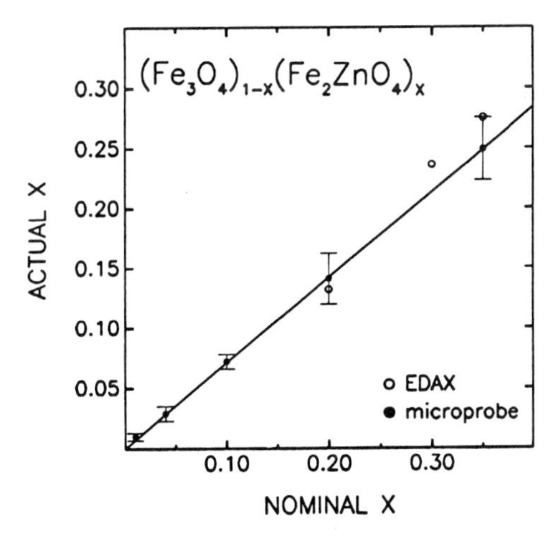

Figure 6. Plot of actual versus nominal zinc concentrations in $(Fe_2O_4)_x\cdot(Fe_2ZnO_4)_{1-x}$ with $x < 0.40$. Error bars indicate variations in zinc content among different single crystals harvested from a single boule.

Individual crystals were found to be single-phase. This was checked on several samples by optical microscopy with polarized light. In standard X-ray powder diffraction analysis there were no indications of any secondary phases. The Laue spots were uniform and sharp in appearance.

Conclusions

The skull-melting technique can be adapted to the growth of zinc ferrites by operating under reduced oxygen pressure, and the oxygen stoichiometry can be adjusted by subsequent annealing. If operated properly one can obtain large $Fe_{3-x}Zn_xO_4$ single crystals of uniform composition in the range $x < 0.4$, that are free from appreciable impurities.

Acknowledgment

The authors wish to acknowledge their indebtedness to Dr. J. E. Keem of Troy, MI, who was instrumental in the initial installation and operation of the equipment; to the late Dr. H. Harrison, and to Drs. D. Buttrey and R. Aragón (presently at the University of Delaware), all of whom developed and perfected the procedures and instituted many improvements; to J. W. Koenitzer, who was always willing to assist with the process of crystal growth and to help in innumerable repairs; and to Mr. C. Hager, who assisted in the microprobe analysis.

This research was supported by the National Science Foundation Grant DMR 86-16533 AO2 and by the Indiana Center for Innovative Superconducting Technology Grant No P-7028/7029.

References

1. H. R. Harrison and R. Aragón, *Mater. Res. Bull.*, **13**, 1097 (1978).
2. R. Aragón, H. R. Harrison, R. J. McCallister, and C. J. Sandberg, *J. Cryst. Growth*, **61**, 221 (1983).
3. H. R. Harrison, R. Aragón, J. E. Keem, and J. M. Honig, *Inorg. Synth*, **22**, 43 (1983).
4. P. Wang, M. A. Wittenauer, D. J. Buttrey, Q. W. Choi, P. Metcalf, Z. Kakol, and J. M. Honig, *J. Cryst. Growth*, **104**, 285 (1990).
5. P. Wang, Q. W. Choi, and J. M. Honig, *Z. Anorg. Allg. Chem.* **550**, 91 (1987).
6. M. Sato, in *Research Techniques for High Pressure and High Temperature*, G. C. Ulmer (ed.), Springer, New York, 1971, pp. 43–99.

28. CONGRUENT GROWTH OF SINGLE-CRYSTAL La_2NiO_4 AND OTHER LAYERED NICKELATES BY RADIOFREQUENCY SKULL MELTING

$$La_2O_3 + NiO \rightarrow La_2NiO_4$$

Submitted by D. J. BUTTREY,* R. R. SCHARTMAN,†‡ and J. M. HONIG†
Checked by BRUCE F. COLLIER† and J. E. GREEDAN§

The physical properties of the substituted lanthanum nickelate solid-solution series $Ln_{2-x}M_xNiO_{4-x/2+\delta}$ (LN = La,Pr,Nd; M = Ca, Sr, Ba) are being extensively investigated both because of interest in their intrinsic behavior and because of their close relationship to the corresponding lanthanum cuprate high-T_c superconductors. For fundamental investigations it is generally desirable, and often essential, to have available well-characteized single crystals of the lanthanide nickelates. We describe here the conditions for growth of large single crystals by congruent RF induction melting, as well as procedures for basic compositional and thermodynamic characterization of the products. We will place primary emphasis on growth and characterization of La_2NiO_4 and comment on related compositions as appropriate.

Growth of single crystals of La_2NiO_4 measuring a few millimeters in length was first reported by Föex et al.[1] in 1960; they used a solar furnace to achieve temperatures approaching 2000°. Small crystals suitable for crystallographic study have also been grown in a plasma furnace[2] and more recently, by a CO_2 laser technique.[3] The first procedure for melt growth of large crystals ($\sim 1\ cm^3$) of La_2NiO_4 was developed by Buttrey and coworkers.[4] More recently, Bassat et al.[5] reported growth of crystals by a floating-zone technique. The growth technique described in this chapter, known as *skull melting*, is a cold-crucible process in which the starting material is inductively coupled to an electromagnetic (RF) field in the presence of a controlled oxygen partial pressure. The melt is then slowly decoupled such that crystals grow in the presence of existing thermal gradients. Following growth, the crystals are removed and annealed separately by subsolidus isothermal equilibration with a selected atmosphere to produce the desired uniform

* Department of Chemical Engineering, University of Delaware, Newark, DE 19716.
† Department of Chemistry, Purdue University, West Lafayette, IN 47907.
‡ Present address: Department of Materials Science and Engineering, University of Wisconsin, Madison, WI 53706.
§ Institute for Materials Science, McMaster University, Hamilton, Ontario, Canada L85 4MI.

concentration of interstitial oxygen defects. In addition to the details concerning operation of the skull melter presented below, further discussion of principles and operational details are discussed elsewhere,[6–11] as well as in a companion chapter[12] in the present volume.

In this synthesis, we focus on the choice and control of conditions for crystal growth, as well as on a method for controlled atmosphere subsolidus annealing. We then briefly discuss the characterization of the crystals.

The starting point for determining optimal growth conditions for La_2NiO_4 is the examination of the features of the binary La_2O_3–NiO phase diagram. Unfortunately, information on this system is incomplete. It is known that there are three stable, congruently melting phases at high temperature: La_2O_3, La_2NiO_4, and NiO, with melting points of approximately 2315, 1900, and 1900°, respectively. A crude representation of the phase diagram is presented in Fig. 1, based on reasonable approximations for the temperatures and compositions of the eutectics; the need for accurate experimental data is obvious. As presented, all compositions are taken to be line phases. Although this is a good approximation for La_2O_3, it is important to recognize that both NiO and La_2NiO_4 exhibit variations in stoichiometry. NiO should really be represented as $Ni_{1-x}O$ ($x < 0.002$ at 1700°), due to the presence of variable concentrations of cation vacancies that depend on the equilibrium oxygen fugacity and temperature. Similarly, for La_2NiO_4 it has been shown that the La:Ni ratio is invariant within experimental error at 2.00 ± 0.01, whereas the total metal to oxygen ratio varies, again depending on the equilibrium oxygen fugacity and temperature.[4] The variation of metal:oxygen ratio

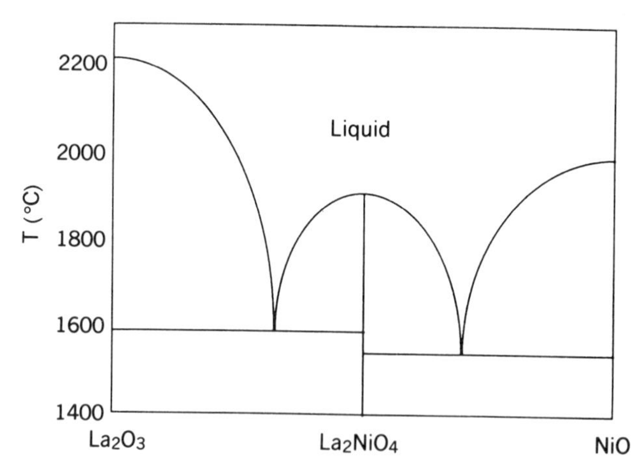

Figure 1. A qualitative representation of the La_2O_3–NiO phase diagram.

results from formation of interstitial oxygen defects, so the nonstoichiometry is represented by writing $La_2NiO_{4+\delta}$ ($0.00 \leq \delta < 0.13$ at 800°). Note that $La_2NiO_{4+\delta}$ behaves as a pseudobinary oxide.

For both synthesis and crystal growth it is essential that the conditions of temperature and atmosphere be adjusted to fall within the stability field of $La_2NiO_{4+\delta}$. In other words, the oxygen partial pressure must remain between the reduction and oxidation boundaries. The reduction boundary may be expected to be similar to that of NiO, since La^{3+} is extremely stable to reduction. The reduction boundary of NiO is described by

$$NiO(s) \leftrightarrows Ni^0(s) + \tfrac{1}{2}O_2(g) \tag{1}$$

for which the equilibrium oxygen fugacity is given by the van't Hoff form[13]

$$\log_{10}\{f_{O_2}(\text{atm})\} = \frac{-24{,}930}{T} + 9.36 \tag{2}$$

where T is in kelvins. For liquid NiO, the reduction boundary is not well characterized, but is expected to be close to that of the solid. For reduction of La_2NiO_4, we find boundary reaction to be

$$La_2NiO_4(s) \leftrightarrows La_2O_3(s) + Ni^0(s) + 1/2O_2(g) \tag{3}$$

for which the associated temperature dependence of the equilibrium oxygen fugacity is given by

$$\log_{10}\{f_{O_2}(\text{atm})\} = \frac{-25{,}040}{T} + 8.07 \tag{4}$$

Unfortunately, the oxidation boundaries of $Ni_{1-x}O$ and $La_2NiO_{4+\delta}$ are not well characterized, although many higher oxides of nickel are known.

The oxygen fugacity in the atmosphere in contact with the melt, equivalent to the partial pressure under the conditions of interest here, must exceed the larger of the two values obtained by extrapolation of Eqs. (2) and (4) to temperatures somewhat above the melting point of $La_2NiO_{4+\delta}$. Since the melt is necessarily superheated, due to the need to establish a stable melt in a cold crucible and because of geometric inhomogeneity of the RF field, we conservatively estimate the excess temperature to be ~300°. From this estimate we conclude that the oxygen partial pressure in contact with the melt should be at least 0.05 atm. An upper limit has been established experimentally at approximately 0.25 atm, above which the crystals diminish in size and frequently fracture on cooling. It is noteworthy that the subsolidus width of the stability field appears to cover many orders of magnitude in oxygen fugacity (although the upper limit remains to be characterized) for $900^\circ \leq T \leq 1300^\circ$, whereas the range suitable for crystal growth is restricted to less than one decade.

An additional complication arises in the cases where lanthanum is replaced by praseodymium or neodymium: at moderate to low temperatures the oxygen defects may segregate into defect-rich and -poor regions, resulting in phase separation.[14] At high temperatures the interstitial defects are randomly dispersed in a single tetragonal phase that crystallizes in the K_2NiF_4 structure for all accessible defect concentrations. Following crystal growth, the cooldown to room temperature frequently results in shattering of the crystals if phase separation occurs. This can be avoided if the defect concentration range throughout a given crystal falls within one of the low-temperature phase composition ranges. Since these single-phase composition ranges are narrow and the two-phase (segregated) compositional ranges are broad, it is difficult to obtain intact single crystals. Furthermore, because of nonuniformities in temperature distributions, the necessary conditions can at best be met in only limited portions of the boule. Homogeneous defect concentrations within the single-phase compositional ranges can, in principle, be established by subsolidus annealing under appropriate oxygen fugacity an temperature ranges as described for $La_2NiO_{4+\delta}$; however, the required conditions for successful growth and single-phase subsolidus annealing are as yet only partially characterized.[15] Although phase separation has also been reported for $La_2NiO_{4+\delta}$,[16] it seems to be kinetically slow, so that no significant segregation of interstitials occurs during a reasonable rapid cooldown (less than 30 min between 800 and 100°).

Some alkaline-earth-substituted analogs of the Ln_2NiO_4 phases have also been grown as single crystals. The phase diagrams remain uncharacterized; however, as expected from the nature of this heterovalent substitution (divalent for trivalent), higher oxygen fugacities are required relative to the unsubstituted members with a given lattice oxygen content for synthesis and crystal growth.

Procedure

The RF source used was a Lepel dual-frequency 50 kW generator with operating frequency ranges centered about 300 kHz and 2.8 MHz. Low-frequency operation is usually best suited for materials with ambient resistivities below 0.01 $\Omega \cdot$cm, whereas the higher-frequency mode is used in case where higher resistivity charges are to be melted. The lanthanide nickelates fall into the latter category, so operation at 2.8 MHz is selected. Details of the cold-crucible design and controlled atmosphere chamber are described in detail elsewhere.[4,10,12] The desired oxygen partial pressure is obtained by using CO_2 as the diluent, since monoatomic gases result in arcing.

In preparation for the crystal growth it is best to prereact the starting materials, dehydrated La_2O_3 and NiO, to obtain a less hygroscopic material for loading into the skull crucible and to facilitate the initial coupling. Since $La_2NiO_{4+\delta}$ prepared at $\sim 1200°$ in air is semiconducting with an ambient resistivity near 1 $\Omega \cdot$cm, it couples inductively much more readily than do the insulating precursors, La_2O_3 and NiO. Even using $La_2NiO_{4+\delta}$ as the charge material for the skull crucible, a small disk-shaped graphite susceptor ($\sim$0.5 g) is used to initiate coupling. As the susceptor begins to burn, forming CO_2, the charge is heated and direct coupling follows rapidly. Using a large susceptor (2–3 g) it is possible to start with La_2O_3 and NiO as the initial charge; however, during burn-off of the susceptor it is likely that Ni^0 will be irreversibly produced from local reduction of NiO. Reduction of nickel near the susceptor occurs when the rate of mass transfer of oxygen is not sufficient to prevent direct reduction of the charge material. Also, since NiO is more readily reduced than La_2NiO_4 (or Pr_2NiO_4) reduction of nickel is more of a problem with charges that have not been prereacted. Placement of the susceptor at the surface of a prereacted charge allows coupling to be initiated with no reduction of nickel. In either case, it is best to use a charge of at least 1.2 kg, as smaller charges generally result in small crystals; with less than $\sim$0.9 kg the melt is difficult to stabilize.

As the melt begins to form, the volume of material in the crucible is substantially decreased and more charge must be added. This is accomplished, without breaking the controlled atmosphere environment, by using a powder hopper positioned above the crucible from which additional material can be sifted in by remote control. Once all material has been loaded and the melt has stabilized, the crucible is lowered at 15 mm/h from the stationary RF coil. As the decoupling occurs upward from the bottom, crystals begin to grow in a Bridgman-like fashion. Toward the end of the run, radiative losses from the surface of the melt becomes sufficient to freeze over the surface, and crystals begin growing from the top down. As coupling diminishes, a "zone of last freezing" is left in which small crystals form. Using a crucible of 7 cm diameter, a 1.4 kg charge results in a boule that a slightly over a 5 cm in height with a density close to that of La_2NiO_4, $\sim$7 g/cm^3. The crystals obtained from below the zone of last freezing are typically 1.5 cm in height, and may exceed 2 cm in some instances. The crystallographic orientation is typically such that one of the $\langle 100 \rangle$ directions of the simple tetragonal K_2NiF_4-type unit cell is vertical. The [001] direction is least favored during growth. Cross-sectional areas vary from crystal to crystal and between runs; crystal volumes may exceed 1 cm^3. The cross-sectional area may be enhanced be loading the bottom of the charge with crystals from a previous run to provide seeding.

Properties

Selected crystals were analyzed to determine the overall La/Ni atomic ratios. A wet-chemical technique, described below, was used for the majority of crystals tested; however, a few specimens were also commercially analyzed by an independent laboratory using atomic absorption analysis; the atomic ratio was reported as 2/1 within experimental error.

We determined the cation ratio by EDTA titration using Xylenol Orange as the indicator. The pH was buffered to the 5.0–6.5 range using sodium acetate/acetic acid to optimize indicator effectiveness. Each crystal was weighed and dissolved in dilute HCl from which two aliquots were taken. One was treated directly with EDTA to obtain the total metal concentration according to the reaction

$$yLa^{3+} + Ni^{2+} + (1 + y)EDTA^{2-} \rightarrow yLa(EDTA)^{+} + Ni(EDTA) \qquad (5)$$

Excess ammonium hydroxide was added to the second aliquot in order to precipitate lanthanum as the hydroxide, while retaining nickel in solution as an ammonia complex:

$$6NH_4OH + yLa^{3+} + Ni^{2+} \rightarrow yLa(OH)_3(s) + Ni(NH_3)_6{}^{2+} + 3yH^{+} + 3(2 - y)H_2O. \qquad (6)$$

The precipitated $La(OH)_3$ was filtered, redissolved in HCl, and titrated with EDTA to obtain total lanthanum content. Combining this result with the first titration Eq. (5), the La/Ni ratio, y, was found to be 2.00 ± 0.01, in agreement with the commerical analysis.[14]

The oxygen:metal ratio was determined by the standard technique of iodometric titration of "active oxygen." $La_2NiO_{4+\delta}$ was dissolved in the presence of excess KI using 1*M* HCl. Any interstitial (active) oxygen, corresponding to $\delta > 0$, must result in excess oxidation potential. This may be described in terms of the formation of Ni^{3+} generated by the electroneutrality constraint, although any altenative mechanism for charge compensation is equally appropriate. Furthermore, if the excess oxidation potential were due to formation of cation vacancies rather than to interstitial oxygen, we could easily reformulate the present analysis to determine the vacancy content. We will justify the defect representation in terms of oxygen excess at the close of this section. The net reaction describing the dissolution and oxidation of I^- may be written as

$$La_2NiO_{4+\delta} + 2\delta I^- + 2(4+\delta)H^+ \rightarrow 2La^{3+} + Ni^{2+} + \delta I_2 + (4+\delta)H_2O \quad (7)$$

The resulting iodine (or triiodide if we account for complex formation in the presence of excess I^-) can then be titrated with sodium thiosulfate using starch as the endpoint indicator

$$2S_2O_3^{2-} + I_2 \rightarrow S_4O_6^{2-} + 2I^- \quad (8)$$

An interference occurs as a result of the presence of dissolved molecular oxygen in the acidic solution:

$$4I^- + O_2 + 4H^+ \rightarrow 2I_2 + 2H_2O \quad (9)$$

To avoid this interference, the titrations must be carried out with deaerated solutions in either a glovebox or in a glovebag free of appreciable levels of oxygen. As a check for interference, a blank should always be used. If the titration is well set up, very little or no titratable color should appear in the blank.

The level of oxygen excess, δ, determined for as-grown crystals depends on the oxygen content of the growth atmosphere and, to a lesser extent, on the region of the boule in which a given crystal grew, falling in the range $0.04 \leq \delta \leq 0.08$ for growth atmospheres with $0.05 \leq f_{O_2}$ (atm) ≤ 0.25.

The initial characterization of the nature of the oxygen nonstoichiometry in terms of an interstitial oxygen defect was accomplished by direct measurement of the density of single crystals of known excess oxidation potential and lattice parameters. Since the La/Ni ratio is fixed at 2.00, regardless of the magnitude of excess oxidation potential, two simple defect schemes can be envisioned. Either (1) two La vacancies are generated for every Ni vacancy, as represented by $(La_2Ni)_{1-x}O_4$, or (2) excess oxygen is incorporated interstitially into the lattice, as represented by the formula $La_2NiO_{4+\delta}$. Clearly, the formula weight, and therefore the density for fixed unit-cell volume, is lower in the cation vacancy scheme than in the interstitial defect case. Measuring the density of single crystals by the Archimedes method, the unit cell volume from different data, and the excess oxidation potential from iodometric titration, the correct model was determined to be that involving interstitial oxygen defects.[17] Neutron diffraction evidence has since provided evidence placing the interstitial defect site near the $(\frac{1}{4}, \frac{1}{4}, \frac{1}{4})$ position in a $\sqrt{2}a \times \sqrt{2}b \times 1c$ supercell relative to the simple K_2NiF_4 structure for specimens possessing large defect concentrations.[16]

As discussed in the introductory remarks, specific, homogeneous concentrations of interstitial oxygen defects may be established after crystal growth by isothermal subsolidus controlled atmosphere annealing. This requires knowledge of the relationship between defect concentration, temperature and oxygen fugacity, that is the functional dependence $\delta = \delta\ (T,\ f_{O_2})$. This relationship has been characterized experimentally for $T = 1000$, 1100, and 1200° by isothermal thermogravimetric studies.[18] Mass changes were monitored under isothermal conditions while varying the oxygen fugacity of the environment surrounding the crystal. As the oxygen fugacity is steadily decreased, the reduction boundary is ultimately reached, below which La_2O_3 and Ni^0 form as indicated by Eq. (3). The oxygen content relative to the reduction boundary is then measured as a function of f_{O_2} at fixed temperature. If we assume that $\delta = 0$ at the reduction boundary, δ is proportional to the mass difference relative to the boundary. This assumption is consistent with the absence of measurable excess oxidation potential in specimens equilibrated very near the reduction boundary, but one cannot rule out substoichiometric compositions. Since charge compensation is substoichiometric specimens would require the existence of Ni^+ to satisfy the electroneutrality constraint, and monovalent nickel is known to be quite unstable, it is likely that the substoichiometric regime is either compositionally very narrow or nonexistent. Crystals prepared near the reduction boundary are also transparent amber in color when less than approximately 100 μm in thickness,[19] suggesting that they are insulating, consistent with the absence of charge carriers ($\delta = 0$).

Atmospheres with controlled oxygen fugacities in the range $-3.0 \leq \log_{10}\{f_{O_2}\,(\mathrm{atm})\} \leq 0$ are achieved by simply mixing appropriate flow rates of O_2 and Ar. For $\log_{10} f_{O_2}\,(\mathrm{atm}) \leq -3.0$, the fugacity may be controlled using gaseous buffers such as CO/CO_2, H_2/CO_2, and H_2/H_2O for which thermodynamic data are readily available.[20] Since some oxides may form carbonates or hydroxides, the choice of buffer must be carefully considered. In the present case, CO/CO_2 buffers work well. This buffer is described by the reaction

$$CO_2 \leftrightarrows CO + \tfrac{1}{2}O_2 \tag{10}$$

for which the equilibrium constant, $K(T)$, is given by

$$K(T) = \frac{f_{CO} f_{O_2}^{1/2}}{f_{CO_2}} \infty \frac{\dot{F}_{CO}}{\dot{F}_{CO_2}} f_{O_2}^{1/2} \tag{11}$$

where $\dot{F}_{CO}$ and $\dot{F}_{CO_2}$ are the volumetric flow rates of CO and CO_2, respectively. Oxygen contamination in the reagent gases, as well as leaks in the

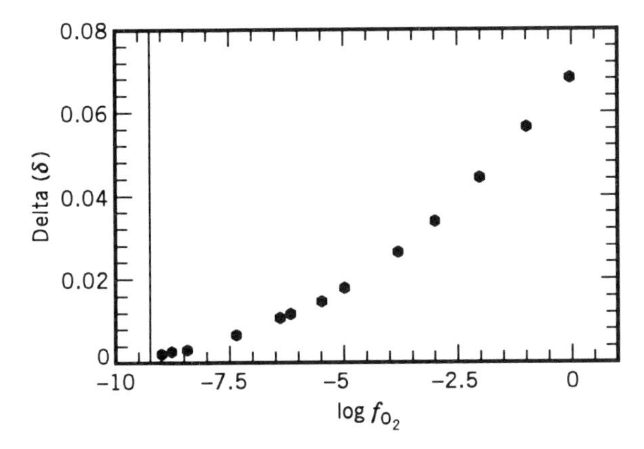

Figure 2. The level of excess (interstitial) lattice oxygen, δ, in $La_2NiO_{4+\delta}$ obtained from thermogravimetric analysis at 1200° is plotted against $\log_{10}\{f_{O_2}\text{ (atm)}\}$.

annealing system, may shift the actual f_{O_2} somewhat; however, if the buffer capacity is reasonably large, the actual f_{O_2} should be close to that expected. As a more reliable determination of the actual f_{O_2}, we have employed an Y_2O_3-stabilized ZrO_2 solid electrolyte cell, calibrated against the NiO/Ni^0 reduction boundary, to make a direct emf measurement of the *in situ* f_{O_2}. This measurement is set up as a isothermal concentration cell, with the outside of a closed-end Y_2O_3-stabilized ZrO_2 tube expected to the annealing environment, while the insides is in contact with pure oxygen at 1 atm pressure. The oxygen fugacity is then determined from the Nernst equation as

$$\log_{10}\{f_{O_2}(\text{atm})\} = \frac{-\varphi}{0.049605T} \tag{12}$$

where φ is in millivolts and T is in kelvins.

The results of a thermogravimetric study performed at 1200° are shown in Fig. 2. These data have been successfully fit to a interstitial oxygen-defect model that is consistent with the previously described density measurements.

Acknowledgments

We wish to acknowledge the assistance of P. Metcalf, K. S. Nanjundaswamy, V. Bhat, and Dr. H. R. Harrison (deceased) of Purdue University, and D. E. Rice of the University of Delaware, for assistance with the growth of single-crystal samples.

References

1. M. Foex, A. Mancheron, and M. Line, *Compt. Rend. Acad. Sci. Paris*, **250**, 3027 (1960).
2. B. Grande, Hk. Muller-Buschbaum, and M. Schweizer, *Z. Anorg. Allg. Chem.*, **428**, 120 (1977).
3. U. Lehmann and Hk. Muller-Buschbaum. *Z. Naturforsch*, **35b**, 389 (1980).
4. D. J. Buttrey, H. R. Harrison, J. M. Honig, and R. R. Schartman, *J. Solid State Chem.*, **54**, 407 (1984).
5. M. Bassat, P. Odier, and F. Gervais, *Phys. Rev. B*, **35**, 7126 (1987).
6. D. Michel, *Rev. Hautes Temp. Refract.*, **9**, 225 (1972).
7. V. I. Aleksandrov, V. V. Osiko, A. M. Prokhorov, and V. M. Tatarinstev, *Vestn. Akad. Nauk SSSR* **12**, 29 (1973).
8. V. I. Aleksandrov, V. V. Osiko, A. M. Prokhorov, and V. M. Tatarinstev, in *Current Topics in Materials Science*, E. Kaldis (ed.), Vol. 1, North Holland Press, Amsterdam 1978, p. 421.
9. J. F. Wenkus, M. L. Cohen, A. G. Emslie, W. P. Menaski, and P. F. Strong, *Design and Fabrication of a Cold Crucible System*, Air Force Cambridge Research Laboratories AFCRL-TR-75-0213 (1975).
10. H. R. Harrison and R. Aragon, *Mater. Res. Bull.*, **13**, 1097 (1978).
11. D. Michel, M. Perez, Y. Jorba, and R. Collongues, *J. Cryst. Growth*, **43**, 546 (1978).
12. M. Wittenauer, P. Wang, P. Metcalf, Z. Kakol, and J. M. Honig, *Inorg. Synth.*, **30**, 1994.
13. G. C. Ulmer (ed.), *Research Techniques for High Pressure and High Temperature*, Springer-Verlag, 1971.
14. D. J. Buttrey, J. D. Sullivan, G. Shirane, and K. Yamada, *Phys. Rev. B*, **42**, 3944 (1990).
15. J. D. Sullivan, D. J. Buttrey, D. A. Cox, and J. Hriljac, in press.
16. J. D. Jorgensen, B. Dabrowski, Shiyou Pei, and D. G. Hinks, *Phys. Rev. B*, **40**, 2187 (1989).
17. D. J. Buttrey, R. R. Schartman, J. M. Honig, C. N. R. Rao, and G. N. Subanna, *J. Solid State Chem.*, **74**, 233 (1988).
18. R. R. Schartman and J. M. Honig, *Mater. Res. Bull.*, **24**, 1375 (1989).
19. T. Freltoft, D. J. Buttrey, G. Aeppli, D. Vaknin, and G. Shirane, *Phys. Rev. B*, **44**, 5046 (1991).
20. P. Deines, R. H. Nafziger, G. C. Ulmer, and E. Woermann, *Temperature–Oxygen Fugacity Tables for Selected Gas Mixtures in the System C–H–O at One Atmosphere Total Pressure*, Bulletin 88 of the Earth and Mineral Sciences Experiment Station, The Pennsylvania State University, University Park, 1974.

29. HYDROTHERMAL SYNTHESIS AND CRYSTAL GROWTH OF POTASSIUM TITANYL ARSENATE, $KTiOAsO_4$

$$TiO_2 + KH_2AsO_4 \xrightarrow[2700\ \text{bar}]{700^\circ} KTiOAsO_4 + H_2O$$

Submitted by MARK L. F. PHILLIPS* and GALEN D. STUCKY‡
Checked by JOHN B. INGS†

The nonlinear optical (NLO) material potassium titanyl arsenate (KTA) has recently become the focus of considerable interest as a medium for second-harmonic generation, particularly at 1.064 μm, and electrooptic (EO) modulation.[1] Crystals used in determining KTA structure and nonlinear optical properties have previously been obtained by fusing TiO_2 with excess KH_2AsO_4 and K_2CO_3 at $\sim 1000^\circ$, then cooling the melt to approximately 800°, allowing crystals to spontaneously nucleate and grow. However, it is reported that elevated synthesis temperatures result in higher concentrations of K_2O vacancy defects in the isomorphous NLO/EO material $KTiOPO_4$ (KTP), and that these defects are associated with lower laser-damage thresholds;[2] it is likely that this relationship holds true for KTA as well. The use of lower growth temperatures is discouraged by increased melt viscosity, which results in either flux inclusions or the need for extremely slow cooling rates. Melt viscosity can be reduced by adding WO_3 to the melt, though crystals again frequently grow with multiple domains, and inclusion of tungsten from the melt imparts a yellowish color to the product.

Hydrothermal synthesis of KTA offers an attractive alternative to flux growth. A review of the hydrothermal synthesis technique is given in ref. 3, and its application to crystal growth of KTP is discussed in Laudise et al.[4] Bierlien and Gier[5] give examples demonstrating the method's utility in preparing crystals with the general formula (K, Rb, Tl)TiO(P, As)O_4. We can view the hydrothermal synthesis and crystal growth of KTA as either a melt growth that uses water to reduce viscosity, or as a solution growth that uses high concentrations of mineralizer to bring the KTA into solution. With the appropriate concentration of water in the mother liquors, their viscosity and temperature can be greatly reduced, improving growth rate and crystal quality without adversely affecting the solubility of the KTA. This work

* Sandia National Laboratories, Albuquerque, NM 87185. (Author to whom correspondence should be addressed.)

† Litton Airtron, 200 East Hanover Ave., Morris Plains, NJ 07950-2496.

‡ Department of Chemistry, University of California, Santa Barbara, CA 93106.

describes the hydrothermal nucleation and growth of KTA crystals potentially suitable for nonlinear optic-electrooptic investigations, or as seeds for gradient growth of larger crystals at lower temperature.

Procedure

■ **Caution.** *Follow all manufacturer's recommendations and limitations in the use of pressure equipment with this synthesis. In particular, use only a pressure vessel rated by the manufacturer to withstand the maximum temperature and pressure to be used in this experiment, and ensure that the equipment is in good condition prior to use. Failure to meet these precautions may result in an explosion.*

A Leco HR-1B-1 Tem-Pres system was used for this synthesis, although any bomb and pressurizing system rated for the synthesis conditions can be employed. The bomb and pressure lines are filled with water, and the reagents are sealed within a collapsible gold tube. This allows the density of the liquors to be determined by externally controlling the pressure within the bomb and prevents the reagents from corroding the interior walls of the vessel and contaminating the solution. Gold is preferred as the container material as it is impervious to H_2 gas, the transport of which would lead to reduction of Ti via the couple $Fe/Fe^{2+}//TiO^{2+}/Ti^{3+}$.

During operation the water in the tubes and bomb is a supercritical fluid, the density of which is independent of temperature and can be selected by either adjusting the initial percent fill or by external pressure control. The density strongly determines the solvent properties of the water in the tubes. To ensure a density of 0.7–0.8 g/cm^3, the bomb is pressurized prior to heating, and as the maximum temperature is approached steam is released until the desired density (measured as pressure) is achieved. Since the crystals in this experiment grow from precipitated nuclei, the reagents must initially be completely dissolved. The bomb temperature is then allowed to slowly cool through all temperature regimes in which significant precipitation of KTA is likely to occur, then drops freely to room temperature. It is not necessary to control pressure externally during the cooling cycle, since a constant solution density is desired during crystal growth.

■ **Caution.** *KH_2ASO_4 is poisonous, and possibly carcinogenic. Avoid contact with this material or its solutions. In particular, use caution when sealing the gold tubes or when opening the bomb at the end of the synthesis, as any leaks will release the arsenic.*

The Au tube used in this procedure (4 in. long $\times \frac{3}{8}$-in.-o.d. $\times$ 0.010-in. wall, Engelhard) is prepared by annealing both ends at a dull red heat using a gas–air flame, then crimped at one end along three evenly spaced points along the diameter such that the folds meet at the center. An oxygen-gas flame is used to weld the tops of the crimp closed, and the seal is checked for holes or gaps under a light microscope. The tube is held upright in a fixture, and 1.26 mL 5.0*N* KOH (6.300 mmol), 0.45 mL H_2O, 0.180 g (2.253 mmol) TiO_2 (anatase) and 2.84 g (15.766 mmol) KH_2AsO_4[6] are added. The top $\frac{3}{4}$ in. of the inner tube wall is cleaned and dried, then crimped in the same fashion as the bottom end. The tube is then suspended in water until the top crimp is approximately $\frac{3}{4}$ in. above the water level, and the crimp is welded. A correct water level is essential to a successful weld; if it is too high, the top will cool and prevent the gold from fusing; a low water level will allow the liquid contents of the tube to boil and perforate the seal during welding. After checking the seal with a light microscope and rewelding if necessary, the tube is weighed.

The sealed tube is placed in the bomb, which is then filled with water. After the bomb stem is screwed on, this also is filled, using a syringe. The bomb is placed in a tube furnace, pressurized to 1000 bar, and heated promptly to 675°, releasing steam as necessary to maintain a final pressure of 2700 bar. The bomb temperature is held at 675° for 8 h, cooled to 475° at 2°/h, to 350° at 6°/h, then allowed to fall to room temperature. The tube is reweighed to determine whether leaks occurred during synthesis and reexpanded by briefly and *cautiously* heating over a gas–air flame. The tube is opened and the contents are filled, rinsed with water, and air-dried. Typical yield is 0.520 g, or 95% of theory. Although many large pieces will be twinned, there will be a number of well-formed single crystals 2–3 mm across. It is worth noting that the crystal size is determined in part by the K_2O/As_2O_5 mole ratio in the mineralizer. For example, a 1:1 ratio gives KTA powder with crystallites not exceeding 0.1 mm in diameter, and a 1.2:1 correspondence produces many well-formed single crystals between 0.1 and 0.5 mm in length.

Powder X-ray diffractometry shows that the lattice is orthorhombic, space group $Pna2_1$, with $a = 13.146(5)$ Å, $b = 6.584(2)$ Å, $c = 10.771(6)$ Å. Typical width of single-crystal X-ray omega (ω) scan peaks is 0.15°, suggesting the absence of domain walls. The techniques available for visualizing ferroelectronic domains in KTP[7,8] (pyroelectric, piezoelectric, and electrooptic) should be applicable to KTA as well.

References and Note

1. J. D. Bierlein, H. Vanherzeele, and A. A. Ballman, *Appl. Phys. Lett.*, **54**, 783 (1989).
2. P. A. Morris, M. K. Crawford, A. Ferretti, R. H. French, M. G. Roelofs, J. D. Bierlein, J. B. Brown, G. M. Loiacono, and G. Gashurov, in *Optical Materials: Processing and Science*, D. B. Poker and C. Ortiz (eds.) Materials Research Society, Pittsburgh, 1989, pp. 95–101.
3. R. A. Laudise, *Chem. Eng. News*, **65**, (39), **30** (1987).

4. R. A. Laudise, R. J. Cava, and A. J. Caporaso, *J. Cryst. Growth*, **74**, 275 (1986).
5. J. D. Bierlein and T. E. Gier, U.S. Patent 3,949,323. Apr. 6, 1976.
6. Potassium dihydrogen arsenate is commercially available through Sigma Chemical Company, PO Box 14508, St. Louis, MO 63178 or Johnson Matthey Electronics, Orchard Road, Royston, Hertfordshire SG8 5HE, England, or may be obtained in sufficient purity by *carefully* reacting stoichiometric quantities of As_2O_5 and KOH in water. The solution is concentrated to near saturation by boiling, then poured into a large excess of methanol and filtered. The product is characterized by titration and/or X-ray powder diffraction.
7. F. C. Zumsteg, J. D. Bierlein, and T. E. Gier, *J. Appl. Phys.*, **47**, 4980 (1976).
8. J. D. Bierlein and F. Ahmed, *Appl. Phys. Lett.*, **51**, 1322 (1987).

30. SYNTHESIS OF $La_2Ta_3Se_2O_8$ SINGLE CRYSTALS

$$La_2Se_3 + Ta_2O_5 + Ta \xrightarrow{1475\ K} La_2Ta_3Se_2O_8 + \text{side products}$$

Submitted by THEODORE D. BRENNAN* and JAMES A. IBERS†
Checked by L. F. SCHNEEMEYER‡

The quaternary rare-earth oxychalcogenides are a relatively rare class of compounds that show interesting physical properties and structures. For example, the $Ln_2Ta_3Q_2O_8$ (Ln = La, Ce, Pr, Nd; Q = Se; Ln = La; Q = S) series[1,2] are mixed-valence compounds as is $La_5V_3O_7S_6$,[3] while the compounds $LaCrOQ_2$ (Q = S, Se)[4] are ferromagnetic. Most known quaternary oxychalcogenides containing a lanthanide have previously been prepared by reacting the lanthanide oxychalcogenide, Ln_2O_2Se or Ln_2O_2S, with the appropriate metal chalcogenide. For example, the compounds $(Ln_2O_2)_m(M_xQ_y)_n$ (M = Cu, Ag, Ga, In, Ge, Sn, As, Sb, Bi; Q = S, Se),[5] $LnCrOQ_2$,[6,7] $La_5V_3O_7S_6$,[3] and $(UOS)_4LuS^8$ have all been prepared in this way. The synthesis presented here for $La_2Ta_3Se_2O_8$ instead uses lanthanide sesquiselenide, tantalum oxide, and tantalum as the starting materials to give single crystals. No attempt has been made to optimize the yield of this synthesis; rather, the objective is to obtain single crystals. Pure $La_2Ta_3Se_2O_8$ powder may be prepared by the reaction of La_2O_3, Ta_2O_5, Ta, and Se (1:1:1:2) at 1475 K for 4 days.[1]

* Department of Chemistry, Morgan State University, Baltimore, MD 21239.
† Department of Chemistry, Northwestern University, Evanston, IL 60208-3113.
‡ AT&T Bell Laboratories, Murray Hill, NJ 07974.

Procedure

The lanthanide sesquiselenides are not commercially available. La_2Se_3 may be prepared by grinding together Se pellets (920 mg, 6 mmol) and La powder (1080 mg, 4 mmol) in an inert atmosphere, and then heating the mixture in an evacuated ($<10^{-3}$-torr) carbon-coated fused-silica tube, 10-mm-i.d., $\approx$100-mm length, at 775 K for 1 day, 1075 K for 1 day, and 1275 K for 1 day.

■ **Caution.** *Attempts to prepare more than 1 g of La_2Se_3 under these conditions have resulted in occasional explosions during heating.*

A fused-silica tube may be coated with carbon through the pyrolysis of an acetone residue in the tube. The La_2Se_3 powder produced is dark red to black in color. It should be kept in an inert atmosphere. Single crystals of $La_2Ta_3Se_2O_8$ are then prepared by grinding La_2Se_3 (257 mg, 0.5 mmol), Ta_2O_5 (221 mg, 0.5 mmol), and Ta (91 mg, 0.5 mmol) powders with an agate mortar and pestle in an inert atmosphere. The reaction mixture is then sealed in an evacuated fused-silica tube that has been carefully cleaned inside and out. It is then heated at 775 K for 2 days, 1075 K for 2 days, and 1475 K for 6 days. At 1475 K the fused-silica tube will soften and collapse slightly. The tube is attacked and mainly black rectangular prisms of $La_2Ta_3Se_2O_8$ are formed along with clear irregular prisms of $La_2Se(SiO_4)$.[9] The latter compound may be removed by washing the mixture with 6*M* HCl. This procedure also works for the synthesis of $M_2Ta_3Se_2O_8$ (M = Ce, Nd, Pr)[1] and for $La_2Ta_3S_2O_8$,[2] where La_2S_3 is used in place of La_2Se_3.

Properties

Crystals of $La_2Ta_3Se_2O_8$ are air-stable and do not dissolve in 6*M* HCl. They crystallize in an orthorhombic cell of dimensions $a = 9.929(4)$, $b = 11.951(4)$, $c = 7.666(3)$ Å at $T = 298$ K. The compound is isostructural with $Pr_2Ta_3Se_2O_8$[1] and $La_2Ta_3Se_2O_8$,[2] whose structures are known from single-crystal studies. Crystals of $La_2Ta_3Se_2O_8$ show low conductivity, about $1 \times 10^{-6}\ \Omega^{-1}\ cm^{-1}$. The magnetic susceptibility of the compound is given by $\chi = C/(T + \theta) + \chi_0$, with $C = 0.0117(4)$ emu K/mol, $\theta = 0.20(2)$ K, and $\chi_0 = 1.6(1) \times 10^4$ emu/mol.

References

1. T. D. Brennan, L. E. Aleandri, and J. A. Ibers, *J. Solid State Chem.*, **91**, 312 (1991).
2. T. D. Brennan and J. A. Ibers, *J. Solid State Chem.*, **98**, 82 (1992).
3. J. Dugué, T. Vovan, and P. Laruelle, *Acta Crystallogr. Sect. C: Cryst. Struct. Commun.*, **41**, 1146 (1985).

4. M. Wintenberger, T. Vovan, and M. Guittard, *Solid State Commun.*, **53**, 227 (1985).
5. M. Guittard, S. Benazeth, J. Dugué, S. Jaulmes, M. Palazzi, P. Laruelle, and J. Flahaut, *J. Solid State Chem.*, **51**, 227 (1984).
6. J. Dugué, T. Vovan, and J. Villers, *Acta Crystallogr., Sect. B: Struct. Crystallogr. Cryst. Chem.*, **36**, 1291 (1980).
7. J. Dugué, T. Vovan, and J. Villers, *Acta Crystallogr., Sect. B: Struct. Crystallogr. Cryst. Chem.*, **36**, 1294 (1980).
8. S. Jaulmes, M. Julien-Pouzol, J. Dugué, P. Laruelle, T. Vovan, and M. Guittard, *Acta Crystallogr., Sect. C: Cryst. Struct. Commun.*, **46**, 1205 (1990).
9. T. D. Brennan and J. A. Ibers, *Acta Crystallogr., Sect. C: Cryst. Struct. Commun.*, **47**, 1062 (1991).

Chapter Four

INTERCALATION COMPOUNDS

31. THE ALKALI TERNARY OXIDES, A_xCoO_2 AND A_xCrO_2 (A = Na, K)

Submitted by C. DELMAS,* C. FOUASSIER,* and P. HAGENMULLER*
Checked by J. F. ACKERMAN†

Reprinted from *Inorg. Synth.*, **22**, 56 (1983)

Several ternary oxides of formula A_xMO_2 (A = Na, K; M = Co, Cr) have been synthesized by solid-state reactions. Depending on the stability and the reducing character of the materials obtained, various synthesis methods may be used. They are described here from the simplest to the most sophisticated.

A. SODIUM COBALT OXIDES: Na_xCoO_2 ($x \leqslant 1$)

$$3x\mathrm{Na_2O_2} + 2\mathrm{Co_3O_4} + (2 - 3x)\mathrm{O_2} \rightarrow 6\mathrm{Na}_x\mathrm{CoO_2}$$

$$(x = 0.60;\ 0.64 \leqslant x \leqslant 0.74;\ 0.77;\ 1).$$

* Laboratoire de Chimie du Solide du CNRS, Université de Bordeaux 1, 351 Cours de la Libération, 33405 Talence Cedex, France.
† General Electric Company, P.O. Box 8, Schenectady, NY 12301.

Procedure

All manipulations are carried out in a dry box. The peroxide, Na_2O_2, can be replaced by the hydroxide NaOH or the oxide Na_2O. The reaction temperatures are the same as for Na_2O_2. Nevertheless, it is advantageous to start the reaction with Na_2O_2 because it is the purest of the three sodium compounds.

Stoichiometric amounts of the two starting oxides (1 g of Co_3O_4, 4.15 mmol, and the corresponding Na_2O_2 amount) are intimately mixed by grinding in an agate mortar. The mixture is introduced into an alumina crucible and then heated for 15 h in an oxygen stream.

As the resulting phases are not very thermally stable, the reaction temperature required is dependent on the value of x. To obtain phases in which $x = 0.60, 0.77$, or 1, the reaction temperature is 550°, while for pure phases in the range $0.64 \leqslant x \leqslant 0.74$, the reaction temperature is 750°.

Properties

The Na_xCoO_2 phases are obtained in the form of black powders that are very sensitive to atmospheric moisture.

The structure is derived from $(CoO_2)_n$ sheets of edge-sharing octahedra. The Na^+ ions are inserted between the slabs in a trigonal prismatic ($x = 0.60$ and $0.64 \leqslant x \leqslant 0.74$) or octahedral ($x = 0.77$ and $x = 1$) environment.[1]

The crystallographic data are summarized in Table 1.

Although the nonstoichiometric compounds have a metallic character, $NaCoO_2$ is a semiconductor.

TABLE I. Crystallographic Data of the Na_xCoO_2 Phases

x	Symmetry	Cell Parameters	Oxygen Packing
0.60	Monoclinic	$a = 4.839$ Å $b = 2.831$ Å $c = 5.71$ Å $\beta = 106.3°$	*AABBCC*
0.64, 0.74	Hexagonal	$a = 2.833$ Å $c = 10.82$ Å	*AABB*
0.77	Monoclinic	$a = 4.860$ Å $b = 2.886$ Å $c = 5.77$ Å $\beta = 111.3°$	*ABCABC*
1	Rhombohedral	$a = 2.880$ Å $c = 15.58$ Å	*ABCABC*

B. POTASSIUM COBALT OXIDE BRONZES: $K_{0.50}CoO_2$, $K_{0.67}CoO_2$

$$6x\text{KOH} + 2\text{Co}_3\text{O}_4 + \frac{4-3x}{2}\text{O}_2 \rightarrow 6\text{K}_x\text{CoO}_2 + 3x\text{H}_2\text{O}$$

Procedure

As commercial potassium hydroxide always contains a few percent of potassium carbonate and water, it is necessary to determine (by acidimetric titration) the potassium content of the starting material.

The procedure is similar to that used for preparing the Na_xCoO_2 phases (Section A). The reaction temperature for preparing either phase is 500°.

Properties

Both phases are obtained in the form of dark blue powders, extremely sensitive to atmospheric moisture.

The K^+ ions are inserted in a trigonal prismatic environment between $(CoO_2)_n$ sheets. The two structural types differ in the oxygen packing: *AABBCC* for $x = 0.50$ and *AABB* for $x = 0.67$.[2] The $K_{0.50}CoO_2$ phase crystallizes in the rhombohedral system. The hexagonal parameters are $a = 2.829$ Å, $c = 18.46$ Å. The $K_{0.67}CoO_2$ phase crystallizes in the hexagonal system with the cell parameters $a = 2.837$ Å, $c = 12.26$ Å.

They are both metallic conductors.

C. POTASSIUM COBALT OXIDE: $KCoO_2$

$$4\text{K}_2\text{O} + \text{KO}_2 + 3\text{CoO}_4 \rightarrow 9\text{KCoO}_2$$

Procedure

The stoichiometric mixture of the starting materials is ground in a dry box in an argon or nitrogen atmosphere whose O_2 and H_2O content is less than 4 ppm. (K_2O is not stable in the presence of oxygen and leads to peroxide, K_2O_2.) As the potassium oxide K_2O is not a commercial product, it is prepared by controlled oxidation of liquid potassium by oxygen diluted with argon according to Rengade.[3] The resulting mixture (K_2O: 0.1609 g, 1.708 mmol; KO_2: 0.0303 g, 0.427 mmol; Co_3O_4: 0.3083 g, 1.281 mmol) is introduced into a gold tube (4 mm diameter, 0.4 mm thickness, 60 mm length). This tube is sealed, weighed, and heated for 15 h.

Depending on the thermal treatment, two allotropic varieties of $KCoO_2$ can be obtained. A synthesis temperature of 500° leads to the low-temperature α form, while the high-temperature β form, which is metastable at room temperature, is obtained by quenching in air from 900°.

As a precautionary measure, the tube is weighed again and opened in a dry box. (A difference in the two weights indicates a crack in the gold tube with resultant reaction with water vapor or oxygen or K_2O volatilization.)

Properties

Whereas α-$KCoO_2$ is a brown powder, β-$KCoO_2$ is black. Both phases are very sensitive to moisture. The structural transition between the α and β forms is reversible; the transition temperature is around 650°. As previously mentioned, the β form can be maintained at room temperature, but is metastable and leads to α-$KCoO_2$ when heated above about 250°.

The α variety crystallizes in the tetragonal system with the cell parameters $\alpha = 3.797$ Å, $c = 7.87$ Å. Its structure is unknown.

The product β-$KCoO_2$ also crystallizes in the tetragonal system. The cell parameters are $a = 5.72$ Å, $c = 7.40$ Å.[2] The structure of β-$KCoO_2$ is related to that of high-temperature cristobalite, the K^+ ion occupying the 12-coordinate site, which is empty in β-SiO_2.

D. POTASSIUM CHROMIUM OXIDE: $KCrO_2$

$$K_2O + Cr_2O_3 \rightarrow 2KCrO_2$$

The compound $KCrO_2$ is used as starting material for the preparation of the K_xCrO_2 phases (Section E).

Procedure

A small amount of potassium is added to a stoichiometric mixture of potassium and chromium oxides (K: 5×10^{-3} g, 0.127 mmol; K_2O: 0.382 g, 4.06 mmol; Cr_2O_3: 0.617 g, 4.06 mmol). After grinding, the powder is introduced into a silver crucible, which is placed in a Pyrex tube. The tube is sealed under vacuum and heated for 15 h at 450°. Theoretically, potassium metal does not participate in the reaction, but it creates a strongly reducing atmosphere in the sealed tube that is necessary for the synthesis of $KCrO_2$. During the reaction, the potassium reacts with the glass and is thus eliminated. After 15 h at 450°, pure $KCrO_2$ is obtained. All manipulations are carried out in a dry box in an oxygen-free argon or nitrogen atmosphere (O_2 and H_2O content less than 5 ppm).

Properties

The product is a green powder, extremely sensitive to moisture or oxygen. At room temperature in air, it is immediately oxidized, giving a mixture of $K_{0.50}CrO_2$ and K_2CrO_4. It crystallizes in the rhombohedral system. The hexagonal parameters are $a = 3.022 \pm 0.004$ Å, $c = 17.762 \pm 0.03$ Å. Its structure consists of $(CrO_2)_n$ layers between which the K^+ ions are inserted in an octahedral environment.[4]

E. POTASSIUM CHROMIUM OXIDE BRONZES: K_xCrO_2

$$0.70 \leqslant x \leqslant 0.77$$

$$0.50 \leqslant x \leqslant 0.60$$

$$KCrO_2 \rightarrow K_xCrO_2 + (1 - x)K \nearrow$$

Procedure

One gram of $KCrO_2$ is put in a gold crucible that is introduced into the apparatus shown in Fig. 1. The temperature is increased while holding the reaction mixture under a vacuum of 10^{-2} torr. At 700° $KCrO_2$ evolves potassium metal which distills and deposits on the cooled part of the Vycor reaction vessel.

After 3 h at 700° a potassium-deficient phase whose formula is K_xCrO_2 ($0.70 \leqslant x \leqslant 0.77$) is obtained. If the temperature is raised to 950°, more potassium volatilizes and a phase more deficient in potassium ($0.50 \leqslant x \leqslant 0.60$) is obtained. All materials must be manipulated in an inert atmosphere.

Direct syntheses of K_xCrO_2 from the oxides K_2O, KO_2, and Cr_2O_3 or K_2O, Cr_2O_3, and CrO_2 cannot be realized by this technique. These reactions give only mixtures of Cr_2O_3 and K_2CrO_4, K_3CrO_4 or K_4CrO_4, depending on the values of x.

Properties

The two phases obtained are dark brown powders, very sensitive to atmospheric moisture. The phase with the higher potassium content is also sensitive to oxygen.

Figure 2 shows the variation of x (as determined by thermogravimetric analysis and X-ray diffraction) versus temperature. The Parts I, III, and V of the curve correspond respectively to $x = 1$, $0.70 \leqslant x \leqslant 0.77$, and $0.50 \leqslant x \leqslant 0.60$.

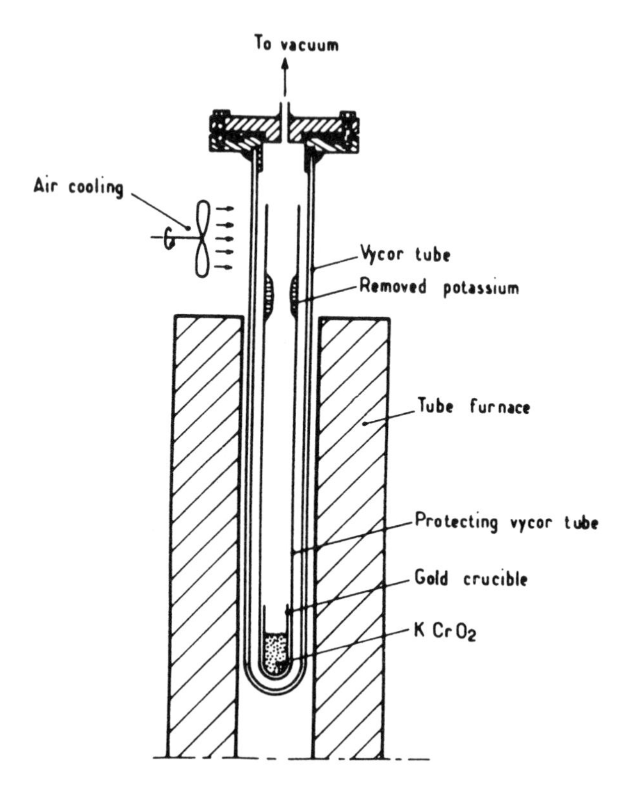

Figure 1. Apparatus for the preparation of K_xCrO_2 phase.

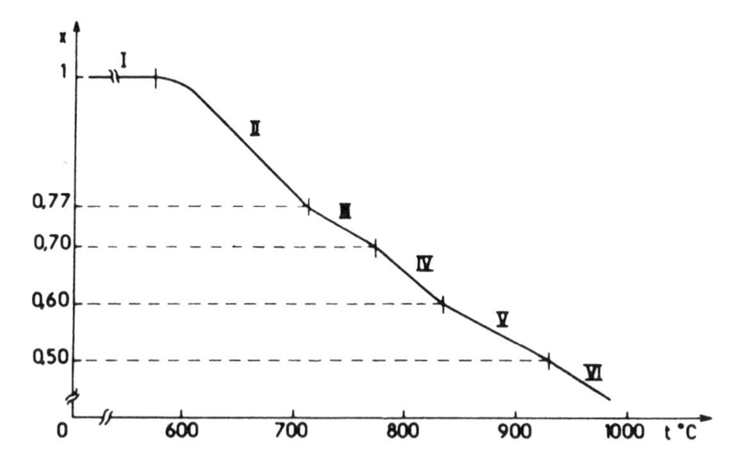

Figure 2. TGA of $KCrO_2$ (heating rate 100 °/h).

The structures of the bronzes are built up of $(CrO_2)_n$ sheets with K^+ ions inserted in distorted octahedral sites for $0.70 \leqslant x \leqslant 0.77$ and trigonal prismatic sites when $0.50 < x < 0.60$.[4] K_xCrO_2 $(0.70 \leqslant x \leqslant 0.77)$ crystallizes in a monoclinic system ($a = 5.062$ Å, $b = 2.986$ Å, $\beta = 112.61°$) and K_xCrO_2 $(0.50 \leqslant x \leqslant 0.60)$ in a rhombohedral system. Its hexagonal parameters are $a = 2.918$ Å, $c = 18.44$ Å.

References

1. C. Fouassier, G. Matejka, J. M. Réau, and P. Hagenmuller, *J. Solid State Chem.*, **6**, 532 (1973).
2. C. Delmas, C. Fouassier, and P. Hagenmuller, *J. Solid State Chem.*, **13**, 165 (1975).
3. E. Rengade, *C. R. Febd. Seances Acad. Sci.* **114**, 754 (1907).
4. C. Delmas, M. Devalette, C. Fouassier, and P. Hagenmuller, *Mater. Res. Bull.*, **10**, 393 (1975).

32. TANTALUM DISULFIDE (TaS_2) AND ITS INTERCALATION COMPOUNDS

Submitted by J. F. REVELLI*
Checked by F. J. DiSALVO†

Reprinted from *Inorg. Synth.*, **19**, 35 (1979)

Over the past 5 years, a considerable amount of research has been devoted to the study of layered transition-metal dichalcogenides and their so called intercalation complexes.[1–5] In particular, the group IV, V, and VI transition-metal dichalcogenides form layered structures with hexagonal, rhombohedral, or trigonal symmetry. The basic layer is composed of a covalently bound X—M—X sandwich (M = transition metal, X = chalcogen), and successive layers are bound together by relatively weak chalcogen–chalcogen var der Waals bonds. Hence, under appropriate conditions, these layers can be "pried" apart, and other chemical species can be inserted between them—much in the same fashion as two decks of cards are mixed together by shuffling. From the point of view of chemical bonding, intercalation can be regarded as a charge-transfer process in which the orbitals of the guest species are mixed together with those of the layered "host."[2–4] In the case of the group VI layered transition-metal dichalcogenides the energy gained in charge transfer is presumably small. Hence, in these materials stable intercalation complexes have been observed only for highly electropositive guest

* Electrical Engineering and Materials Science Departments, Northwestern University, Evanston, IL. Current Address: Xerox Corp., Webster Research Center, Rochester, NY 14600.
† Bell Laboratories, Murray Hill, NJ 07974.

species, such as the alkali metals. For this case a net donation of charge occurs from the alkali metal to the transition-metal dichalcogenide.[4–8] The group V layered dichalcogenides, on the other hand, exhibit the phenomenon of intercalation for a wide range of organic (Lewis base) materials,[2–5] transition and posttransition metals,[9] and the alkali metals.[7] Synthesis techniques of the group V layered compounds and their intercalation complexes are described in the following sections. In particular, TaS_2 and the tantalum sulfide intercalates are used as the primary examples.

Systematic studies of the structural properties of TaS_2 by Jellinek[10] revealed the presence of several polymorphic forms of the compound as a function of temperature. Within a layer, the tantalum atom sits in the holes formed between two layers of sulfur atoms in the S—Ta—S sandwich. The coordination of the tantalum is trigonal prismatic or octahedral, depending on whether the two sulfur layers lie one on top of the other or are rotated by 60°. Thus the various polymorphs observed in TaS_2 result from the variety of stacking sequences of the basic S—Ta—S slabs. Figure 1 shows the various phases that are formed as a functions of temperature. The high-temperature "1*T*" (*T* = trigonal symmetry) phase has octahedral coordination of the

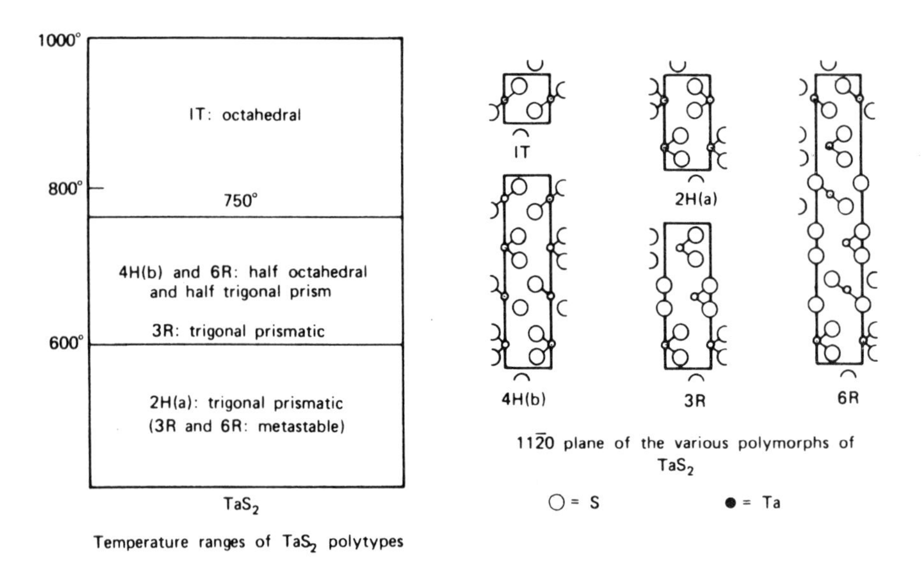

Figure 1. The *a* notation indicates that the tantalum atoms are aligned along the crystallographic *c* axis. *R* indicates rhombohedral symmetry. The 3*R* phase has three trigonal prismatic slabs per unit cell, whereas the 6*R* has six slabs that are alternately trigonal prismatic and octahedral. F. Jellinek, *J. Less Common Met.*, **4**, 9–15 (1962); F. J. DiSalvo et al., *J. Phys. Chem. Solids*, **342**, 1357 (1973).

tantalum atom and one S—Ta—S slab in the unit cell. The room-temperature $2H(a)$ phase (H = hexagonal symmetry) has trigonal prismatic coordination of the tantalum atoms and two S—Ta—S slabs per unit cell (see $11\bar{2}0$ section, Fig. 1). For reasons as yet not fully understood, only those phases that have trigonal prismatic character tend to form intercalation compounds readily. Of these, the $2H(a)$ polytype of TaS_2 seems to be the most favorable host material.

A. POLYCRYSTALLINE $2H(a)$ PHASE OF TaS_2

$$\mathrm{Ta} + 2\mathrm{S} \xrightarrow{850^\circ} 1T\text{-}\mathrm{TaS_2}$$

$$1T\text{-}\mathrm{TaS_2} \xrightarrow{750^\circ} 6R\text{-}\mathrm{TaS_2} \xrightarrow{550^\circ} 2H(a)\text{-}\mathrm{TaS_2}$$

Procedure

The techniques used in the preparation of polycrystalline and single-crystal TaS_2 are much the same as those described for TiS_2, ZrS_2, and SnS_2 and have been described in this series,[12,13] The latter compounds, however, exist in the $1T$ phase structure throughout the entire range from the crystal growth temperature down to room temperature. The compound TaS_2, on the other hand, must undergo two first-order phase transitions as it is cooled from above 750° to room temperature. To assure complete transformation to the $2H(a)$ phase, the TaS_2 must be cooled very slowly through these transition temperatures.

The synthesis of $2H(a)$-TaS_2 powder is carried out in fused quartz or Vycor ampules—typically 10 cm long and 1.7 cm in diameter, with a wall thickness of 1.0–2.0 mm (Fig. 2a). About 1.41 g (0.008 mol) of 0.020-in.-diameter tantalum wire (99.8% purity)* is cut in the form of 3.8-cm lengths, washed in hot dilute HCl to remove iron contamination introduced in the cutting process, rinsed in distilled water, and dried. Sulfur powder, 0.500 g (0.016 mol, 99.9999% purity),† is then added, and the ampule is sealed under a vacuum of about 10^{-3} torr. The sealed ampule is placed into a cold laboratory tube furnace with the tantalum wire charge placed in the end of the ampule towards the center of the oven and elevated about 1 cm (see Fig. 2b). Thus, as the sample is heated, the sulfur melts and remains in the lower, cooler end of the ampule. A thermocouple is placed near this end to ensure that the temperature is at or below 450° (the pressure of sulfur vapor in equilibrium

* Available from Materials Research Corp., Route 303, Orangeburg, NY 10962

† Available from United Mineral and Chemical Corp., 129 Hudson St., New York, NY 10013.

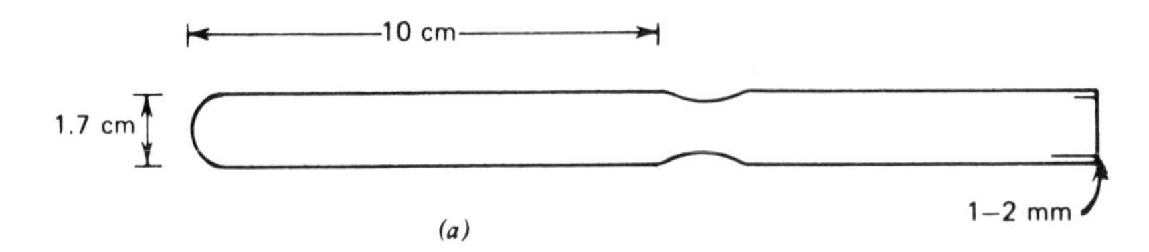

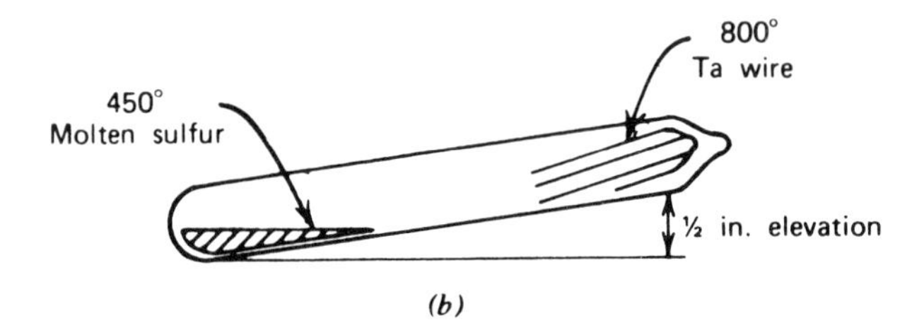

Figure 2. (a) Typical sample tube (Vycor or quartz). (b) Loaded sample tube.

with liquid sulfur is 1 atm at this temperature). The end of the ampule containing the tantalum may be heated to 800 or 900°.

■ **Caution.** *Because the reaction of tantalum and sulfur is exothermic and because of the high vapor pressure of sulfur, it is essential, in order to avoid explosions, that the liquid sulfur be prevented from coming into direct contact with the hot tantalum wire.*

The use of tantalum wire along with the "two-zone" technique ensures a safe reaction rate. When the sulfur has reacted completely with the tantalum wire (usually within 7–10 days) the ampule is removed from the furnace, shaken vigorously to break up clumps of unreacted material, and placed in the center of the furnace. It may be necessary to repeat this procedure several times to ensure that all the wire has reacted.* After 1 or 2 days at a 850–900°, the cooling process may be started. The sample is cooled initially to 750° and annealed for 1 day. It is then annealed at 650° for about a day and at 550° for another 2–3 days. The furnace is then shut off the sample is allowed to cool for 5–6 h to room temperature.

* An alternate method for quickly obtaining smaller (1-g) batches of TaS_2 powder is to place tantalum powder (– 60 mesh) in a 10–12 mm (inside diameter fused) quartz tube, 10 cm long, with a stoichiometric quantity of sulfur. This mixture will react completely in a relatively short while (2–3 days) at 450°. The sample may then be heated uniformly to 850–900° as described for the larger batch.

Properties

The polycrystalline $2H(a)$-TaS_2 should be free-flowing and in the form of black platelets. Any gold-colored material indicates the presence of the $1T$ phase and results from improper annealing. Further, the presence of a fibrous or needle-like material indicates incomplete reaction of the tantalum. This fibrous material is most likely TaS_3, which decomposes above 650°.[14]

$$Ta + 2TaS_3 \xrightarrow{650^\circ} 3TaS_2$$

This material may be removed by reheating the sample to 850°, followed by the same annealing procedure outlined above. The X-ray diffraction pattern for $2H(a)$-TaS_2 may be used for identification. The following d values have been obtained for major low-angle X-ray diffraction lines (and intensities): 6.05(1); 3.025(0.06); 2.8709(0.32); 2.7933(0.07); 2.3937(0.80); 2.3389(0.04) Å. Note that it is difficult to obtain the ideal intensities because of preferred orientation of the crystallites. This material is a superconducting metal with $2T_c = 0.8 \pm 0.05$ K[2].

B. SINGLE-CRYSTAL $2H(a)$PHASE OF TaS_2

$$TaS_2 + 2I_2 \underset{750^\circ}{\overset{850^\circ}{\rightleftharpoons}} TaI_4 + 2S$$

Procedure

Single crystals of $2H(a)$-TaS_2 may be obtained by chemicai transport either from prereacted TaS_2 powder or directly from the elements. If the crystals are to be prepared from the elements, care must be taken to heat the ampule slowly in the manner described previously to avoid explosions. The reactants are loaded into a 20-cm-long, 1.7-cm-diameter fused-quartz ampule. About 5 mg of iodine is added per cubic centimeter of volume of the reaction ampule. This serves as a transport agent according to the equation given above.[15] Both the sulfur and TaI_4 are volatile components. The equilibrium constant is such that at the higher temperature the reaction proceeds to the right, as described, while at the cooler end of the ampule the equilibrium favors TaS_2 and I_2. Thus the iodine is regenerated constantly as the TaS_2 crystals grow in the cooler zone.

Several techniques exist for loading iodine into the reaction vessel, and some are described in Reference 13. The simplest method is to load iodine quickly in air followed by pumping down to a few micrometers of pressure and sealing. The ampule is then heated to 900° (with appropriate precautions taken if the elements are undergoing reaction for the first time). The ampule must then be placed in a thermal gradient of 850–750° such that the charge is at the hot end and the 750° zone extends over a region of 50–75 mm. This may be accomplished by using two tubular furnaces joined together end to end or by a single furnace with a continuously wound filament having taps every 3.8 cm or so. These taps are then shunted with external resistances to achieve the desired temperature profile. One week is usually sufficient to achieve 100% transport of the charge. This transport period must then be following by the annealing procedure described above; however, the annealing times should be prolonged somewhat to ensure that the large crystals transform properly; $1\frac{1}{2}$ days at 650° and 3 days at 550° are usually sufficient. On cooling, the ampule may be cracked open and the crystals are removed and rinsed in CCl_4 and then CS_2 to remove I_2 and S respectively.

■ **Caution.** *The tube should be wrapped in several layers of cloth before it is opened.*

As mentioned in Reference 13, larger crystals can be obtained by using larger ampule diameter (2.5 cm) and/or heating the growth zone above 900° for a few hours (with the charge end at 800° or so) before beginning the transport. This results in "back-transporting" multiple TaS_2 nucleation sites that are in the growth zone.

Properties

The crystals obtained in this fashion have hexagonal symmetry (space group $P6_3/mmc$) with $a = 3.314$ Å and $c/2 = 6.04$ Å ($c/2$ is the basic S—Ta—S slab thickness). The d values given above for the polycrystalline material may be used to check the identity of a crushed crystal.

C. INTERCALATION COMPOUNDS OF TaS_2

Three main categories of intercalation compounds can be formed with layered transition-metal dichalcogenides: Lewis base complexes, alkali-metal complexes, and transition-metal or posttransition-metal "complexes." A representative synthesis for each category is included. For the Lewis bases the tendency to form intercalation compounds increases as the Lewis basicity of the molecule increases and as the molecular size of the base decreases.[2–4] Hence, NH_3 ($pK_a \approx 9$) readily intercalates a wide variety of layered transition metal dichalcogenides (including TiS_2 and ZrS_2), whereas pyridine ($pK_a \approx 5.3$) has been shown to intercalate only $2H(a)$-TaS_2 or NbS_2. Under certain circumstances, the smaller NH_3 molecule may be used to "pry" the layered dichalcogenide apart, making possible subsequent intercalation of a large molecule.[2–4] This "double intercalation" is carried out by preintercalating with NH_3.

1. Procedure 1a: Pyridine Intercalate of Tantalum Disulfide

$$2TaS_2 + C_5H_5N \text{ (excess)} \xrightarrow{200^\circ} 2TaS_2 \cdot C_5H_5N$$

Procedure

In the compound $2TaS_2 \cdot C_5H_5N$ the pyridine rings are normal to the crystallographic *c* axis.[2–4] Two such rings inserted between the layers give the stoichiometric composition $2TaS_2 \cdot C_5H_5N$ (see Fig. 3). Under certain conditions, the complex $4TaS_2 \cdot C_5H_5N$ also may be obtained.[4]

$2H(a)$-TaS_2 powder (1.91 g about 0.008 mol) is placed in a Pyrex tube about 20 cm long and 1 cm in diameter, with a 2-mm wall thickness. An excess of redistilled pyridine is then added. The volume of pyridine should be three or four times that of the TaS_2 powder. (■ **Caution.** *Direct contact with pyridine or pyridine vapor should be avoided.*) The tube is connected to a vacuum system and is quickly pumped down to 15-torr pressure (vapor pressure of liquid pyridine at room temperature). The pyridine is then frozen with liquid nitrogen, and the evacuation is continued to a pressure of 10^{-3} torr. To remove dissolved air, the sample is carried through two or more additional cycles of freezing, pumping on the frozen material, and thawing. The final cycle is followed by sealing the tube under vacuum with the contents frozen at liquid nitrogen temperature. The sample is then heated gradually to 200° with stirring (an oil bath with magnetic stirring is adequate). After 15 min to 1 h at this temperature (if the TaS_2 powder is off good

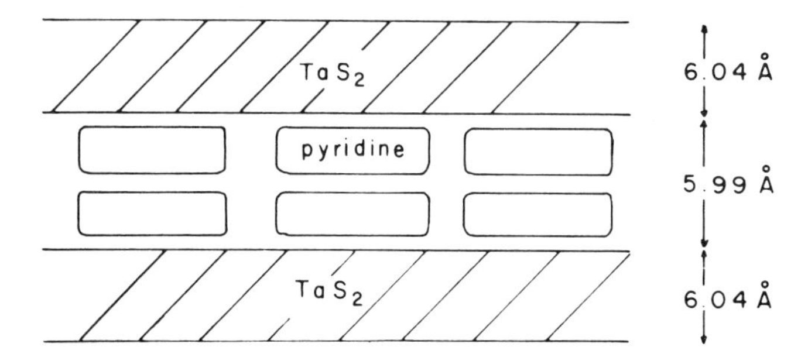

Figure 3. Schematic representation of tantalum disulfide–pyridine intercalate.

quality), the powder swells to nearly twice its original volume as the pyridine intercalate forms.

Intercalation of single crystals of $2H(a)$-TaS_2 may be carried out in a similar manner, although it is possible that, if the crystals are too large in surface areas intercalation may proceed only around the edges. It should be noted that the time for full intercalation of crystals increases with crystal dimensions.

Properties

The pyridine intercalate is blue-black. Examination of the individual platelet under a microscope reveals a characteristic exfoliated appearance. Hexagonal symmetry is retained with $a = 3.325$ Å and $c/2 = 12.03$ Å. Comparison with the slab thickness for the parent $2H(a)$-TaS_2 shows a c-axis expansion, δ, of 5.99 Å. The following d values have been obtained for long-angle X-ray diffraction lines 12.03, 6.015, 4.010, 3.008, 2.880 and 2.895 Å. $2TaS_2 \cdot C_5H_5N$ is also super conducting with a transition temperature of 3.5 ± 0.3 K[2,3]

To check the composition of the complex thus synthesized, a weight-gain analysis may be carried out. The Pyrex tube is scribed carefully and broken open after being wrapped in a cloth. The contents are filtered through a tared fitted glass filter funnel (medium- or fine-porosity filter). The Pyrex tube is rinsed with CH_2Cl_2 to remove any material left clinging to its walls. The excess pyridine liquid may take on a brownish color. This coloration is due to a small amount of sulfur being extracted from the TaS_2 by the pyridine and can be avoided by the addition of some sulfur to the pyridine before intercalation.[16] After suction filtration of its contents, the funnel is stoppered and the $2TaS_2 \cdot C_5H_5N$ powder is vacuum-dried (10–50μ) at room temperature and carefully weighed. The weight of intercalated pyridine is determined by difference between the original TaS_2 charge and the $2TaS_2 \cdot C_5H_5N$.

2. Procedure 1b: Ammonia Intercalate of Tantalum Disulfide as an Intermediate

$$TaS_2 + NH_3 \rightarrow TaS_2 \cdot NH_3$$
$$2(TaS_2 \cdot NH_3) + B \rightarrow 2TaS_2 \cdot B + 2NH_3$$

Dried gaseous NH_3 is condensed into a Pyrex combustion tube (2-mm wall thickness, 10-mm i.d.) to which the host material has already been added. The NH is frozen with liquid nitrogen and the tube is evacuated and sealed.

■ **Caution.** *The vapor pressure of liquid NH_3 at room temperature is 8 atm. The seal on the Pyrex tube should be carefully annealed so that the tube* *will not explode. The sealed tube always should be kept behind a protective shield or in a protective metal pipe.*

After a short while ($\frac{1}{2}$ to several hours) at room temperature the layered host material reacts fully to yield the stoichiometry $MX_2 \cdot NH_3$. The excess NH_3 may turn light blue because of the extraction of a slight amount of sulfur from the TaS_2, but here again, as in the case of pyridine, a small amount of excess sulfur may be added to the Pyrex tube (i.e., a few milligrams) before it is sealed. The Pyrex tube is then removed from its bomb, cooled to refreeze the NH_3, wrapped in a protective cloth, and cracked open. The intercalated powder is quickly transferred to a flask containing refluxing liquid of the molecules to be intercalated. (■ **Caution.** *The Pyrex tube should be opened in a hood to avoid inhalation of* NH_3.) Care should be taken to avoid prolonged exposure of the $TaS_2 \cdot NH_3$ powder to water vapor in the air during transfer.

Other intercalation techniques involve melting or dissolving solid organic materials in solvents such as benzene to obtain a mobile species of the intercalate. Table I gives other organic materials that have been intercalated in $2H(a)$-TaS_2 along with the reaction times and temperatures. The intercalation complexes retain hexagonal symmetry and the crystallographic a and c parameters are listed with the expansion of the c-axis because of the inclusion of the organic molecule (δ). Also included in the table are the onset temperatures of superconductivity.

3. Procedure 2: Sodium Intercalate of Tantalum Disulfide

$$TaS_2 + x\text{Na} \rightarrow \text{Na}_x\text{TaS}_2$$
$$(0.4 \leqslant x \leqslant 0.7)$$

Omloo and Jellinek[7] have described the synthesis and characterization of intercalation compounds of alkali metals with the group V layered transition-metal dichalcogenides. Typically, these types of intercalation complexes are sensitive to moisture and must be handled in dry argon or nitrogen atmospheres. The alkali metal atoms occupy either octahedral or trigonal prismatic holes between X—M—X slabs. There are two principal means by which these compounds may be prepared.

Procedure 2a

A 1.911-g quantity (0.008 mol) of $2H(a)$-TaS_2 powder (or an appropriate mixture of the elements; see Section A) is loaded into a quartz ampule such as the one described in Fig. 2*a*. The ampule is then transferred to a dry box with an argon or nitrogen atmosphere, and 74–129 mg (0.0032–0.0056 mol) of

TABLE 1. Other Molecules that Intercalate TaS_2

Intercalate	Time (days)	Temp (°C)	*a* (Å)	*c* (Å)	*δ* (Å)	T_{onset} (°K)
Amides						
Butyramide	21	150	—	11.0	5.0	3.1
Hexanamide	21	150	—	11.2	5.2	3.1
Stearamide (I)	10	150	—	57.0	51.0	3.1
Thiobenzamide	8	160	—	11.9	5.9	3.3
Phenylamines						
Aniline	16	150	—	18.15	12.11	3.1
N,N-Dimethylaniline	3	170	—	12.45	6.41	4.3
N,N,N′,N′-Tetramethyl-*p*-phenylenediamine	13	200	3.335	2 × 9.66	3.62	2.9
Cyclic amines						
4,4′-Bipyridyl	21	160	3.316	2 × 12.08	6.04	2.5
Quinoline	6	160	—	12.08	6.04	2.8
Pyridine *N*-oxide	8 hr	100	3.335	11.97	5.93	2.5
Pyridinium chloride	4	170	3.329	9.30	3.26	3.1
Hydroxides						
CsOH	30 min	25	3.330	2 × 9.28	3.24	3.8
LiOH	30 min	25	3.330	2 × 8.92	2.88	4.5
NaOH	30 min	25	3.326	2 × 11.86	2.82	4.8
Triton B[a]	30 min	25	3.331	2 × 11.98	5.94	5.0
Alkylamines						
Ammonia	3	25	3.328	2 × 9.22	3.17	4.2
Methylamine	30	25	3.329	2 × 9.37	3.32	4.2
Ethylamine	30	25	3.334	2 × 9.58	3.53	3.3
Propylamine	30	25	3.330	2 × 9.66	3.61	3.0
Butylamine	30	25	—	2 × 9.73	3.68	2.5
Decylamine	30	25	—	14.6	8.5	—
Dedecylamine	30	25	3.325	34.4	28.3	—
Tridecylamine	30	25	3.322	40.5	36.4	2.5
Tetradecylamine	30	25	3.325	46.2	40.1	2.4
Pentadecylamine	30	25	3.328	45.1	39.0	2.8
Hexadecylamine	30	25	—	39.7	33.6	—
Heptadecylamine	30	25	—	48.5	43.4	2.7
Octadecylamine	30	25	—	55.8	49.7	3.0
Tributylamine	7	200	—	2 × 10.28	4.23	3.0
Miscellaneous						
Ammonium acetate	1 hr	150	3.330	9.08	3.04	2.0
Hydrazine	10 min	10	3.334	9.16	3.12	4.7
Potassium formate	1 hr	200	3.334	9.05	3.01	4.7
Guanidine	3	25	—	—	—	—

[a] Tetrasodium (ethylenedinitrilo) tetraaetate.

Source: From F. J. DiSalvo, Ph.D. dissertation, Stanford University 1971.

freshly cut sodium metal is added. Very low moisture and oxygen concentrations must be maintained in the dry box to prevent attack of the sodium. The techniques for maintaining pure inert atmospheres are described elsewhere.[17] The specified amount of sodium produces a product within the range of $0.4 \leqslant x \leqslant 0.7$[7] over which the intercalation compound exists. The ampule is stoppered, removed from the dry box, attached to a vacuum pump, quickly evacuated to 10^{-3} torr, and sealed under vacuum. It is then heated to 800° for 1 day (again, caution must be exercised if the sample is prepared from the elements; see Section A) and cooled slowly to room temperature. The resulting powder is black or gray and, as mentioned earlier, is very sensitive towards moisture. The crystal structure exhibits hexagonal symmetry with $a = 3.337 \pm 1$ and $c/2 = 7.30 \pm 1$Å (corresponding to $\delta \approx 1.2$ Å) for $x = 0.7$. When the amount of sodium is decreased below $x = 0.7$, δ increases and the a axis decreases slightly as x decreases toward 0.4.[7] If the amount of sodium added to the ampule initially corresponds to less than 0.4 mol per mole of TaS_2, a phase separation into unreacted TaS_2 and $Na_{0.4}TaS_2$ occurs. On the other hand, if more than 0.7 mol of sodium is added per mole of TaS_2, free sodium remains to attack the quartz reaction vessel. For identification purposes, the prominent low-angle powder X-ray diffraction lines of $Na_{0.7}TaS_2$ are 7.295, 3.648, 2.890, 2.835, 2.687, and 2.484 Å. Two polymorphic forms of Na_xTaS_2 are reported: η phase and δ phase. The η phase is obtained by heating $1T$-TaS_2 and sodium at 500°, while the δ phase is obtained when $2H(a)$ TaS_2 is used as the starting material.[7]

Alternate Procedure 2b

In this method the metallic sodium is dissolved in a solvent such as liquid NH_3 or tetrahydrofuran and the resulting solution is used to intercalate $2H(a)$-TaS_2. This technique was used by Cousseau[18] and Trichet et al.[19] in preparation of group IV layered transition-metal dichalcogenide–alkali-metal intercalation compounds. The advantage of this method is that it is carried out at room temperature and, consequently, there is less likelihood of reaction between sodium and the reaction vessel. On the other hand, this method is more difficult in that it involves the use of liquid NH_3. Furthermore, undesirable side reactions may occur if the NH_3 is not dried thoroughly or if the reaction vessel is not clean. For example,

$$Na + 2NH_3 \rightarrow NaNH_2\downarrow + H_2(g)$$

is a competing reaction that can occur under "dirty" conditions and is evidenced by the formation of a white precipitate (sodium amide) in the Na–NH_3 solution. Cousseau[18] describes a technique that employs a Pyrex reaction vessel such as the one shown in Fig. 4. A 1.911-g quantity (0.008 mol) of 2*H*(*a*)-TaS_2 is placed in branch *A* of the vessel along with a small sealed glass ampule containing a known weight of distilled sodium (between 74 and 129 mg, as described earlier). Dried ammonia gas is condensed (using liquid nitrogen) into this branch, and the vessel is sealed off under vacuum. The sodium ampule is then broken by briskly shaking the vessel (■ **Caution.** *This should be done behind a protective barrier.*), and a characteristic blue solution results as the sodium is dissolved. The blue coloration of the liquid NH_3 solution disappears as the TaS_2 is intercalated by the sodium. The liquid NH_3 is then poured off into branch *B* and cooled with liquid nitrogen. The temperature gradient causes any residual NH_3 to condense in branch *B*. This branch is removed by sealing at point R_1. Residual NH_3 in the Na_xTaS_2 may be removed by heating branch *A* gently (about 200°), while simultaneously cooling branch *C* to liquid nitrogen temperature. This branch is removed by sealing at point R_2 after the NH_3 has been frozen in *C*. The final product should be removed from the Pyrex tube only in an argon- or nitrogen-filled dry box as described in the preceding procedure.

Other layered materials intercalated with alkali metals have been prepared and are given in Table II. Only the maximum alkali metal concentration is listed for the various A_xMX_2.

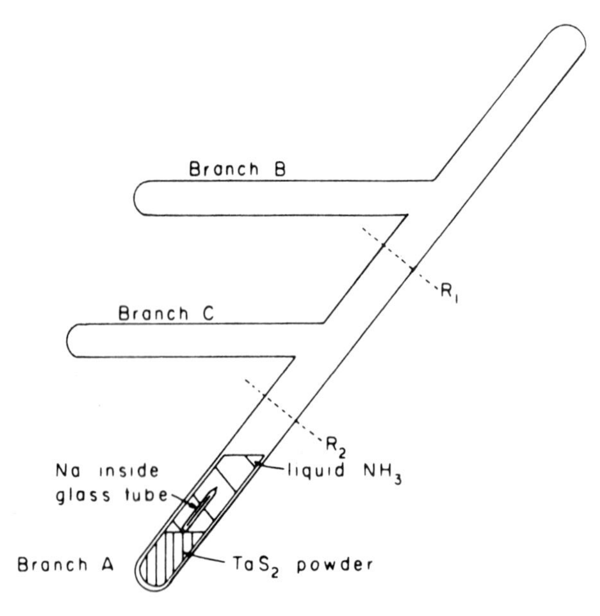

Figure 4. Adapted from J. Cousseau, Ph.D. dissertation, Université de Nantes, France, 1973.

TABLE II. Summary of Intercalation Data for A_xMX_2 Complexes (*Maximum* Alkali-Metal Concentration)

Group IVb[a]			Group Vb[b]			Group VIb[a]		
Compount	c Axis (Å)	δ (Å)	Compound	c Axis (Å)	δ (Å)	Compound	c Axis (Å)	δ (Å)
TiS_2	5.67	—	TaS_2	2×6.05	—	MoS_2	2×6.15	—
$Li_{0.6}TiS_2$	6.16	0.49	$Li_{0.7}TaS_2$	2×6.45	0.40	$Li_{0.8}(NH_3)_{0.8}MoS_2$	2×9.5	3.35
$Na_{0.8}TiS_2$	3×6.68	1.01	$Na_{0.7}TaS_2$	2×7.27	1.22	$Na_{0.6}MoS_2$	2×7.5	1.35
$K_{0.8}TiS_2$	3×7.56	1.89	$K_{0.7}TaS_2$	2×8.10	2.05	$K_{0.6}MoS_2 \cdot$	2×8.1	1.95
$Cs_{0.6}TiS_2$	3×8.36	2.69	—	—	—	$Cs_{0.5}MoS_2$	2×8.89	2.74
$TiSe_2$	6.01	—	$TaSe_2$	2×6.36	—	$MoSe_2$	2×6.46	—
$Na_{0.95}TiSe_2$	3×6.99	0.98	$Na_{0.7}TaSe_2$	2×7.70	1.34	—	—	—
—	—	—	$K_{0.7}{}^{TaSe}{}_{2}$	2×8.52	2.16	$K_{0.5}MoSe_2$	2×8.57	2.21

[a] Alkali-metal–NH_3 solution intercalation. Data from W. Rudorff, *Chimia.*, **19**, 496 (1965).

[b] Direct heating intercalation. Data from W. P. F. A. M. Omloo and F. Jellinek, *J. Less-Common Met.*, **20**, 121 (1970).

Properties

Table II gives the c-axes and slab expansions δ. These materials react with air and moisture. No superconductivity has been found in the group V alkali-metal complexes.

4. Procedure 3: Tin Intercalate of Tantalum Disulfide

$$TaS_2 + Sn \xrightarrow{880^\circ} SnTaS_2$$

DiSalvo et al.[9] have carried out a systematic survey of intercalation compounds of $2H(a)$-TaS_2 with posttransition metals. In particular, the system Sn_xTaS_2 was found to exist in two composition domains, $0 < x \leqslant \frac{1}{3}$ and $x = 1$. The following discussion briefly describes the techniques used by DiSalvo to synthesize the compound $SnTaS_2$. Syntheses of other transition and posttransition-metal intercalation complexes with the layered transition-metal dichalcogenides are discussed in References 9 and 20–24.

Procedure

$2H(a)$-TaS_2 is synthesized according to the procedure outlined in Section A. A 1.91-g sample of TaS_2 powder (0.008 mol) is loaded into a fused quartz ampule such as the one shown in Fig. 2*a*, along with 1.19 g of tin powder (about 0.010 mol; ~50-mesh powder; 99.5% purity).* The tube is then evacuated to 10^{-3} torr, sealed, and heated in a small temperature gradient ($\Delta T \sim 20^\circ$) with TaS_2 at the hot end (850°). The excess tin not intercalated sublimes to the cooler end of the tube over a period of several weeks. An alternate technique involves pressing a pellet of a mixture of TaS_2 and Sn powders (1.91 g TaS_2 and 0.950 g Sn) at 40,000 psi. This pellet is then sealed in a fused quartz ampule under vacuum, fired to 600° for a week or so, cooled, and then removed from the ampule. Regrinding, pressing, and refiring several times will ensure a homogeneous sample.

Properties

The $SnTaS_2$ has hexagonal symmetry with lattice parameters $a = 3.28$ Å and $c/2 = 8.7$ Å.[9] For identification purposes, the prominent low-angle X-ray powder d spacings and (relative intensities) are: 8.6(mw), 4.33(ms), 2.85(m), 2.81(s), 2.71(s), and 2.56(m). $SnTaS_2$ undergoes a superconducting transition at $T_c = 2.95$ K.[9]

* Available from Alfa Products, Ventron Corp., P.O. Box 299, Danvers, MA, 01923.

References

1. A. Weiss and R. Ruthardt, *Z. Naturforsch.*, **24**, 256, 355 and 1066 (1969).
2. F. R. Gamble, F. J. DiSalvo, R. A. Klemm, and T. H. Geballe, *Science*, **168**, 568 (1970).
3. F. R. Gamble, J. H. Osiecki, and F. J. DiSalvo, *J. Chem. Phys.*, **55**, 3525 (1971).
4. F. R. Gamble, J. H. Osiecki, M. Cais, R. Pisharody, and F. J. DiSalvo, *Science*, **174**, 493 (1971).
5. F. R. Gamble and T. H. Geballe, in *Treatise on Solid State Chemistry*, Vol. 3, N. B. Hannay (ed.), Plenum Press, 1976. This is a comprehensive survey of intercalation in a wide variety of host materials.
6. J. V. Acrivos, W. Y. Liang, J. A. Wilson, and A. D. Yoffe, *J. Phys. Chem.*,**4**, 6–18 (1971).
7. W. P. F. A. M. Omloo and F. Jellinek, *J. Less-Common Met.*, **20**, 121 (1970).
8. W. Rudolf and H. H. Sich, *Angew. Chem.*, **71**, 724 (1959).
9. F. J. DiSalvo, G. W. Hull, Jr., L. H. Schwartz, J. M. Voorhoeve, and J. V. Waszczak, *J. Chem. Phys.*, **59** (4), 1922 (1973).
10. F. Jellinek, *J. Less-Common Met.*, **4**, 9 (1962).
11. J. A. Wilson and A. D. Yoffe, *Adv. Phys.*, **18** (73), 193–335 (1969). This article offers a complete discussion of the polymorphic phases found in the various layered transition metal dichalcogenides.
12. R. C. Hall and J. P. Mickel, *Inorgan. Synth.*, **5**, 82 (1957).
13. L. E. Conroy and R. J. Bouchard, *Inorg. Synth.*, **12**, 158 (1970).
14. G. Brauer, *Handbook of Preparative Inorganic Chemistry*, Vol. 1, Academic Press, New York, 1963, p. 1328.
15. H. Shaefer, *Chemical Transport Reactions*, Academic Press, New York, 1964.
16. A. H. Thompson, *Nature*, **251**, 492 (1974).
17. D. F. Shriver, *The Manipulation of Air-Sensitive Compounds*, McGraw-Hill, New York, 1969, pp. 164, 193.
18. J. Cousseau, Thèse de Doctorat de Spécialité, Université de Nantes France, 1973.
19. L. Trichet, J. Cousteau, and J. Rouxel, *Compt. Rend. Acad. Sci.*, **273**, 243 (1971).
20. F. Hulliger, *Struct. Bonding*, **4**, 421 (1968).
21. J. M. Voorhoeve, nee van den Berg, and M. Robbins, *J. Solid State Chem.*, **1**, 132 (1970).
22. K. Trichet and J. Rouxel, *Compt. Rend. Acad. Sci.*, **267**, 1322 (1969); **269**, 1040 (1969).
23. M. S. Whittingham, *Chem. Commun.*, **1974,** 328 (1974).
24. G. V. Subba Rao and Ji. C. Tsang. *Mater. Res. Bull.*, **9**, 921 (1974).

33. AMMONIA SOLVATED AND NEAT LITHIUM–TITANIUM DISULFIDE INTERCALATES

$$0.22\text{Li} + y\text{NH}_3 + \text{TiS}_2 \xrightarrow{\text{NH}_{3(l)}} \text{Li}_{0.22}^{+}(\text{NH}_3)_y\text{TiS}_2^{0.22-}$$

$$\text{Li}_{0.22}^{+}(\text{NH}_3)_y\text{TiS}_2^{0.22-} \xrightarrow{250^\circ} \text{Li}_{0.22}^{+}\text{TiS}_2^{0.22-} + y\text{NH}_3$$

Submitted by M. J. McKELVY* and W. S. GLAUNSINGER†
Checked by V. B. CAJIPE‡

The lithium–ammonia and lithium intercalates of titanium disulfide are of considerable practical and fundamental interest. Lithium–ammonia intercalates are of interest as intermediates in the synthesis of the lithium intercalates of TiS_2, which are the cathode materials formed during the discharge–charge cycles of Li-TiS_2 high-energy-density batteries.[1,2] They are also of substantial fundamental interest in the investigation of intercalant solvation phenomena in transition-metal disulfide hosts.[3,4] Intercalates with < 0.22 mol Li/mol TiS_2, prepared by pouring a lithium–ammonia solution onto TiS_2, also contain cointercalated NH_4^+ due to ammonia–host redox reactions.[5] These compounds are best described by the formula $\text{Li}_x^+(\text{NH}_4^+)_{y'}(\text{NH}_3)_{y''}\text{TiS}_2^{(x+y')-}$, where the total cation concentration is charge compensated such that $x + y' = 0.22 \pm 0.02$. Herein, we describe the synthesis and characterization of the simpler intercalate $\text{Li}_{0.22}^+(\text{NH}_3)_{0.64}\text{TiS}_2^{0.22-}$ and its ammonia deintercalation to form the intermediate high-energy-density battery cathode material $\text{Li}_{0.22}^+\text{TiS}_2^{0.22-}$.

Although the lithium intercalates of TiS_2, Li_xTiS_2, can also be prepared by reacting TiS_2 in *n*-butyllithium solution, electrochemically with lithium, and in lithium halide melts in the presence of H_2S,[6] the liquid-ammonia technique has several advantages. These include relatively short equilibrium times to form nonstoichiometric intercalates, the absence of a difficult to separate matrix from which the product has to be isolated for further investigation, and the ability to perform the synthesis using standard glovebox and high-vacuum techniques without the need for an inert-atmosphere electrolytic cell.

Synthesis of the lithium–ammonia intercalate requires about 6 days to complete and involves about 12 h of lab time. Preparation of the lithium

* Center for Solid State Science, Arizona State University, Tempe, AZ 85287-1704.
† Department of Chemistry and Biochemistry, Arizona State University, Tempe, AZ 85287-1604.
‡ Institut des Matériaux de Nantes, CNRS-UMR 110, Nantes, France.

intercalate from its lithium–ammonia precursor can be accomplished in about 5–6 h.

Procedure

All the glassware associated with the following syntheses is cleaned with a solution consisting of 45 mL of 48% HF, 165 mL of concentrated HNO_3, 200 mL of H_2O, and 10 g of Alconox detergent.

■ **Caution.** *Aqueous hydrogen fluoride solutions are highly corrosive and cause painful, long lasting burns. Rubber gloves, eye protection, and a lab coat should be worn, and the solution should be handled in a well ventilated hood.*

After a 3 min exposure to the cleaning solution, the glassware is rinsed at least 25 times with distilled water and subsequently rinsed 5 additional times with megaohm-cm water that has been filtered to remove organic contaminants. The glassware is then dried at 110°.

■ **Caution.** *The reagents used in this synthesis are air-sensitive. The synthesis and associated sample handling should be carried out under rigorous inert conditions. This requires the use of an inert-atmosphere glovebox (He atmosphere, < 1 ppm total H_2O and O_2) (Vacuum Atmospheres Corp.) and a liquid–nitrogen trapped vacuum line ($< 10^{-4}$ torr).*

The preparation of the TiS_2 used as the host for these syntheses was previously described in this volume.[7] In addition to using high purity starting materials, minimizing the amount of excess Ti in the van der Waals gap of the host is desirable, since it can retard or prevent intercalation and affect both the intercalate formed and its properties.[8]

The ammonia used (99.9%) (Matheson Gas Products) is dried by condensation onto and subsequent storing over sodium (99.9%) [ROC/RIC] prior to use. About a quarter gram of sodium is loaded into a Pyrex container that is sealed with a high-vacuum stopcock in the glovebox. The container is evacuated to $\leq 10^{-4}$ torr. Ammonia (typically 25–50 mL) is cryopumped into it directly from an ammonia cylinder using a liquid nitrogen bath. A typical setup for this procedure is shown in Fig. 1. Care should be taken to avoid filling the bulb to more than half full and to keep the ammonia pressure well below atmospheric pressure during the transfer. The mixture is then evacuated with the high-vacuum stopcock and with the stopcock closed, allowed to warm to form a sodium–ammonia solution in a dry ice/alcohol bath ($-78°$) and stored in this fashion. Just prior to use the solution should

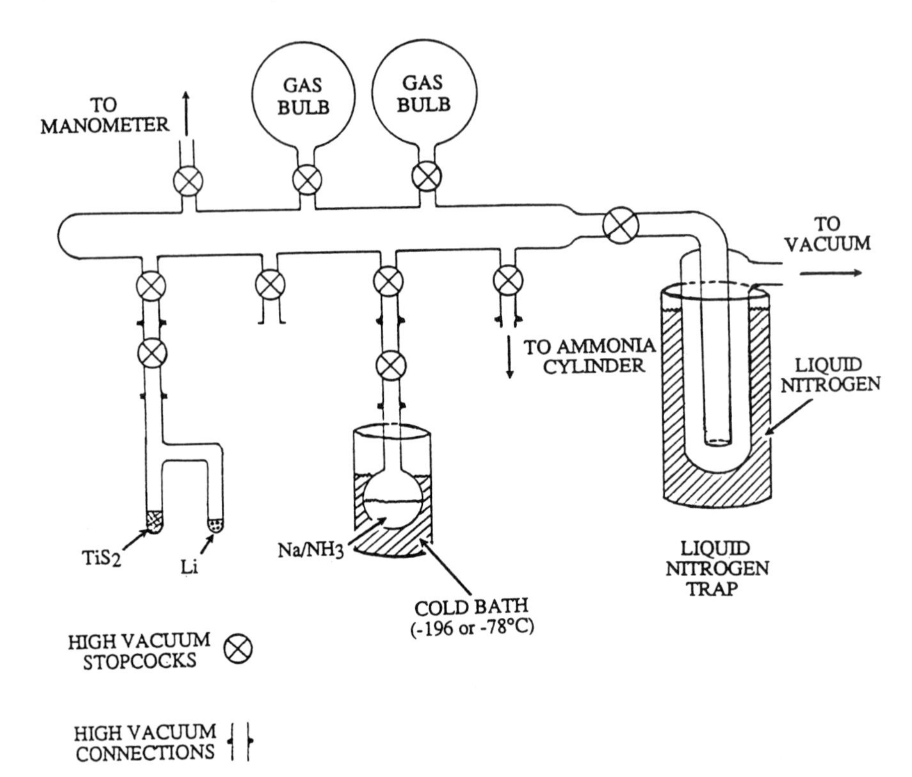

Figure 1. A typical high-vacuum setup for $Li^{+}_{0.22}(NH_3)_yTiS_2^{0.22-}$ synthesis. The manometer pressure gauge (either a capacitance manometer or a Hg manometer, preferably with a thin layer of ultra-low-vapor-pressure oil on top of the Hg column to suppress Hg volatility) is used to monitor the gas pressure and ensure the NH_3 pressure does not exceed ambient pressure at any time.

be frozen using a liquid–nitrogen bath followed by pumping out any noncondensibles present to $\leq 10^{-4}$ torr.

■ **Caution.** *Liquid ammonia has a vapor pressure of about 10 atm at 20°. Gaseous ammonia is both toxic and corrosive. The sodium–ammonia solution should be stored behind a clear explosion shield in a well-ventilated hood and kept at −78°. If it is allowed to warm, an explosion may result. A face shield, lab coat, and thick gloves should always be worn when handling the solution.*

The h-tube shown in Fig. 2 is evacuated to $\leq 10^{-4}$ torr and flamed to 200–500° to remove adsorbed water. TiS_2 (1 g, 8.9 mmol) with a particle size ≤ 0.5 mm, is placed in one leg of the h-tube, with lithium (13.6 mg,

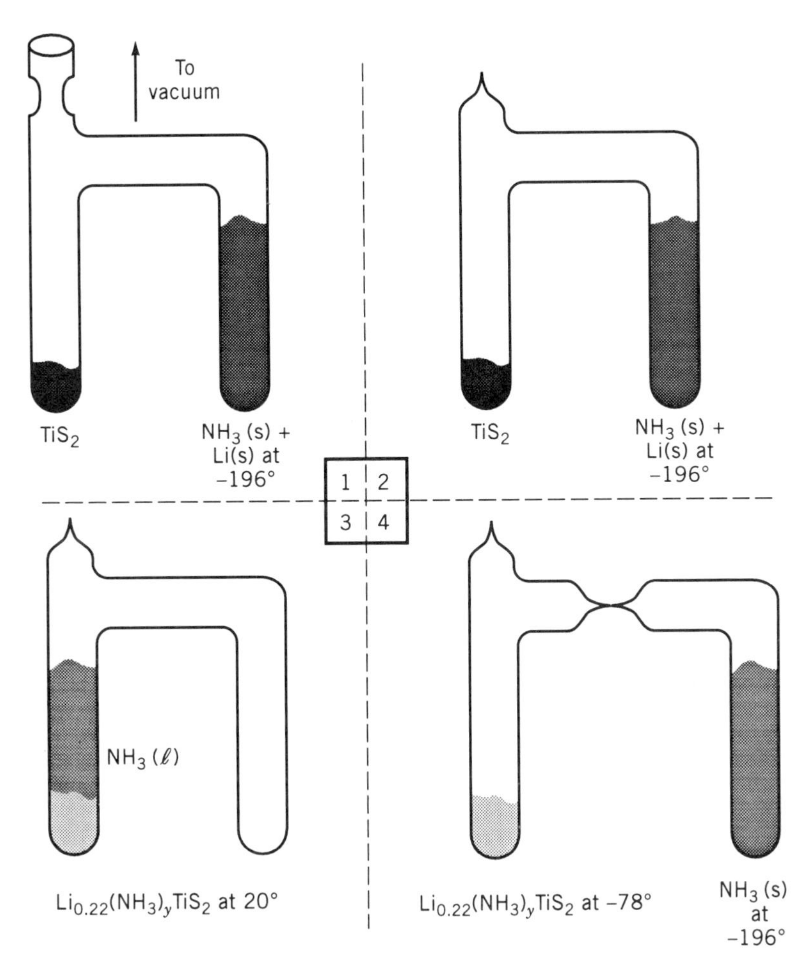

Figure 2. The synthesis of lithium–ammonia TiS_2: (1) condensation of NH_3(g) onto the lithium in the h-tube; (2) sealing of the h-tube to form an n-tube; (3) reaction of the Li–NH_3 solution with TiS_2 to form the intercalate; (4) separation of the intercalate from the excess ammonia. The h-tube typically is made from 12-mm-o.d. Pyrex tubing with $\geq$1-mm wall thickness. The legs of the h-tube are about 8 cm long, with about 7 cm separating them. Slight constriction of the glass seal-off regions of the reaction vessel prior to synthesis facilitates seal-off. Important: any increase in the external dimensions or decrease in the wall thickness of the h-tube will weaken it, possibly resulting in explosion.

1.96 mmol) in the other leg. (If difficulty is encountered loading the h-tube, an H-tube should be used with TiS_2 and Li being loaded through the separate openings.) The h-tube is then evacuated to $\leq 10^{-4}$ torr and enough NH_3(g) is slowly cryopumped ($-196°$) from the Na–NH_3 solution into the Li leg of the tube to ensure the presence of a layer of $NH_{3(l)}$ over the intercalate throughout the intercalation process. The NH_3 gas can be directly cryopumped onto the Li or a precalculated volume and pressure of NH_3 gas can be collected in the gas bulbs shown in Fig. 1 and subsequently condensed onto the Li. The h-tube is flame-sealed to form an n-tube, with the seal being carefully annealed to remove residual stress, as shown in Fig. 2. The Li–NH_3 is then slowly warmed to ambient temperature and the resulting solution is poured onto the TiS_2 with vigorous agitation.

■ **Caution.** *Sealed reaction vessels containing $NH_{3(l)}$ at ambient temperature are potentially explosive. Care should be taken to avoid reaction vessels having scratched glass, as this may weaken them allowing an explosion to occur. The reaction vessel should be allowed to warm to ambient temperature in a well-ventilated hood behind a clear explosion shield. A face shield, lab coat, and thick gloves should be worn when handling the vessel or any of its subsequently sealed-off parts. Both the vessel and its ammonia containing parts should be stored in an explosion proof container in the hood when not in use.*

In less than a minute, the characteristic blue color of the lithium–ammonia solution disappears, indicating complete lithium intercalation. The reaction vessel is then stored at ambient temperature for 2 days to ensure homogeneous Li intercalation.

When the reaction is complete, the $NH_{3(l)}$ is decanted into the opposite leg of the n-tube from the intercalate. The ammonia may be colored as a result of polysulfide formation.[9] Any residual polysulfides in the $NH_{3(l)}$ remaining around the intercalate may be removed by repeatedly distilling the ammonia back onto the intercalate, using an ice-water bath, and decanting it until the ammonia over the intercalate is essentially colorless. The ice-water bath is then transferred from the intercalate leg to the $NH_{3(l)}$ leg to remove the remaining $NH_{3(l)}$ from around the intercalate. Once the intercalate is dry, care must be taken to avoid significant NH_3 deintercalation into the NH_3 leg by cryopumping during sample isolation. This can be accomplished by the following procedure. The intercalate and $NH_{3(l)}$ legs are simultaneously placed in ice-water and dry-ice alcohol baths, respectively, for half an hour. The $NH_{3(l)}$ is then frozen using a liquid nitrogen trap, while the intercalate is cooled in the dry-ice alcohol bath. After allowing an hour to ensure ammonia condensation, the intercalate leg is sealed off from the ammonia leg of the n-tube, as shown in Fig. 2. Again, the seal-off region should be carefully flame annealed to remove residual stress. Then the intercalate-containing tube is transferred to the glovebox, scratched with a glass knife, wrapped with parafilm and carefully opened. The sudden change in ammonia pressure over the sample on opening the vessel results in substantial structural disorder.[3] The equilibrium, well-ordered intercalate can be obtained by placing the intercalate in a previously evacuated and flamed Pyrex storage vessel, cooling the intercalate to $-196°$, pumping out the He introduced in the glovebox, flame-sealing the vessel, and allowing the intercalate to anneal at ambient temperature for 2 days.[3]

The primary limitation to scaling up the synthesis is being able to safely contain the liquid-ammonia solution in the glass reaction vessel. The quantity of $Li^{+}_{0.22}(NH_3)_yTiS_2^{0.22-}$ to be synthesized can be increased as long as the reaction vessel can reliably withstand the vapor pressure of liquid ammonia at ambient temperature. However, the time needed to attain equilibrium intercalates under $NH_{3(l)}$ may be longer.

Thermogravimetric analysis (TGA) and vapor-pressure measurements (VPM) coupled with mass-spectrometric evolved-gas analysis (EGA) and X-ray powder diffraction (XPD) are the primary techniques used for product characterization. Intercalated ammonia can be thermally deintercalated quantitatively at 250° to yield the lithium intercalate.[5] The ammoniate is transferred to a previously evacuated ($\leq 10^{-4}$ torr) and flamed glass reaction vessel in the glovebox. The intercalate is chilled to $-196°$ and the residual glovebox He evacuated to $\leq 10^{-4}$ torr. The intercalate is then warmed to 250° under static vacuum conditions while monitoring the evolved-gas pressure, slowly evacuated to $\leq 10^{-4}$ torr, and subsequently annealed at 300° under dynamic vacuum for 2–3 h. For optimum structural order, the product should be annealed at 300° for several days in a sealed, evacuated ampule. Mass-spectrometric analysis of the gas evolved during the thermal deintercalation process showed almost exclusively ammonia was evolved, with the only other detectable gas being a trace ($<1\%$) of hydrogen sulfide. VPM of this process and/or TGA are used to quantify the ammonia composition. TGA demonstrates that the thermal deintercalation of ammonia is a simple one-step process.[3,5,10]

■ **Caution.** *The intercalate must be handled under rigorous inert conditions during analysis, as during synthesis. Only leakproof, airtight containers should be used.*

Properties

X-ray powder diffraction of $Li^{+}_{0.22}(NH_3)_{0.64}TiS_2^{0.22-}$ showed it to be a single-phase, $3R$ intercalate, with $a = 3.423(3)$ Å and $c = 26.75(1)$ Å.[10] Rietveld refinement of neutron powder diffraction data for a similar deuterated intercalate, $Li^{+}_{0.23}(ND_3)_{0.63}TiS_2^{0.23-}$, combined with ammonia compositional analysis, strongly suggests the presence of discrete two-dimensional ammonia-solvation complexes of Li^+ for these intercalates, where the coordination number for lithium is 3.[3] Magnetic susceptibility and charge compensation measurements demonstrate that guest-to-host charge transfer is accomplished through the donation of one e^-/Li to the host conduction band for both product intercalates, with no significant charge transfer being associated with intercalant ammonia.[5]

X-ray powder diffraction of the product $Li^{+}_{0.22}TiS_2^{0.22-}$ was used to verify homogeneous and complete lithium intercalation. Li_xTiS_2 ($0.00 < x < 1.00$) has a single-layer trigonal structure where the c lattice parameter is a strong function of x in the region of interest.[11,12] The observed value of c for the product was 6.011(3) Å, in good agreement with the values of 5.995 Å[11] and 6.015 Å[12] reported previously.

Acknowledgment

This work was supported by National Science Foundation grants DMR-8605937 and DMR-8801169.

References

1. M. S. Whittingham, *Science*, **192**, 112 (1976).
2. M. S. Whittingham, *J. Solid State Chem.*, **29**, 303 (1979).
3. V. G. Young Jr., M. J. McKelvy, W. S. Glaunsinger, and R. B. Von Dreele, *Chem. Mater.*, **2**, 75 (1990).
4. E. W. Ong, M. J. McKelvy, L. A. Dotson, and W. S. Glaunsinger, *Chem. Mater.*, **4**, 14 (1992).
5. M. McKelvy, L. Bernard, W. Glaunsinger, and P. Colombet, *J. Solid State Chem.*, **65**, 79 (1986).
6. J. Rouxel, in *Intercalated Layered Materials*, F. Levy (ed.), Reidel, Dordrecht, 1979, p. 201.
7. M. J. McKelvy and W. S. Glaunsinger, *Inorg. Synth.*, **30**, 28 (1994).
8. M. J. McKelvy and W. S. Glaunsinger, *Solid State Ionics*, **34**, 211 (1989).
9. A. H. Thompson, *Nature*, **251**, 492 (1974).
10. G. W. O'Bannon, M. J. McKelvy, W. S. Glaunsinger, and R. F. Marzke, *Solid State Ionics*, **32/33**, 167 (1989).
11. J. R. Dahn and R. R. Haering, *Solid State Commun.* **40**, 245 (1981).
12. B. G. Silbernagel and M. S. Whittingham, *J. Chem. Phys.*, **64**, 3670 (1976).

34. INTERCALATION COMPOUNDS WITH ORGANIC π-DONORS: $FeOCl(TTF)_{1/8.5}$, $FeOCl(TSF)_{1/8.5}$, AND $FeOCl(Perylene)_{1/9}$

Submitted by J. F. BRINGLEY† and B. A. AVERILL*
Checked by DAVID A. CLEARY†

The layered transition-metal oxyhalides (MOX: M = Fe, Ti, V; X = Cl, Br) are able to undergo intercalation reactions that involve the reversible insertion of a guest species (i.e., atoms or molecules) between the two-dimensional sheets of layered inorganic hosts (e.g., FeOCl), thus forming intercalation compounds of the type $MOX(guest)_{1/n}$. On intercalation, the host layers move apart to accommodate the size and steric packing constraints of the guest molecules, but the structure of the individual layers of the host itself remain unchanged.[1] Guest molecules can be introduced between the host layers by a variety of methods: by neat reaction of the guest with the host, treating the host with a solution containing the guest molecules dissolved in a suitable solvent, electrochemical intercalation, or ion-exchange reaction.[2] The rate of intercalation depends on a number of factors, including the available host surface area, the solubility of the guest species, the basicity of solvent, and the purity of the host.

Intercalation compounds may have potential use as catalytic materials, cathodes for lightweight batteries, ammonia getters, and ion-exchange materials. In addition, the present approach is intended to demonstrate the use of layered intercalation hosts as macroanionic acceptors for the synthesis of organic conductors.[3, 4] This strategy involves the oxidative insertion of the organic donor molecules used to form organic conductors between the layers of inorganic hosts, forming stacks of potentially conducting, organic donors separated by inorganic layers. A large number of organic donors can be intercalated into FeOCl ([e.g., tetrathiafulvalene (TTF),[3] tetraselenafulvalene (TSeF),[5] and bis(ethylenedithio)tetrathiafulvalene)[6]]. The syntheses of the TTF[3] and perylene[4] intercalates are represented here.

A. SYNTHESIS OF $FeOCl(TTF)_{1/8.5}$

$$FeOCl + \tfrac{1}{8.5}TTF \rightarrow FeOCl(TTF)_{1/8.5}$$

* Department of Chemistry, University of Virginia, McCormick Road, Charlottesville, VA 22901.
† Department of Chemistry, Washington State University, Pullman, WA 99164-4630.

Procedure

A thick-walled Teflon-stoppered reaction flask (~10 cm × 3 cm) (Fig. 1) containing a teflon-coated stir bar is charged with 0.100 g (0.931 mmol) of finely divided microcrystalline FeOCl[7] and 0.300 g (1.47 mmol) of TTF (Aldrich Chemical Co.). Finely divided FeOCl is obtained by pulverizing FeOCl in a ball mill or by sonication in dry toluene for 20 min at a power of 50 W. The flask is then fitted with a rubber septum, evacuated, and purged with dry nitrogen several times using a Schlenk line. Care should be taken in all steps of the synthesis since both the starting materials and products are slightly H_2O sensitive. Dimethoxyethane (DME), 10.0 mL, degassed and freshly distilled from CaH_2, is then added by syringe. The Teflon stopper is then replaced, and while stirring, the flask is carefully evacuated to the vapor pressure of the solvent and sealed with the Teflon stopper. The flask is then wrapped in aluminum foil (TTF is light-sensitive) and placed in an oilbath whose temperature is carefully controlled at 70° using a variable resistor. The oilbath is placed on a magnetic stirrer so that both the reaction mixture and

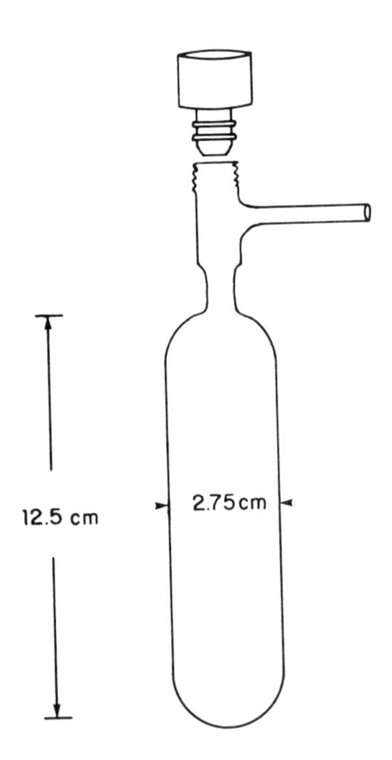

Figure 1. Schematic drawing of the reaction flask used for intercalation reaction.

the oil can be stirred simultaneously. (■ **Caution**. *The thermal treatment should be carried out in a hood since a sealed system is being heated.*) After 7 days at this temperature, the contents are allowed to cool to room temperature and then filtered over a fine glass frit. The black solid is then washed with DME until the eluent is colorless and dried *in vacuo*, 25° and 10^{-2} torr.

Properties

The product is characterized by X-ray powder diffraction; diffraction lines d(Å) = 13.099(s), 6.511(m), 3.474(s), 3.065(m), 2.8212(m), 2.4842(s), 2.4604(w), 2.3419(m), 2.1880(w), 1.8937(m), 1.8748(mw), 1.8048(mw), 1.6706(mw), 1.6568(w), 1.5056(m), (s = strong, m = medium, w = weak); a = 3.3412 Å, b = 3.7874 Å, c = 26.02 Å. FT-IR: strong absorptions at 490, 684, 745, 820, and 1332, $\nu(\text{cm}^{-1})$. (In addition to the broad band due to the FeOCl host at 470–490 cm^{-1}.) The absence of the strong characteristic interlayer reflection of FeOCl, d = 7.917 Å is used to confirm complete intercalation. In cases where intercalation is determined to be incomplete, the product may be treated with a fresh solution of the intercalant, TTF. Elemental analyses are consistent with the stoichiometry $FeOCl(TTF)_{1/8.5}$.

TSF, the selenium analog of TTF, can be intercalated into FeOCl by the same procedure using the same molar amounts, resulting in the isomorphous compound $FeOCl(TSF)_{1/8.5}$.[8,9] X-ray powder diffraction: d(Å) = 13.47(s), 6.761(w), 3.485(vs), 3.101(vw), 2.834(w), 2.516(w), 2.455(m), 2.344(w), 1.891(m), 1.871(w), 1.832(m), 1.662(m), 1.650(m), 1.574(w), 1.512(m), 1.249(m), 1.243(m); FT-IR (ν, cm^{-1}): 640(m), 658(m), 1075(m), 1301(s), 1.457(m), 1355(sh).

B. SYNTHESIS OF $FeOCl(Perylene)_{1/9}$

$$FeOCl + \tfrac{1}{9}\text{perylene} \rightarrow FeOCl\ (\text{perylene})_{1/9}$$

Procedure

The materials and methods used are identical to those described above for $FeOCl(TTF)_{1/8.5}$. A flask is charged with 0.200 g (1.80 mmol) FeOCl and 0.400 (1.6 mmol) perylene (Aldrich Chemical Co.). Freshly distilled dimethoxyethane 20.0 mL is then added by syringe. Because of the poor solubility of perylene in the solvent, much of the intercalant is suspended in the solvent at this point. The flask is then evacuated and purged with dry nitrogen several times and sealed at the vapor pressure of the solvent. The flask is then wrapped in Al foil, and heated in an oilbath at 85° for 30 days. After this time, the solid is collected on a glass frit, washed with 300 mL CH_2Cl_2 in 30 mL portions, and 100 mL DME, and dried *in vacuo*.

Properties

The product is characterized by X-ray powder diffraction: diffraction lines d(Å) = 16.68(s), 8.332(m), 2.718(w), 1.889(s), 1.835(w), 1.658(w), 1.244(w): $a = 3.321(1)$ Å, $b = 3.782(2)$ Å, $c = 33.38(9)$ Å. FT-IR 490(vs), 696(m), 819(s), 1186(s), 1348(m), 1541(s, broad). Elemental analysis consistent with stoichiometry $FeOCl(perylene)_{1/9}$. In many cases DME is found to cointercalate with perylene into FeOCl, resulting in the compound $FeOCl(perylene)_{1/9}$ $DME_{1/20}$; chemical ionization mass spectrum $m/z = 253$ (perylene) and $m/z = 91$ (DME). However, DME deintercalates at lower temperatures (~160–200°) than does perylene (~300–400°) and can be selectively removed from the interlayer region of the host by heating *in vacuo* (190°, 10^{-2} torr, 2 days) with no significant changes in structural or physical properties.

References and Note

1. T. R. Halbert, in *Intercalation Chemistry*, M. S. Whittingham and A. J. Jacobson (eds.), Academic Press, Orlando, 1982, Chapter 12.
2. M. S. Whittingham and A. J. Jacobson (eds,), *Intercalation Chemistry*, Academic Press, Orlando, 1982.
3. S. M. Kauzlarich, J. F. Ellena, P. D. Stupic, W. M. Reiff, and B. A. Averill, *J. Chem. Soc.*, **109**, 4561 (1987).
4. J. F. Bringley and B. A. Averill, *Chem. Mater.*, **2**, 180 (1990).
5. J. F. Bringley, B. A. Averill, and J.-M. Fabre, *Mol. Cryst. Liq. Cryst.*, **170**, 215 (1988).
6. J. F. Bringley, J.-M. Fabre, and B. A. Averill, *J. Am. Chem. Soc.*, **112**, 4577 (1990).
7. S. Kikkawa, F. Kanamuru and M. Koizumi, *Inorg. Synth.* **22**, 86 (1983).
8. J. F. Bringley, J.-M. Fabre, and B. A. Averill, *Chem. Mater.*, **4**, 522 (1992).
9. Tetraselenafulvalene is not commercially available. There are several reported syntheses in the literature, but most utilize highly toxic and malodorous CSe_2 or H_2Se as reagents. Probably the safest and most convenient published procedure is that of Cava et al.,[10] which in our hands gives low and variable yields (10–30%) of TSF.
10. Y. A. Jackson, C. White, V. V. Lakshmikantham, and M. P. Cava, *Terahedron Lett.*, **28**, 5635 (1987).

35. LAYERED INTERCALATION COMPOUNDS

Submitted by S. KIKKAWA,* F. KANAMARU,* and M. KOIZUMI*
Checked by SUZANNE M. RICH† and ALLAN JACOBSON†

Reprinted from *Inorg. Synth.*, **22**, 86 (1983)

The layered compound iron(III) chloride oxide FeOCl absorbs pyridine derivative molecules into its van der Waals gap, forming the intercalation compounds FeOCl(pyridine derivative)$_{1/n}$. Iron(III) chloride oxide and the pyridine derivatives act, respectively, as a Lewis acid and base in the reaction with a partial transfer of the pyridine derivative electrons to the host FeOCl layer. Electrical resistivity of the host FeOCl decreases from 10^7 Ω·cm to 10^3 Ω·cm with intercalation, and the interlayer distance expands to almost twice the original value.

Methanol and ethylene glycol, which are not directly intercalated into FeOCl itself, will enter the expanded interlayer region of FeOCl(4-amino pyridine)$_{1/4}$. These molecules attack the chloride ions which are weakly bound to the previously intercalated 4-aminopyridine (APy) and cause the elimination of the 4-aminopyridine. They substitute for Cl in FeOCl layer and are grafted to the FeO layer without the reconstruction of the host FeO layer.

The layered oxide $KTiNbO_5$ will form intercalation compounds, although it does not intercalate organic amines directly. If its interlayer potassium is replaced by protons through treatment with HCl, the product, $HTiNbO_5$, will intercalate organic amine molecules.

Herein are described the preparations of the charge-transfer-type intercalation compound FeOCl (pyridine derivative)$_{1/n}$; of grafted-type intercalation compounds $FeO(OCH_3)$ and $FeO(O_2C_2H_4)_{1/2}$; and of some organic intercalates of $HTiNbO_5$.

A. CHARGE-TRANSFER-TYPE INTERCALATION COMPOUNDS: FeOCl(PYRIDINE DERIVATIVE)$_{1/n}$

$$\text{FeOCl} + 1/n\ (\text{pyridine derivative}) \rightarrow \text{FeOCl (pyridine derivative)}_{1/n}$$

* The Institute of Scientific and Industrial Research, Osaka University, Osaka 567, Japan.
† Exxon Research and Engineering Company, Corporate Research, P.O. Box 45, Linden, NJ 17036.

Procedure

Iron(III) chloride oxide is prepared by sealing a mixture of 100 mg α-Fe_2O_3 (0.63 mmol) and 220 mg $FeCl_3$ (1.36 mmol) in a Pyrex glass tube 20 cm long and 2.0 cm in diameter with a wall thickness of 2.0 mm and heating at 370° for 2–7 days.[1] The product is washed with water to remove excess $FeCl_3$. The red-violet blade like crystals of FeOCl which are obtained are then washed with acetone, dried, and stored in a desiccator with silica gel. Prolonged exposure to moist air will cause hydrolysis of the FeOCl and render it incapable of forming intercalation compounds.

Iron(III) chloride oxide (100 mg) is introduced into a Pyrex tube $\approx$10 cm long, 0.8 cm in diameter, 2.0 mm wall thickness, which has been sealed at one end. To this is added 4 mL of reagent-grade pyridine (Py) or 2,4,6-trimethylpyridine (TMPy) which has been previously dried over 3A molecular sieves. The tube is sealed and heated at 100° for 1 week; then the black product is collected by filtration and washed several times with dry acetone.[2] (To prepare the APy derivative 4 mL of a 1*M* acetone solution of APy, dried as above, is used and the reaction temperature is maintained at 40°.)

Anal. Calcd. for $FeOCl(py)_{1/4}$: C, 11.80; N, 2.75; H, 1.00. Found: C, 13.25; N, 3.09; H, 1.11. Calcd. for $FeOCl(tmpy)_{1/6}$: C, 12.55; N, 1.83; H, 1.45. Found: C, 12.26; N, 1.81; H, 1.63. Calcd. for $FeOCl(apy)_{1/4}$: C, 11.47; N, 5.35; H, 1.16. Found: C, 12.67; N, 5.32; H, 1.94.

Properties

The *d* values corresponding to the interlayer distances are 7.92 Å for FeOCl, and 13.27(P), 13.57(APy), and 11.9(TMPy), for the intercalated FeOCl. Iron(III) oxide chloride is a semiconductor with a resistivity of 10^7 Ω·cm at room temperature. The intercalated complexes are still semiconductive but exhibit improved electrical conductivities along their *c* axes. The electrical resistivities at room temperature are 10, 10^3, and 10^3 Ω·cm, respectively, for Py, APy, and TMPy intercalated FeOCl.

B. GRAFTED-TYPE COMPOUND FROM FeOCl

1. Methanol-Grafted Compound $FeO(OCH_3)$

$$FeOCl(APy)_{1/4} + CH_3OH \rightarrow FeO(OCH_3) + HCl + 1/4AP$$

Procedure

Direct reaction of FeOCl with methanol does not occur up to a temperature of 100°. In contrast, the reaction of $FeOCl(APy)_{1/4}$ is a facile one, due primarily to the expanded interlayer distance which permits penetration of the methanol molecules.

The intercalation compound $FeOCl(APy)_{1/4}$, 80 mg, is soaked in 2 mL of methanol (previously dried over 3A molecular sieves for 3 days) in a sealed Pyrex glass tube, which is typically 10 cm long and 0.8 cm in diameter, with a wall thickness of 2.0 mm.[3] The tube is maintained at 100° and shaken twice daily for 10 days. (If difficulty is encountered in obtaining intercalation, it may be helpful to add a small amount of APy.) The resulting brown solid is collected by filtration, washed with several volumes of dry methanol, then stored in a desiccator with silica gel.

Anal. Calcd. for $FeOOCH_3$: Fe, 54.3; C, 11.7; H, 2.94; N, 0.00; Cl, 0.00. Found: Fe, 55.9; C, 9.29; H, 2.43; N, 0.33; Cl, 4.20.

Properties

The interlayer distance in $FeO(OCH_3)$ is 9.97 Å. The infrared spectrum shows a C—O stretching vibration around 1050 cm^{-1}, but no O—H stretching vibration. A study of the Mössbauer effect at room temperature indicates that the isomer shift and the quadrupole splitting are 0.37 and 0.60 mm/s, respectively.

2. Ethylene Glycol-Grafted Compound $FeO(O_2C_2H_4)_{1/2}$

$$FeOCl(APy)_{1/4} + 1/2C_2H_4(OH)_2 \rightarrow FeO(O_2C_2H_4)_{1/2} + HCl + 1/4(APy)$$

Procedure

Eighty milligrams of $FeOCl(APy)_{1/4}$ is well ground and placed in a sealed Pyrex tube, 10 cm long and 0.8 cm in diameter, with a wall thickness of 2.0 mm,[4] which contains 2 mL ethylene glycol and approximately 20 mg of APy. (The addition of APy is important to drive the reaction to completion.) The reaction mixture is heated at 110° for 1 week. This produces a brown crystalline material which is collected by filtration, then washed with dry acetone until the silver chloride test is negative.

Anal. Calcd. for $FeO(O_2C_2H_4)_{1/2}$: Fe, 54.8; C, 11.8; H, 1.98. Found: Fe, 54.2; C, 11.6; H, 2.21.

Properties

The basal spacing of $FeO(O_2C_2H_4)$ is 14.5 Å when the product is sill in ethylene glycol, but it shrinks to 10.98 Å on washing with acetone.

Mössbauer isomer shift and quadrupole splitting are respectively 0.38 and 0.59 mm/s at room temperature.

C. THE ORGANIC INTERCALATES OF $HTiNbO_5$

$$KTiNbO_5 + HCl \rightarrow HTiNbO_5 + KCl$$

$$HTiNbO_5 + b \rightarrow H_{1-1/m}\,(bH)_{1/m}\,TiNbO_5$$

where b is an amine

Procedure

White solid $KTiNbO_5$ is prepared by heating an intimate mixture of K_2CO_3, TiO_2, Nb_2O_5 in the molar ratio 1:2:1 at 1100° overnight.[5] The reaction product (~3 g) is treated with 2*M* HCl (200 mL) at 60° for 1 h to produce $HTiNbO_5$.[6] The sample is washed with distilled water several times, until a flame test shows no detectable potassium. Approximately 20 mg of the sample is sealed in a Pyrex tube, 15 cm long and 1.5 cm in diameter with a wall thickness of 2.0 mm, which contains 3 mL of aliphatic amines $CH_3(CH_2)_nNH_2$ and the mixture warmed at 60° for 5 h.[7] The products are filtered and washed with acetone three times. The products have composition $(H^+]_{1/2}\,[CH_3(CH_2)_nNH_3^+]_{1/2}\,[TiNbO_5^-]$.

Properties

Both parent compounds ($KTiNbO_5$ and $HTiNbO_5$) and amine-intercalated $HTiNbO_5$ belong to the orthorhombic system; $a = 6.44$, $b = 3.78$, $c = 17.52$ Å for $HTiNbO_5$. The $(TiNbO_5)^-$ layer lies in the *ab* plane and intercalates amine in its protonated form. The interlayer distance $c/2$ expands on intercalation: $c = 9.02$ Å for ammonia, 22.96 Å for methanamine, 26.78 Å for ethanamine, 35.34 Å for 1-propanamine, 36.82 Å for 1-butanamine.

References

1. M. D. Lind, *Acta Cryst.*, **B26**, 1058 (1970).
2. S. Kikkawa, F. Kanamaru, and M. Koizumi, *Bull. Chem. Soc. Jpn.*, **52**, 963 (1979).
3. S. Kikkawa, F. Kanamaru, and M. Koizumi, *Inorg. Chem.* **15**, 2195 (1976).
4. S. Kikkawa, F. Kanamaru, and M. Koizumi, *Inorg. Chem.*, **19**, 259 (1980).
5. A. D. Wadsley, *Acta Cryst.*, **17**, 623 (1964).
6. H. Rebbash, G. Desgardin, and B. Raveau, *Mater. Res. Bull.*, **14**, 1125 (1979).
7. S. Kikkawa and M. Koizumi, *Mater, Res. Bull.*, **15**, 533 (1980).

36. LITHIUM INSERTION COMPOUNDS

Submitted by D. W. MURPHY* and S. M. ZAHURAK*
Checked by C. J. CHEN† and M. GREENBLATT†

Reprinted from *Inorg. Synth.*, **24**, 200 (1986)

Many inorganic solids are capable of undergoing insertion reactions with small ions such as H^+, Li^+, and Na^+. The host solid in these reactions undergoes reduction in order to those of the respective hosts, with the inserted cation occupying formerly empty sites of the host.

The examples presented here illustrate a variety of reagents for the insertion or removal of lithium ions from inorganic solids. A variety of reagents exhibiting a range of redox potentials allow access to intermediate stoichiometries and control of side reactions. Four reagents are used in these syntheses: butyllithium, a strongly reducing source of lithium; lithium iodide, a mild reducing source; ethanol, a mild oxidant; and iodine, a stronger oxidant. A discussion of the redox levels of these and other reagents may be found elsewhere.[1]

A. VANADIUM DISULFIDE

$$2LiVS_2 + I_2 \xrightarrow{CH_3CN} 2VS_2 + 2LiI$$

The class of layered transition-metal dichalcogenides has been of great interest because of their varied electronic properties and chemical reactions. Most compounds of this may be prepared by stoichiometric reactions of the elements above 500°. However, the highest vanadium sulfide that can be made in this manner is V_5S_8. An amorphous VS_2 has been prepared by the metathetical reaction of Li_2S and VCl_4.[2] The method presented here allows preparation of polycrystalline VS_2 with the CdI_2 structure.[3]

Procedure

The $LiVS_2$[4] is prepared from an intimate mixture of V_2O_5 (0.094 g, 0.05 mol) and Li_2CO_3 (3.694 g, 0.05 mol). The mixture is placed in a vitreous carbon boat inside a quartz tube in a tube furnace. The tube is connected to a two-way inlet valve for argon (or another inert gas) and hydrogen sulfide. Gas

* AT&T Bell Laboratories, Murray Hill, NJ 07974.
† Department of Chemistry, Rutgers University, The State University of New Jersey, New Brunswick, NJ 08903.

exits through a bubbler filled with oil. An H_2S flow of ~100 ml/min is maintained, and the furnace is heated to 300°, then to 700° at the rate of 100°/h. (■ **Caution.** *Hydrogen sulfide is a highly poisonous gas. The reaction should be carried out in a well-ventilated fume hood.*) Water and sulfur are deposited on the tube downstream. The reaction is held at 700° for 12 h. The reaction mixture is cooled under H_2S and then flushed with argon. The boat is removed into a jar filled with argon and placed in a good dry box (a dry atmosphere is sufficient). The mixture is reground and refired in H_2S at 700° for another 16 h. The $LiVS_2$ prepared in this way in actually $Li_{0.90-0.95}VS_2$. The stoichiometry is adjusted to $Li_{1.0}VS_2$ by treatment with a dilute (0.05*N*) solution of butyllithium in hexane. See the preparation of Li_2ReO_3 below for details of butyllithium reactions.

A solution of ~0.1*M* I_2 is prepared from freshly sublimed I_2 and acetonitrile distilled from P_4O_{10}. The solution is standardized by titration with a standard aqueous thiosulfate solution to the disappearance of the I_2 color. Addition of iodide and/or starch gives no color enhancement in acetonitrile.

The solid polycrystalline $LiVS_2$ (4.670 g, 38.3 mmol) is placed in a 300 mL round-bottomed flask under argon. The flask is fitted with a serum cap. A solution of I_2 in CH_3CN (225 mL, 0.091*M*, 20.5 mmol of I_2) is added to the $LiVS_2$ using a transfer needle through the serum cap. The iodine color rapidly dissipates as solution is added. The heterogeneous reaction mixture is stirred using a magnetic stirrer. After 16 h the mixture is filtered in air, and the solid product is washed with acetonitrile. The filtrate and washings are titrated with standard aqueous thiosulfate to determine the unreacted excess I_2 (1.35 mmol).

The yield is quantitative.

Combustion Anal. Calcd. for VS_2: V, 44.3. Found, 44.5. Atomic absorption for Li gives 180 ppm. X-ray fluorescence shows no iodine.

Properties

The VS_2 produced in this manner is a shiny metallic gray powder. The compound is hexagonal (CdI_2 type) with $a = 3.217$ Å and $c = 5.745$ Å. Sulfur loss occurs in air or inert gas above 300°.

B. LITHIUM DIVANADIUM PENTOXIDE

$$2LiI + 2V_2O_5 \xrightarrow{CH_3CN} 2LiV_2O_5 + I_2$$

Ternary alkali-metal vanadium oxide bronzes are well known, including γ-LiV_2O_5.[5] It was recognized that some other composition or structure was formed from the combination of lithium and V_2O_5 at room temperature through electrochemical or butyllithium reactions.[61] It is possible to prepare the low-temperature δ-LiV_2O_5 with butyllithium,[6,7] although irreversible overreduction is difficult to avoid. The use of LiI as reductant avoids any overreduction.[8]

Procedure

The divanadium pentoxide is prepared from reagent-grade NH_4VO_3 by heating in air first at 300° for 3 h and then at 550° for 16 h. Anhydrous LiI (Alfa Products) is dried under vacuum at 150° prior to use. A solution of ~1.5*M* LiI in acetonitrile is prepared using acetonitrile freshly distilled from P_4O_{10}. Care is taken to exclude moisture and oxygen in the preparation and storage of this solution. The V_2O_5 powder (5.005 g, 27.5 mmol) is placed in a flask fitted with a serum cap, and an excess of the LiI solution (25.0 mL of 1.44*M*, 36 mmol) is added via syringe. The supernatant rapidly develops a dark yellow-brown color characteristic of iodine. The reaction mixture is stirred at room temperature for ~24 h, using a magnetic stirrer. The color of the solid changes from yellow to green to blue-black over several hours. The product is isolated by filtration in air and is washed with acetonitrile. Titration of the filtrate and washings with standard aqueous thiosulfate determines that 13.59 mmol of I_2 is formed in the reaction.

Properties

The LiV_2O_5 formed in this way is dark blue and is stable in air for moderate lengths of time. Long-term storage in a desiccator is satisfactory. The compound is orthorhombic[7,8] with $a = 11.272$ Å, $b = 4.971$ Å, and $c = 3.389$ Å. A reversible first-order structural transformation occurs at 125° to ε-LiV_2O_5,[8] which is also orthorhombic with $a = 11.335$ Å, $b = 4.683$ Å, and $c = 3.589$ Å. Above 300° the structure changes irreversibly to that of the thermally stable γ-LiV_2O_5.

C. LITHIUM RHENIUM TRIOXIDES: Li_xReO_3 ($x \leq 0.2$)

$$ReO_3 + 2BuLi \rightarrow Li_2ReO_3 + \text{octane}$$

$$2Li_2ReO_3 + 2EtOH \rightarrow 2LiReO_3 + 2LiOEt + H_2$$

$$ReO_3 + \text{excess LiI} \rightarrow Li_{0.2}ReO_3 + LiI + I_2$$

Rhenium trioxide has one of the simplest extended structures. Octahedral (ReO_6) units share oxygen atoms between units such the Re—O—Re bonds are linear. The symmetry is cubic, and each cell contains one Re and one empty cubeoctahedral cavity. This lattice serves as a starting point for the generation of a number of other structures, including perovskites and shear compounds. Since ReO_3 is the simplest of this large family of compounds, an understanding of its behavior with lithium is intrinsic to understanding the class as a whole.

Three phases have been identified in the Li_xReO_3 system.[10] For $x < 0.35$ the structure remains cubic. A line phase at $x = 1.0$ is rhombohedral, as is a phase at $1.8 \leq x \leq 2.0$.

1. Dilithium Rhenium Trioxide

■ **Caution.** *Concentrated butyllithium (n-BuLi) (~2.0–2.0M in hexane) is extremely air- and moisture-sensitive and must be handled in an inert atmosphere. Any amount that must be handled in air should be diluted with hexane or other inert solvent before exposure.*

Procedure

Reaction and filtration operations are carried out in a helium-filled glovebox. However, the procedures are easily adapted to the use of Schlenk techniques. The hexane used should be distilled from sodium, and the concentrated *n*-BuLi solution (Alfa Products) standardized by total base titration. If the solution is cloudy as received, it can be filtered in the glovebox.

To a flame-dried 100 mL round-bottomed flask is added 4.08 g (17.42 mmol) of ReO_3, along with sufficient hexane (~10.0 mL) to cover the ReO_3. A small excess of an *n*-BuLi solution (12.5 mL, 38.24 mmol, 3.059*N*) is slowly added via a 20.0-mL gastight syringe. The reaction is exothermic, and, depending on the particle size of the ReO_3 and on the *n*-BuLi concentration, addition of *n*-BuLi may cause boiling of the solvent. The product is purer (by X-ray powder diffraction) when boiling is avoided. The flask is capped with a serum stopper and stirred with a magnetic stirrer for ~24 h at room temperature. The original red-bronze ReO_3 turns to a dark red-brown color on completion of the reaction.

The reaction mixture is filtered in the glovebox and the Li_2ReO_3 collected on a medium-porosity fritted-glass filter. The product is washed several times with hexane to ensure removal of any excess *n*-BuLi. The yield is quantitative. The filtrate and washings are then removed from the glovebox after sufficient dilution with hexane and titrated for excess *n*-BuLi.

The titration consists of addition of a few milliliters of distilled water and an excess of standard HCl (0.1*N*), followed by back-titration with standard NaOH solution (0.1*N*). The lithium stoichiometry is calculated on the basis of the titration results. It has been shown that this total base titration gives the same results as are obtained with active lithium reagent and atomic absorption analysis.

Complete lithiation of the limiting lithium stoichiometry of $Li_{2.0}ReO_3$ may require more than one *n*-BuLi treatment. This can be due in part to dilution of *n*-BuLi as the reaction proceeds. On titrating the initial *n*-BuLi reaction solution, 10.491 mmol of Li remains from an original 3.059*N* *n*-BuLi solution containing 12.5 mL (38.238 mmol) of *n*-BuLi in hexane and 4.080 g (17.421 mmol) of ReO_3. This indicates 1.59 mmol of Li per millimole of ReO_3. Further lithiation and subsequent titration results in $Li_{2.1}ReO_3$. X-ray powder diffraction data indicates the lithium composition, in excess of two Li per ReO_3, is due to impurities in the ReO_3.

Confirmation of the lithium stoichiometry is determined by an iodine reaction that yields the amount of lithium removed from the structure. A titration (described in Section A) performed after reaction of Li_xReO_3 with a standard iodine solution affords the stoichiometry $Li_{1.99}ReO_3$.

Properties

The compound Li_2ReO_3 is a dard red-brown solid that is reactive with atmospheric moisture, forming LiOH and H_2. The X-ray powder pattern data given the following hexagonal crystallographic parameters: $a = 4.977$ Å, $c = 14.793$ Å, $V = 52.88 \times 6$ Å^3.[10]

Neutron diffraction powder profile analysis establishes rhomohedral structure in the space group $R3c$,[11] $a = 4.9711(1)$, $c = 14.788(1)$, $Z = 6$. Both Li and Re atoms occupy type (00*z*) positions, and oxygen atoms the general position (*x*,*y*, *z*) with six formula units per cell. The host lattice, ReO_3, undergoes a twist that creates two octahedral sites from the cubeoctahedral cavity of ReO_3. These are the sites occupied by lithium.[11]

2. Lithium Rhenium Trioxide

Procedure

In a 60.0 mL Schlenk filter within a helium glovebox is placed 4.1891 g (17.37 mmol) of $Li_{2.0}ReO_3$. The filter is equipped with a serum cap and stir bar and removed from the box. The serum cap is secured with a twist of wire, and ~40 mL of absolute ethanol is transferred into the flask, under argon, using a double-edged transfer needle. Evolution of hydrogen is noted on this addition, and the system is kept under argon flow using an oil bubbler. The mixture is stirred the system is kept under argon flow using an oil bubbler. The mixture is stirred ~ 24 h or until gas evolution has ceased, and it is then filtered, using the same Schlenk filter along with a 250 mL receiving flask. The product is rinsed 3 times with 5.0 mL aliquots of absolute ethanol under positive argon pressure using a transfer needle. Care must be taken so as not to overrinse, as more Li can be removed from the structure. The dark red product is isolated in quantitative yield by filtration and is then vacuum-dried. The filtrate containing lithium ethoxide is titrated using a standard acid–base method with phenolphthalein as the indicator.

Reaction of 4.189 g of Li_2ReO_3 together with excess absolute ethanol ($\sim$55.0 mL total) forms 19.45 mmol of lithium ethoxide. Therefore, 1.0 mmol of Li is removed from the structure, leaving $Li_{1.0}ReO_3$. Removal of all of the lithium using the standard iodine solution technique confirms the lithium stoichiometry.

Properties

The compound $LiReO_3$ is very similar in appearance and properties to Li_2ReO_3. Both are red-brown hygroscopic solids; however, $LiReO_3$ is less moisture-sensitive and can be exposed to air for brief periods without damage. The X-ray powder pattern shows that $LiReO_3$ is single-phase with hexagonal lattice parameters of $a = 5.096$ Å, $c = 13.400$ Å, and $V = 50.23 \times 6$ Å^3.[10,11] The structure of $LiReO_3$, according to neutron-diffraction powder profile analysis studies, shows $a = 5.0918(3)$, $c = 13.403(1)$, $z = 6$ in the $R3c$ space group.[11] The ReO_3 skeleton has undergone the same twist as in Li_2ReO_3. The lithium atoms order in half the octahedral sites. The compound is isostructural with $LiNbO_3$. Both $LiReO_3$ and Li_2ReO_3 exhibit temperature-independent Pauli paramagnetism (0.8×10^{-6} and 3.3×10^{-6} emu/g, respectively).[10]

3. Lithium(0.2) Rhenium Trioxide

Procedure

As in the preparation of Li_2ReO_3, reaction and filtration operations take place in a helium-filled glovebox. Rhenium(VI) oxide (10.381 g, 44.33 mmol) together with 70.0 mL of a 0.5*M* LiI solution in acetonitrile are added to a flame-dried 100 mL round-bottomed flask equipped with a stirring bar. The flask is stoppered, and the reaction is monitored for completeness by periodic sampling of the solid using X-ray powder diffraction analysis. A total of three successive treatments with 70.0 mL aliquots of LiI in CH_3CN, with stirring over 7 days, are needed to obtain the homogeneous single-phase $Li_{0.2}ReO_3$ (smaller-scale reactions[12] are complete with a single treatment). Each treatment is followed by filtration of the total solution and two washings with acetonitrile. The filtrates are removed from the glove box and titrated for iodine content using standard aqueous $Na_2S_2O_3$ solution as the titrant. Reaction of 10.381 g of ReO_3 with LiI forms 8.87 meq of iodine, indicating a final stoichiometry of $Li_{0.2}ReO_3$. Flame-emission analysis of Li confirms this result.

References

1. D. W. Murphy and P. A. Christian, *Science*, **205**, 651 (1979).
2. R. R. Chianelli and M. B. Dines, *Inorg. Chem.*, **17**, 2758 (1978).
3. D. W. Murphy, C. Cros, F. J. DiSalvo, and J. V. Waszozak, *Inorg. Chem.*, **16**, 3027 (1977).
4. B. van Laar and D. J. W. Ijdo, *J. Solid State Chem.*, **3**, 590 (1971).
5. J. Galy and A. Hardy, *Bull. Soc. Chim. Fr.*, **1964**, 2808.
6. M. S. Whittingham, *J. Electrochem. Soc.*, **123**, 315 (1976).
7. P. G. Dickens, S. J. French, A. T. Hight, and M. F. Pye, *Mater. Res. Bull.*, **14**, 1295 (1979).
8. D. W. Murphy, P. A. Christian, F. J. DiSalvo, and J. V. Waszczak, *Inorg. Chem.*, **18**, 2800 (1979).
9. G. Brauer (ed.), *Handbook of Preparative Inorganic Chemistry*, 2nd ed., Academic Press, New York, 1965, p. 1270.
10. D. W. Murphy, M. Greenblatt, R. J. Cava, and S. M. Zahurak, *Solid State Ionics*, **5**, 327 (1981).
11. R. J. Cava, A. J. Santoro, D. W. Murphy, S. M. Zahurak, and R. S. Roth, *J. Solid State Chem*, **42**, 251 (1982).
12. R. J. Cava, A. J. Santoro, D. W. Murphy, S. M. Zahurak, and R. S. Roth, *J. Solid State Chem.*, **50**, 121 (1983).

Chapter Five

OXIDE SUPERCONDUCTORS AND RELATED COMPOUNDS

37. SYNTHESIS OF SUPERCONDUCTING OXIDES

Submitted by J. J. KRAJEWSKI*
Checked by S. C. CHEN,† J. YAN,† J. LEE,† K. V. R. CHARY,† E. B. JONES,† Z. ZHANG,† Z. S. TEWELDEMEDHIN,† and M. GREENBLATT†

Since the discovery of superconducting $(La, Ba)_2CuO_4$ in 1986 by Bednorz and Muller,[1] many copper oxide and several non-copper oxide superconductors with high-transition temperatures (T_c) have been discovered. All these copper oxides contain layers of copper–oxygen squares, pyramids, sticks, and/or octahedra as their electronically active structural components. Because of copper's versatility with coordination geometry and valence states, many combinations of elements as well as various dopant levels can be achieved using alkaline earth, rare earth and transition metals, forming a tremendous array of compounds and possibilities. Using standard as well as innovative solid-state chemistry techniques, we describe the detailed procedure for synthesizing six superconducting compounds.

The examples presented here show a wide variety of structures from the simple perovskite in $(Ba, K)BiO_3$ to the more complex orthorhombic $Pb_2Sr_2(Ln, Ca)Cu_3O_8$, but all the compounds can be made using simple

* AT & T Bell Laboratories, Murray Hill, NJ 07974.
† Department of Chemistry, Rutgers University, The State University of New Jersey, New Brunswick, NJ 08903.

starting materials, box furnaces, and laboratory fume foods. All except two of the compounds are copper-based; the others are based on lead and bismuth. A detailed discussion of the magnetic, resistivity, and other properties for each family of superconductors can be found elsewhere.[2]

A. $La_{2-x}Sr_xCuO_4$

$$\frac{(2-x)}{2}La_2O_3 + xSrCO_3 + CuO \rightarrow La_{2-x}Sr_xCuO_4 + xCO_2$$

Procedure

The substitution of Sr for La in the insulating La_2CuO_4 oxidizes some of the copper to Cu^{3+}, resulting in a mixed-valence superconductor. The optimal superconducting composition for $La_{2-x}Sr_xCuO_4$ is $x = 0.15$.[3] This bulk superconductor has planes of CuO_6 octahedra sharing corners alternating with (La, Sr)O layers (Fig. 1*a*). It is prepared from a stoichiometric mixture of La_2O_3 (3.0138 g, 0.00925 mol), $SrCO_3$ (0.2214 g, 0.0015 mol) and CuO (0.796 g, 0.01 mol) powders, heated in air at 1000° in alumina crucible for a few days with intermittent grindings until single phase is obtained as determined by powder X-ray diffraction. Prior to use, La_2O_3 should be freshly dried at 1000° for 8 h; $SrCO_3$ should be dried at 300° for a few hours and stored in a dessicator. High-purity $La(OH)_3$ (3.5132 g, 0.0185 mol) can also be used as a source of lanthanum oxide. However, $La(OH)_3$ should be weighed out in a glove dry box, because of its hydroscopic nature and ability to absorb CO_2 from air. Pellets pressed at 365 kg/cm^2 annealed in flowing O_2 at 1050–1100° for one day, then cooled to room temperature in flowing O_2 over several hours, are useful for physical properties measurements. Annealing at higher temperatures in air results in a lower transition temperature and increased resistivity.

Properties

$La_{1.85}Sr_{0.15}CuO_4$ has a superconducting transition temperature of 38.5 K with a midpoint at 36.2 K. At $x = 0.15$, a single-phase tetragonal material is formed with lattice parameters of $a = 3.78$ Å and $c = 13.23$ Å.

B. $LnBa_2Cu_4O_8$ (Ln = Er, Ho)

$$Ln(NO_3)_3 \cdot 5H_2O + 2Ba(NO_3)_2 + 4Cu(NO_3)_2 \cdot 2.7H_2O \rightarrow$$
$$LnBa_2Cu_4O_8 + 15NO_2 + 3.5O_2 + 7.7H_2O$$

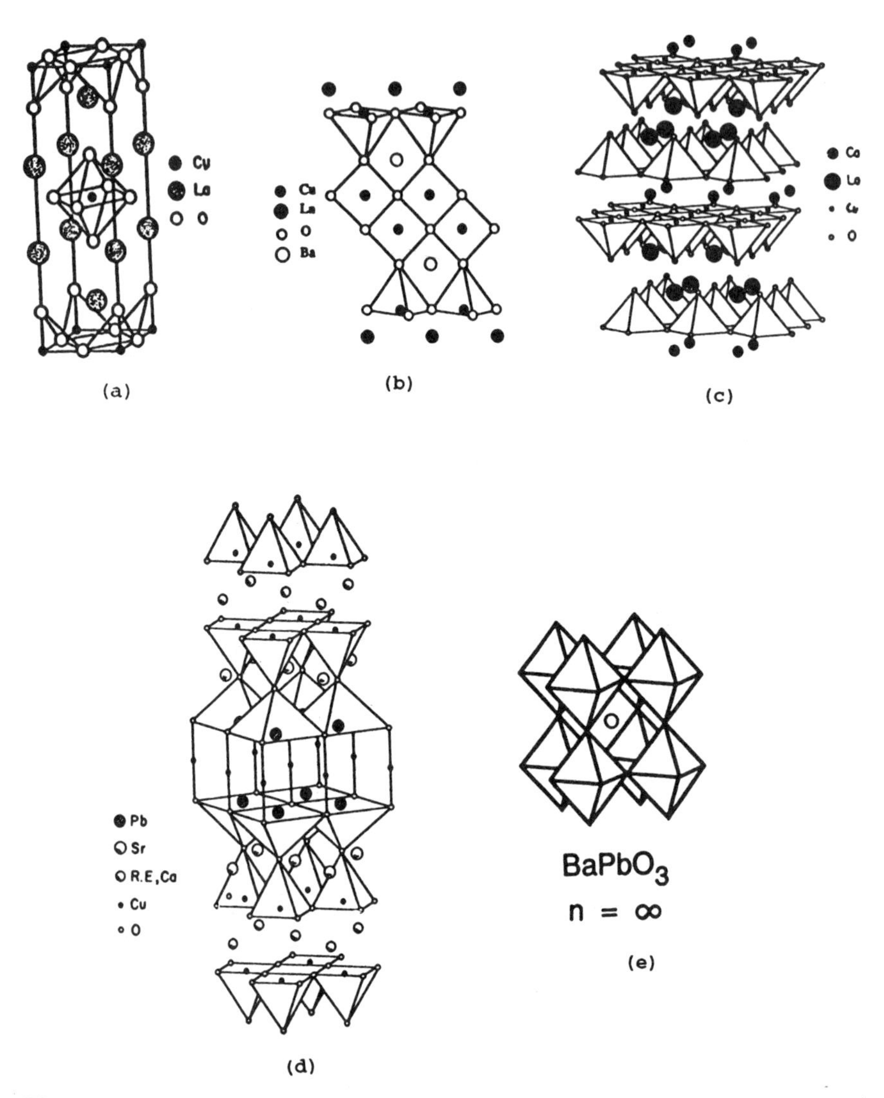

Figure 1. Structures of oxide superconductors: (a) $La_{2-x}Sr_xCuO_4$, (b) $LnBa_2Cu_4O_8$, (c) $La_{2-x}Sr_xCaCu_2O_6$, (d) $Pb_2Sr_2Y_{0.5}Ca_{0.5}Cu_3O_8$, (e) $BaPbO_3$.

Procedure

The synthesis of this double copper–oxygen chain superconductor (Fig. 1*b*) can be accomplished at one atmosphere (1 atm) oxygen pressure and is carried out in two steps.[4] In step 1, the correct molar proportions of each starting material, $Er(NO_3)_3{\cdot}5.5H_2O$ (4.43 g, 0.01 mol), or $Ho(NO_3)_3{\cdot}5.5H_2O$

(4.41 g, 0.01 mol), $Ba(NO_3)_2$ (5.228 g, 0.02 mol), $Cu(NO_3)_2{\cdot}2.7H_2O$ (9.448 g, 0.04 mol), were weighed, intimately mixed in an agate mortar and pestle, and then heated to 325° in air for 15–30 min in an oversized glazed porcelain crucible. Using rare earths other than Er and Ho results in multiphase samples.

■ **Caution.** *This procedure should be carried out in a well-ventilated hood due to the release of nitrogen dioxide, a red-brown toxic gas. Do all grindings in the hood to avoid breathing toxic dust particles.*

The correct amount of water in the starting materials is critical for proper stoichiometry, and the water content may vary depending on the source of the starting materials and storage conditions. For example, using Johnson Matthey Alfa Products Chemicals, the range for the water in the copper nitrate, $Cu(NO_3)_2{\cdot}xH_2O$, can vary from $2 < x < 6$, but we measured $x = 2.7$ on the material used. The water content of each starting material is determined directly by weight loss on thermal decomposition to the simple oxides. At 350°, the $LnBa_2Cu_4O_8$ hydrate–nitrate mixture is first molten but rapidly dehydrates. Because of this molten nature, oversized glazed crucibles are used to facilitate complete recovery of the material and reduce material loss due to splashing caused by evolution of gases. The time and temperature of the first step are kept at a minimum to prevent the formation of carbonates from atmospheric CO_2.

Step 2 consists of grinding the pre-reacted material and placing the light gray powder on 10–20-mil silver foil or in a dense alumina crucible and heating to 825° in flowing oxygen. This alumina crucible or the silver foil is loaded into a fused-silica tube, which is closed at one end and purged with flowing oxygen for several minutes from the other end. The gas input and output are located at the same end of the fused silica to permit the removal of the tube from the furnace while maintaining the CO_2-free environment. The entire assembly is inserted into a precalibrated tube furnace and heated at 825° for 3–5 days. During this period, daily grindings are performed, after the tube assembly has been removed from the furnace and cooled to room temperature. The final cooling period (after 3–5 days) is done at 300°/h with the fused-silica tube still in place. The product is a black, chunky material that grinds to a fine-grained powder.

Properties

$ErBaCu_4O_8$ and $HoBaCu_4O_8$ formed by this procedure are single-phase at 825° but degrade to a multiphase powder at 850°, and by 875° substantial $LnBa_2Cu_3O_{7-\delta}$ is formed. The Ho compound is *A*-centered orthorhombic with $a = 3.847$ Å, $b = 3.872$ Å, and $c = 27.278$ Å. The same structure for the Er compound has lattice parameters of $a = 3.855$ Å, $b = 3.874$ Å, and $c = 27.295$ Å. The onset of the superconducting transition occurs at 80 K for both materials, although the bulk of the transition is near 70 K.

C. $La_{2-x}Sr_xCaCu_2O_6$

$$\frac{(2-x)}{2}La_2(C_2O_4)_3{\cdot}10H_2O + xSr(NO_3)_2 + Ca(NO_3)_2{\cdot}H_2O + 2CuO + 0.6O_2$$

$$\downarrow$$

$$La_{2-x}Sr_xCaCu_2O_6 + (11-5x)H_2O + (6-3x)CO_2 + (2x+2)NO_2$$

Procedure

The synthesis of this simple double-layer superconductor at $x = 0.4$ (Fig. 1*c*) can be accomplished in two steps.[5] In step 1, the correct molar proportions of each starting material, $La_2C_2O_4{\cdot}10H_2O$ (5.7762 g, 0.008 mol), $Sr(NO_3)_2$ (0.8465 g), 0.004 mol), $Ca(NO_3)_2{\cdot}4H_2O$, (2.362 g, 0.01 mol), and CuO (1.591 g, 0.02 mol) are weighed, mixed, and ground in an agate mortar and pestle, then dehydrated at 400° for one hour in air in oversized glazed porcelain crucibles, as detailed in Section A.

■ **Caution.** *This procedure should be carried out in a well-ventilated hood as is detailed in Section A.*

The powders are slowly heated to 900° in flowing oxygen for at least 24 h with daily mechanical grindings. The powders are then removed, cooled, and pressed into 1-cm-diameter pellets under 700 kg/cm^2 and reacted for another 1–3 days at 925° in flowing oxygen. Step 2 consists of a moderately high-oxygen-pressure treatment since one atmosphere of flowing oxygen yields lower superconducting transition temperatures and nonsuperconducting impurities. The high-temperature and high-pressure treatment involves wrapping pellets of each sample in thin (10–20 mils) platinum foil and sealing in a thick-walled fused-silica tube 14 mm o.d. and 8 mm i.d. and 150 mm long), containing enough liquefied oxygen to produce a pressure of 20 atm at 970°. The tube is sealed at one end and the platinum-wrapped pellet is placed in this tube. The tube is connected to a vacuum manifold via thick-wall Tygon tubing and hose clamps. The amount of oxygen gas at room temperature was first measured on the vacuum line with a known volume by using the Boyle–Charles gas law. The measured oxygen was then condensed into the bottom of the tube with liquid N_2. (Only the lower half of the tube is immersed in liquid N_2.) While still in the liquid N_2, the tube is sealed with a torch. Platinum is used to protect the sample from reaction with the fused silica at high pressures and temperatures.

■ **Caution.** *This technique produces a possible explosion hazard. Be sure that no organic matter contacts liquid oxygen. Overfilling the tubes or poor seals may lead to tube failure. Tubes should be kept in liquid nitrogen while transferring to the furnace, and protective gloves, clothing, and face shields should be worn at all times. Steel pipes or insulating sheets are placed around each reaction tube so that should an explosion occur, damage to the furnace or other quartz tubes will be minimized.*

Samples are annealed for 2 days at 970°, with subsequent gradual cooling and 5-periods of annealing of 850, 750, 650, and 500°. When the samples are at room temperature, they are removed one at a time from the furnace and immersed in liquid nitrogen until the unreacted oxygen condenses in the tube. At this moment, the tube is removed quickly and wrapped in a heavy cloth. the end of the tube is gently broken with a hammer. If done properly, the risk of injury is minimized from glass shards and the sample is easily recovered. The resulting pellets have a brownish coating where they have partially reacted with the platinum, but this film is easily removed with light sanding with commercial sandpaper leaving a blue-black dense sample. Samples show a few percent impurities by X-ray diffraction after heating at 1 atm, but are single-phase after the high-pressure treatment.

Properties

$La_{1.6}Sr_4CaCu_2O_6$ formed by this procedure is body-centered with $a = 3.825$ Å and $c = 19.428$ Å. The highest superconducting transition temperature observed is 60 K for the composition $x = 0.4$.

D. $Pb_2Sr_2Y_{0.5}Ca_{0.5}Cu_3O_8$

$$2PbO + 2SrCO_3 + 0.25Ln_2O_3 + 0.5CaCO_3 + 3CuO \rightarrow$$
$$Pb_2Sr_2Ln_{0.5}Cu_3O_8 + 2.5CO_2 + 0.625O_2$$
$$(Ln = La, Pr, Nd, Sm, Eu, Gd, Dy, Ho, Tm, Yb, Lu)$$

Procedure

The general formula that describes this phase is $Pb_2Sr_2LnCu_3O_{8+\delta}$ (Fig. 1*d*) with superconductivity imparted by partial substitution of a divalent ion

(Sr or Ca) for Ln or by addition of excess oxygen.[6] The compound must be synthesized under mildly reducing conditions to maintain Pb in the 2^+ state. The preparative conditions are more complicated than other copper-based superconductors. Attempts at direct synthesis of this compound from oxides and carbonates results in the formation of the very stable $SrPbO_3$-based perovskite; thus two steps are necessary. Step 1 is to prepare a precursor without Pb. The starting materials for the precursor are $SrCO_3$ (2.952 g, 0.02 mol), $CaCO_3$ (1.001 g, 0.01 mol), Y_2O_3 (2.258 g, 0.01 mol) and CuO (2.387 g, 0.03 mol). These materials are combined in the appropriate ratios, calcined for 16 h in dense Al_2O_3 crucibles, at 920–980° in air with one intermediate grinding. Step 2 involves adding the stoichiometric amount of PbO (4.464 g, 0.02 mol) to the precursor, grinding, and pressing a small pellet.

■ **Caution.** *Do all grindings and heatings in a well-ventilated hood to avoid inhalation of lead particles that are highly toxic, and very volatile. OSHA-approved breathing masks are highly recommended. Lead is easily absorbed by the respiratory tract and is a cumulative poison.*

The optimal conditions or $Pb_2Sr_2Y_{0.5}Ca_{0.5}Cu_3O_8$ are heating these pellets at 865° for 12 h in a slightly reducing gas environment of 1% oxygen in nitrogen and cooling in the same gas mixture to room temperature in 15 min. It should be noted that a gas environment of 1% oxygen in nitrogen can be achieved by using a tube furnace and a fused-silica insert with the input and output positioned together so that the tube can be removed from the hot zone without exposing the sample to air. Using higher temperatures, higher Ca content, or higher partial pressures of oxygen results in the intergrowth of a 123-type $YSr_2(Pb,Cu)_3O_x$ in addition to the desired compound or the formation of an $SrPbO_3$-based second phase. Single crystals of the superconducting compounds have been grown from PbO- and CuO-rich melts using the precursor technique.[7]

Properties

The bulk-phase superconductors have a transition temperature of 70–85 K with a 20% or higher diamagnetic Meissner fraction. The lattice parameters are $a = 5.37$ Å, $b = 5.42$ Å, and $c = 15.79$ Å for orthorhombic, superconducting $Pb_2Sr_2Y_{0.5}Ca_{0.5}Cu_3O_8$.

E. $Ba_{1-x}K_xBiO_3$

$$(1 - x)BaO_2 + xKO_2 + 0.5Bi_2O_3 \rightarrow Ba_{1-x}K_xBiO_3 + 0.25O_2$$

Procedure

The superconductivity in this compound occurs within the framework of a three-dimensionally connected bismuth–oxygen array (Fig. 1*e*). Synthesis of this non-copper-based superconductor at $x = 0.4$ can be accomplished in two steps.[8] Step 1 consists of weighing and grinding the correct molar ratios of the starting materials: BaO_2 (1.0160 g, 0.006 mol), KO_2 (0.5688 g, 0.008 mol) and Bi_2O_3 (2.3298 g, 0.005 mol) in a dry box. A 100% excess of KO_2 was used to allow for loss of potassium during heating.

■ **Caution.** *KO_2 reacts vigorously with water and in large quantities may react explosively. This is a powerful oxidizing material. Avoid organics and readily oxidizable substances. To avoid atmospheric moisture all grindings are done in a glove bag or dry box. Avoid breathing toxic barium dust particles. Use OSHA-approved breathing apparatus.*

The optimal conditions for the generation of the highest-transition-temperature single-phase perovskite of stoichiometry $Ba_{0.6}K_{0.4}BiO_{3-\delta}$ are as follows. The sample is loaded into 0.25 in. silver tube 6 in. in length, sealed by rolling the ends and pressing the ends in a vise, and the metal tube is sealed in an evacuated fused silica tube and heated at 675° for 3 days. The powder that is removed from the tube is dark red or brown in color. This sample is ground in agate mortar and pestle, placed on silver foil, and heated in flowing oxygen at 450–475° for 45 min and then quickly cooled to room temperature in oxygen. The resulting powder is deep blue-black in color. Oxygen anneals at temperatures above 475° or reaction times longer than 45 min greatly increase the loss of potassium from the compound. The excess potassium is not detectable by X-ray diffraction but is clearly present as a transparent white coating on the compound. This coating is completely gone after the oxygen anneal.

Properties

This single-phase material with $x = 0.4$ has a cubic perovskite structure with a lattice parameter of $a = 4.293$ Å and has a superconducting transition temperature of 29.8 K.

F. $BaPb_{1-x}Sb_xO_3$

$$BaCO_3 + 0.33(1 - x)Pb_3O_4 + 0.5xSb_2O_3 + \frac{(4.955 + 0.18\,x)}{2}O_2 \rightarrow BaPb_{1-x}Sb_xO_3 + CO_2$$

Procedure

$BaPb_{1-x}Sb_xO_3$ is a perovskite (Fig. 1*e*) and superconducts at 3.5 K. It is synthesized at $x = 0.25$ using the correct molar ratios of the starting materials, $BaCO_3$ (1.974 g, 0.01 mol). Pb_3O_4 (1.714 g, 0.0025 mol) and Sb_2O_3 (0.3644 g, 0.00125 mol).[9]

■ **Caution.** *See Section D for lead dust and vapor warnings.*

These compounds are mixed and ground mechanically for at least 30 min in an agate mortar and pestle. The powder is transferred to dense Al_2O_3 crucible, heated to 825° for 12 h in flowing O_2. The powders are then removed from the crucible and ground mechanically for another 30 min and heated for 5 h at 825° in flowing O_2. After this treatment, a third grinding is performed; then the powder is pressed into a pellet, buried in the powder of its own composition, and fired again for 5 h in flowing O_2 at 825°. Grinding for shorter periods or omitting the grinding after each heat treatment results in a multiphase material. The burying procedure helps prevent volatilization of Pb. The resultant pellets are not dense and are very fine-grained. The best samples are produced when the cooling rate is rapid, 10°/min to room temperature. Longer times or lower temperature O_2 anneal results in a multiphase sample.

Properties

Superconductivity with the highest transition in the $BaPb_{1-x}Sb_xO_3$ occurs at $x = 0.25$ and T_c is 3.5 K. The symmetry is tetragonal with $a = 6.028$ Å and $c = 8.511$ Å.

References

1. J. G. Bednorz and K. A. Muller, *Z. Phys. B*, **64**, 189 (1986).
2. T. A. Vanderah (ed.), *Chemistry of Superconducting Materials*, Noyes Publications, Park Ridge, NJ, 1992.
3. R. J. Cava, R. B. van Dover, B. Batlogg, and E. A. Rietman, *Phys. Rev. Lett.*, **58**(4) 408–410 (1987).
4. R. J. Cava, J. J. Krajewski, W. F. Peck, Jr., B. Batlogg, and L. W. Rupp, *Physica C*, **159**, 372–374 (1989).
5. R. J. Cava, B. Batlogg, R. B. van Dover, J. J. Krajewski, J. V. Waszczak, R. M. Fleming, W. F. Peck, Jr., L. W. Rupp, P. Marsh, A. C. W. P. James, and L. F. Schneemeyer, *Nature*, **345** (6276), 602–604 (1990).
6. R. J. Cava, B. Batlogg, J. J. Krajewski, L. W. Rupp, L. F. Schneemeyer, T. Siegrist, R. B. van Dover, P. Marsh, W. F. Peck, Jr., P. K. Gallagher, S. H. Glarum, J. H. Marshall, R. C. Farrow, J. V. Waszczak, R. Hull, and P. Trevor, *Nature*, **336**, 211 (1988).

7. L. F. Schneemeyer, R. J. Cava, A. C. W. P. James, P. Marsh, T. Siegrist, J. V. Waszczak, J. J. Krajewski, W. F. Peck, Jr., R. L. Opila, S. H. Glarum, J. H. Marshall, R. Hull, and J. M. Bonar, *Chemistry of Materials*, **1**, 548 (1989).
8. R. J. Cava, B. Batlogg, J. J. Krajewski, R. Farrow, L. W. Rupp, A. E. White, K. Short, W. F. Peck, Jr., and T. Kometani, *Nature*, **332**, 814 (1988).
9. R. J. Cava, B. Batlogg, G. P. Espinosa, A. P. Ramirez, J. J. Krajewski, W. F. Peck, Jr., L. W. Rupp, and A. S. Copper, *Nature*, **339**, 291 (1989).

38. SYNTHESIS OF $Tl_2Ba_2Ca_{n-1}Cu_nO_{2n+4}$ ($n = 1, 2, 3$), (Tl, M)$Sr_2CuO_{5\pm\delta}$ (M = Bi, Pb), AND (Tl, Pb)$Sr_2CaCu_2O_{7\pm\delta}$

Submitted by M. GREENBLATT,* L. E. H. McMILLS,* S. LI,* K. V. RAMANUJACHARY,* M. H. PAN,* and Z. ZHANG*
Checked by THOMAS E. SUTTO,† JIAN-MING ZHU,† and BRUCE, A. AVERILL†

The thallium-based superconductors with the general formula $Tl_mA_2Ca_{n-1}Cu_nO_{2n+m+2}$, where $m = 1$ or 2; $n = 1$–5; A = Ba,Sr, are of interest because of their high superconducting temperatures. Unlike their rare-earth-based relatives, the Tl-based copper oxides are thermally unstable phases and can be difficult to prepare in the pure phase.

Sheng and Hermann were the first to synthesize the thallium-based compounds.[1] Their method consisted of heating a Ba—Ca—Cu—O precursor oxide with Tl_2O_3 powder at ~ 880–910° for approximately 3–5 min in flowing oxygen. However, the major disadvantage of this method is that the reaction temperatures needed to produce the pure Tl phase can lead to Tl deficiency due to the volatile nature of thallium. The closed system, first used by researchers at DuPont, removes this problem of excess thallium loss.[2] In this method, the starting materials are sealed in a gold tube and heated at ~900° for a period of 0.5–3 h. Both pure powder and single crystals can be prepared since longer reaction times can be used under closed reaction conditions, however, gold tubes are expensive. Sealed fused-silica tubes have also been used for the preparation of these compounds.[3,4] The disadvantages of this method are that the tubes can be attacked by thallium oxide vapor, resulting in a deviation from desired stoichiometry, and there is also the possibility of explosion at high temperatures due to the high vapor pressure of thallium oxide.

* Department of Chemistry, Rutgers, The State University of New Jersey, New Brunswick, NJ 08903.
† Department of Chemistry, University of Virginia, Charlottesville, VA 22901.

■ **Caution.** *Thallium and its compounds are extremely toxic and must be handled with care.*[5] *Tl_2O_3, one of the starting materials used in the preparation of these materials, is highly volatile even below its melting point of 717 ± 5°, therefore, all preparations should be done in a hood. Gloves and an appropriate breathing mask should be used as the dust can be inhaled and thallium, being water-soluble, can penetrate unbroken skin.*

The following preparation conditions have been used successfully in our laboratory. In all cases unless otherwise specified, 1–2 g samples were fired in high-density alumina boats.

A. $Tl_2Ba_2CuO_{6\pm\delta}$ (2201)

$$Tl_2O_3 + 2BaO_2 + CuO \rightarrow Tl_2Ba_2CuO_6 + O_2$$

Procedure

Stoichiometric amounts of Tl_2O_3 (Morton Thiokol, Inc. 99.99%), BaO_2 (J. T. Baker Co., ~ 99.9%), and CuO (Aldrich Chemical Co., 99.5%) are intimately mixed in an agate mortar and heated at 600° for 6 h. The product is then reground, pelletized, and subjected to various heat treatments, the results of which are shown in Table I.

TABLE I. $Tl_2Ba_2CuO_{6\pm\delta}$ Preparation and Physical Properties[6]

Preparation Conditions	Parameters (Å)	T_c^{onset} (K)	T_c^{zero} (K)
860°, 20 min, quenched in air	a = 5.459 (2), b = 5.474 (1), c = 23.290 (5)	84	70
860°, 20 min, furnace cooled in air	a = 5.454 (1), b = 5.503 (2), c = 23.24 (1)	20	—
860°, 20 min, furnace cooled in air, Ar annealed 450°, 1.5 h	a = 5.462 (4), b = 5.474 (20), c = 23.26 (1)	56	48
Pellet in gold foil, sealed in fused-silica tube; 860°, 20 min; furnace-cooled	a = 5.46 (2), b = 5.51 (4), c = 23.241 (3)	—	—

Properties

The structure of this material ranges from pseudotetragonal to orthorhombic (space group *Abma*) as seen in Table I. From X-ray diffraction, essentially single-phase samples are obtained under all preparation conditions indicated in Table I, however, small amounts (~1–2%) of $BaCuO_2$ impurity are often present. Increasing either the temperature or heating period will result in a significant amount of this impurity phase. Deviations from tetragonal symmetry are smaller for quenched samples (lowest oxygen content) than furnace-cooled ones. Samples prepared in fused-silica tubes (highest oxygen content) show greatest splitting of the (020) and (200) reflections. Therefore the oxygen content has a large influence on the structural distortions of 2201. Post-annealing the 20 K superconducting and nonsuperconducting samples in argon atmosphere at 400–450° for 1–2 h results in superconductivity at ~50 K. A reversible structural phase transition from pseudotetragonal to orthorhombic symmetry accompanied by oxygen uptake/loss occurs in controlled oxygen atmosphere at ~300°.

B. $Tl_2Ba_2CaCuO_2O_{8\pm\delta}$ (2212) and $Tl_2Ba_2Ca_2Cu_3O_{10\pm\delta}$ (2223)

$$Tl_2O_3 + 2BaO_2 + CaO_2 + 2CuO \rightarrow Tl_2Ba_2CaCu_2O_8 + 1.5O_2$$

$$Tl_2O_3 + 2BaO_2 + 2CaO_2 + 3CuO \rightarrow Tl_2Ba_2Ca_2Cu_3O_{10} + 2O_2$$

Procedure

Stoichiometric amounts of Tl_2O_3, BaO_2, and CaO_2 (Pfaltz and Bauer, 99%) are ground in an agate mortar, pressed into pellets, wrapped with very thin (0.025-mm-thick) silver or gold foil, and sintered in air. The samples are then quenched to room temperature.

Properties

In the case of $Tl_2Ba_2CaCu_2O_{8\pm\delta}$ preparations, either silver or gold foil can be used to protect against severe thallium loss. Single-phase materials of 2212 can be prepared in the temperature range 860–870° for a period of 30–90 min. The best T_c values (T_c onset = 106–111 K) are obtained when the sample is sintered at 870° for approximately 90 min. A study of the effect of sintering temperature and reaction time showed that T_c remains relatively constant with increasing temperature. The preparation of $Tl_2Ba_2Ca_2Cu_3O_{10\pm\delta}$ is not as straightforward as that of 2212. In general, the 2212 phase is the most common impurity found in multiphase 2223. Often, the 2212 phase will be the dominant phase in the material when silver foil is used or if the seal fails during the sintering step. Therefore it is necessary to use gold foil to wrap the pellet as the 2223 material tends to be more reactive with silver foil than 2212. The optimum conditions to prepare single-phase 2223 are to sinter the sample at 860–870° for 50–60 min. If lower sintering temperatures are used to prepare 2223 samples, single-phase samples can be obtained, however, the T_c values will be lower (T_c onset = 102–106 K) than that reported in the literature. Using higher sintering temperatures will often result in small amounts of 2212 contamination, however, the T_c values are higher (T_c onset = 112–118 K). Annealing samples in argon or air does not appear to have any effect on the T_c value. Annealing 2223 in oxygen for 4 h at 450° only slightly increases the T_c value (± 2 K). Both 2212 and 2223 are tetragonal with space group *I4/mmm.* The lattice parameters for 2212 are $a = 3.8850(6)$ Å and $c = 29.318(4)$ Å, whereas for 2223 they are $a = 3.8503(6)$ Å and $c = 35.88(3)$ Å[2].

C. $(Tl, Bi)Sr_2CuO_{5\pm\delta}$

Procedure

Stochiometric amounts of Tl_2O_3, Bi_2O_3 (Johnson Matthey, 99.99%), SrO (Morton Thiokol, 99.5%), and CuO are ground in an agate mortar and are preheated to 650–700° for 1 h. The mixture is then reground, pelletized, sealed in a silver bag (a self-sealing bag is made by folding over a piece of 0.025-mm silver foil and crimping with pliers), and sintered for 12 h at 850° with several intermediate grindings. As the value of x is increased, it is necessary to sinter for longer periods of time.

Properties

Pure-phase material $Tl_{1-x}Bi_xSr_2CuO_{5\pm\delta}$ can be made in the solid solution range $0.20 \leq x \leq 0.50$.[7] $Tl_{1-x}Bi_xSr_2CuO_{5\pm\delta}$ is tetragonal with space group *P4/mmm.* Superconductivity with a T_c^{onset} at 45 K has been found in samples of $0.2 < x < 0.3$ that are quenched from 850° into liquid nitrogen and then annealed at 400° for 1 h in air.[8] The synthetic conditions and properties of $Tl_{1-x}Bi_xSr_2CuO_{5\pm\delta}$ are summarized in Table II.

D. $(Tl, Bi)Sr_2CaCu_2O_{7\pm\delta}$

Procedure

The precursor $Sr_4Tl_2O_7$ is prepared from a stoichiometric mixture of $Sr(NO_3)_2$ (Aldirch, 99%) and Tl_2O_3 that is sintered in an open crucible at

TABLE II. Synthetic Conditions and Properties of $Tl_{1-x}Bi_xSr_2CuO_{5\pm\delta}$

No.	x value	Synthetic Conditions	Cell Parameters (Å)	Properties
1	0.2	850°, 4 h, Ag bag[a]	$a = 3.729(1)$, $c = 9.043(1)$	Semiconducting
2	0.2	850°, 4 h, Ag bag[a] 400°, 1 h, air[b]	$a = 3.735(1)$, $c = 9.017(1)$	Superconducting $T_c^{onset} = 45$ K
3	0.2	850°, 4 h, Ag bag[a] Furnace-cool to 500°, 1.5 h, air[c]	$a = 3.736(1)$, $c = 9.014(1)$	Metallic
4	0.25	850°, 5 h, Ag bag[a]	$a = 3.732(1)$, $c = 9.037(1)$	Semiconducting
5	0.25	Sample (4) annealed in air at 400° for 1 h[b]	$a = 3.741(1)$, $c = 9.018(2)$	Superconducting $T_c^{onset} = 50$ K
6	0.25	850°, 5.5 h, Ag bag,[a] FC–500°-1.5 h, O_2[c]	$a = 3.742(1)$, $c = 9.013(2)$	Metallic
7	0.3	850°, 4 h, Ag bag[a]	$a = 3.740(1)$, $c = 9.040(2)$	Semiconducting
8	0.3	Sample (7) annealed in air at 400° for 1 h[b]	$a = 3.747(1)$, $c = 9.025(1)$	Superconducting $T_c^{onset} = 45$ K
9	0.3	850°, 4 h, Ag bag[a] Furnace-cool to 500°, 1.5 h, air[c]	$a = 3.749(1)$, $c = 9.016(1)$	Metallic
10	0.4	850°, 9 h, Ag bag[a]	$a = 3.743(1)$, $c = 9.063(2)$	Semiconducting
11	0.4	Sample (10) annealed in air at 400° for 3.5 h[b]	$a = 3.752(1)$, $c = 9.043(1)$	Metallic
12	0.5	850°, 13 h, Ag bag[a]	$a = 3.747(1)$, $c = 9.052(2)$	Semiconducting
13	0.5	Sample (12) annealed in air at 400° for 3.5 h[b]	$a = 3.754(1)$, $c = 9.039(1)$	Metallic

[a] The pelletized samples were sealed in a Ag bag in air and fired at 850° for the indicated time, followed by quenching into liquid nitrogen temperature.

[b] After quenching in N_2(l), the sample was removed from the Ag bag room temperature (RT), exposed to air and heated at the temperature and for the time indicated, and then quenched to room temperature in air.

[c] The sample was quenched in N_2(l) removed from the Ag bag (RT); sample then heated to 850° in air or oxygen as indicated and cooled from 850° to the temperature and for the time indicated. FC is furnace cooled at cooling rate of ~5 min.

600° for 1–2 h[9]. Then stoichiometric amounts (Tl:Bi ratio 1:1) of $Sr_4Tl_2O_7$, Bi_2O_3, CaO_2 and CuO are mixed and preheated at 800° for 30 min. The powder samples are then ground, pressed into pellets, sealed in a silver bag and sintered at 850° in air for 2 h. Single-phase samples can also be prepared by wrapping the pellet in gold foil and sealing it in a fused-silica tube in air. If a higher sintering temperature such as 900° or a time period of >3 h is used, severe decomposition of the phase will occur. If a sintering temperature of

$<850°$ is used, the reaction will proceed very slowly and multiphase samples will result. $Sr_4Tl_2O_7$ is used as the thallium source as it is less volatile than Tl_2O_3 and therefore minimizes thallium loss.

Properties

The resulting Tl:Bi ratio in the product is close to one but is not known exactly because of the volatile nature of thallium at high reaction temperatures. Singnificant variation of the Tl:Bi ratio from 1.0 results in the formation of impurity phases. The $(Tl, Bi)Sr_2CaCu_2O_{7\pm\delta}$ phase can be indexed with a tetragonal unit cell ($P4/mmm$) of $a = 3.796(1)$Å and $c = 12.113(2)$Å[10]. The T_c^{onset} is ~100 K and T_c^{zero} is ~90 K.

E. $(Tl, Pb)Sr_2CuO_{5\pm\delta}$

Procedure

Stoichiometric amounts of Tl_2O_3, PbO (Fisher), SrO, and CuO are combined in an agate mortar and preheated at 600° for 2 h in flowing oxygen with two intermediate grindings. The powder samples are then ground, pressed into pellets, sealed in a silver bag or fused-silica tube, and subjected to various heat treatments. Preparations of $(Tl_{1-x}Pb_x)Sr_2CuO_{5\pm\delta}$ with $x = 0$–0.3 were

TABLE III. $Tl_{0.5}Pb_{0.5}Sr_2CuO_{5\pm\delta}$ Preparation and Physical Properties

Preparation Conditions	Physical Properties	Cell Parameters (Å)
900°, 1.5 h, Ag bag; quench in liquid N_2	Semiconducting	$a = 3.726(1)$, $c = 9.055(3)$
850°, 1.5 h, Ag bag; quench to room temp.	Superconducting $T_c^{onset} = 60$ K	$a = 3.736(1)$, $c = 9.022(3)$
850°, 1.5 h, Ag bag; quench to room temp.	Superconducting $T_c^{onset} = 40$–60 K	$a = 3.730(9)$, $c = 9.032(3)$
850°, 1 h, Ag bag; then wrap in Au foil, seal in fused-silica tube (in O_2) 850°, 15 min; furnace-cool to 500° for 9 hr; quench to room temp.	Metallic	$a = 3.737(1)$, $c = 9.009(1)$
850°, 1.5 h in air (on Ag foil), quench to room temp.	Metallic	$a = 3.734(1)$, $c = 9.014(1)$

multiphase with $Sr_4Tl_2O_7$ as the major impurity. Table II summarizes the preparation conditions and properties for single-phase $Tl_{0.5}Pb_{0.5}Sr_2CuO_{5\pm\delta}$. When heating periods of 30 min are used, $Sr_4Tl_2O_7$ is the major impurity. If the heating periods longer than 90 min are used, the 1201 X-ray diffraction peaks become broadened and impurity peaks are present.

Properties

There are varying reports on the properties of single-phase $Tl_{0.5}Pb_{0.5}Sr_2CuO_{5\pm\delta}$.[11-14] The properties of the sample are highly dependent on the preparation conditions as seen in Table III. The observed reflections for $Tl_{0.5}Pb_{0.5}Sr_2CuO_{5\pm\delta}$ can be indexed with a tetragonal unit cell (*P*4/*mmm*) and are preparation-dependent.

Acknowledgment

This work was supported in part by the office of Naval Research and the Nation Science Foundation-Solid State Chemistry Grant DMR-87-14072.

References

1. Z. Z. Sheng and A. M. Hermann, *Nature*, **332**, 55, 138 (1988).
2. (a) C. C. Torardi, M. A. Subramanian, J. C. Calabrese, J. Gopalakrishnan, K. J. Morrissey, T. R. Askew, R. B. Flippen, U. Chowdhry, and A. W. Sleight, *Science*, **240**, 631 (1988); (b) M. A. Subramanian, J. C. Calabrese, D. D. Torardi, J. Gopalakrishnan, T. R. Askew, R. B. Flippen, K. J. Morrissey, U. Chowdhry, and A. W. Sleight, *Nature*, **332**, 420 (1988).
3. For example see S. S. P. Parkin, V. Y. Lee, A. I. Nazzal, R. Savoy, T. C. Huang, G. Gorman, and R. Beyers, *Phys. Rev. B*, **38**, 6531 (1988).
4. For example see M. Hervieu, A. Maignan, C. Martin, C. Michel, J. Provost, and B. Raveau, *J. Solid State Chem.*, **75**, 212 (1988).
5. G. E. Myers, *Superconductor Industry*, 13 (1988).
6. K. V. Ramanujachary, S. Li, and M. Greenblatt, *Physica C*, **165**, 377 (1990).
7. M. Greenblatt, S. Li, L. E. H. McMills, and K. V. Ramanujachary, in *Studies of High Temperature Superconductors*, A. V. Narlikar (ed.), Nova Science Publishers, New York, 1990, 143–165.
8. M.-H. Pan and M. Greenblatt, *Physica C*, **184**, 235 (1991).
9. R. V. von Schenck and H. Muller-Buschbaum, *Zei Anorg. Allg. Chem.*, **396**, 113 (1973).
10. S. Li and M. Greenblatt, *Physica C*, **157**, 365 (1989).
11. C. Martin, D. Bourgault, C. Michel, J. Provost, M. Hervieu, and B. Raveau, *Eur. J. Solid State Inorg. Chem.*, **26**, 1 (1989).
12. J. B. Shi, M. J. Shieh, T. Y. Lin, and H. C. Ku, *Physica C*, **162–164**, 721 (1989).
13. J. C. Barry, Z. Iqbal, B. L. Ramakrishna, R. Sharma, H. Eckhardt, and F. Reidinger, *J. Appl. Phys.*, **65**, 5207 (1989).
14. V. Manivannan, A. K. Ganguli, G. N. Subbanna, and C. N. R. Rao, *Solid State Commun.*, **74**, 87 (1990).

39. PREPARATION OF $Ba_4BiPb_2TlO_{12}$, A 10.7 K SUPERCONDUCTOR

$$Tl_2O_3 + Bi_2O_3 + 2PbO_2 + 8BaO \rightarrow Ba_4BiPb_2TlO_{12}$$

Submitted by THOMAS E. SUTTO* and BRUCE A. AVERILL*
Checked by M.-H. PAN† and Z. TEWELDEMEDHIN†

The discovery of high-temperature superconductivity in the thallium and bismuth cuprates,[1] as well as reports of high-temperature superconductivity in $Ba(Bi, Pb)O_3$[2] and $(Ba, K)BiO_3$,[3] prompted a reinvestigation of the cubic perovskite system, $BaMO_3$, where M = Bi, Pb, Tl. Superconductivity in $(Ba, K)BiO_3$ ($T_c = 32$ K)[3] and in $Ba(Bi, Pb)O)_3$ ($T_c = 13.5$ K)[2] is attributed to interaction of the 6*s* energy level of the metal species with the 2*p* level of the oxygen.[2] Therefore, Tl substitution into the $BaMO_3$ system was attempted since Tl possesses similar chemistry to the other 6*s* valence metals, Bi and Pb.

Because the known thallium cuprate systems are two-dimensional materials,[4] only stoichiometric combinations that would give rise to a layered analog of a cubic perovskite structure containing three different metals were examined, corresponding to the empirical formula $Ba_xBi_aPb_bTl_cO_{3x}$, in which a, b, and c are whole numbers with $x = a + b + c = 4$. Thus, the following three syntheses were attempted: $Ba_4Bi_2PbTlO_{12}$, $Ba_4BiPbTl_2O_{12}$, and $Ba_4BiPb_2TlO_{12}$. The last of these materials proved to be a new superconductor with a T_c of 10.5 K,[5] whose synthesis is described in detail herein. An apparently identical phase has also been prepared independently in a study of the effect of doping in $BaBi_{1-x}Tl_xO_3$.[6]

Procedure

Stoichiometric amounts of 99.999% Bi_2O_3 (233 mg, 0.50 mmol), 99% PbO_2 (478 mg, 2.0 mmol), and 99.999% BaO (614 mg, 4.0 mmol), and a 33% excess of 98% Tl_2O_3 (304 mg, 0.66 mmol) are weighed out, loaded into a plexiglass container with a plexiglass grinding ball, and pulverized in a Wig-L Bug shaker with a stainless-steel vial for 4 min. (All chemicals were obtained from Aldrich Chemical Co.)

■ **Caution.** *Tl_2O_3 is highly toxic and is volatile at temperatures well below its melting point of ~720°; gloves and a filter mask should be used in its*

* Department of Chemistry, University of Virginia, McCormick Road, Charlottesville, VA 22903.
† Department of Chemistry, Rutgers University, Piscataway, NJ 08855-0939.

handling, and the reaction should be carried out in a fume hood. Glassware or other equipment that has contacted thallium should be carefully wiped, rinsed, and cleaned in a fume hood, and the waste disposed of in an appropriate fashion.

The mixture is then loaded into a sintered aluminum oxide crucible (10 mL, 20 × 12 mm) and placed in a 60-cm fused-silica tube with an inner diameter of 28 mm. The tube is inserted into a 1-in. tube furnace and rapidly heated (20 min) to 850° under flowing argon (2–5 cm^3/min). After one hour, the material is quenched in flowing argon by removal of the tube from the oven. The prereacted black material is then ground with a mortar and pestle for 15 min and pressed into a pellet using a $\frac{1}{2}$-in. pellet press at ~2 kbar of pressure. The pellet is then placed in a sintered aluminum oxide boat (100 × 20 mm) that is inserted into a fused-silica tube which is then placed in a tube furnace at room temperature. The furnace, equipped with a calibrated thermocouple, is heated rapidly (~15 min) to 800° under flowing argon and held at 800° for approximately one minute. The temperature of the furnace is then allowed to drop to 450° over 10–15 min, the argon flow is replaced by oxygen at a flow rate of 2–5 cm^3/min, and the temperature is lowered to 425° and maintained there for 1.25 h. The furnace is then turned off, and the oven is allowed to cool to room temperature.

Properties

The resulting blue-black pellet is characterized by X-ray powder diffraction.[7] The indexed data are listed below:

h	k	l	Intensity	d_{obs}
1	0	1	2	4.9590
0	0	2	6	4.2742
0	2	0	100	3.0238
2	2	0	4	2.1578
0	0	4	10	2.1431
1	1	4	8	1.9151
2	0	4	31	1.7509
3	1	2	3	1.7506
0	2	4	27	1.7473
1	3	2	27	1.7473
2	2	4	6	1.5143
0	4	0	13	1.5135
3	3	0	3	1.4298
1	1	6	11	1.3521
4	0	4	4	1.2368
0	4	4	4	1.2360

These data are consistent with an orthorhombic unit cell with $a = 6.0769(15)$ Å, $b = 6.0588(17)$ Å, and $c = 8.5580(17)$ Å. Chemical analysis gives 12.76% Tl, versus 13.05% Tl calculated for a stoichiometry of $Ba_4BiPb_2TlO_{12}$. Superconductivity is observed magnetically in a field of 5 G using an S.H.E. Squid. A Meissner signal of 90%, indicating the single phase nature of the product, is characteristically observed with an onset temperature of 10.7 K, followed by a very sharp increase in the diamagnetic signal. Resistivity data indicate a remarkably narrow transition width of less than 0.5 K, with zero resistance observed at 9.8 K with an onset of 10.2 K.

References

1. R. M. Hazen, L. W. Finger, R. J. Angel, C. Prewitt, N. L. Ross, C. E. Hadidiacos, P. J. Heaney, D. R. Veblen, and Z. Z. Sheng, *Phys. Rev. Lett.*, **60**, 1657 (1988).
2. A. W. Sleight, *Sol. State Commun.*, **17**, 27 (1975).
3. R. J. Cava, B. Batlogg, J. J. Krajewski, R. Farow, L. W. Ripp, Jr, A. E. White, K. Short, W. F. Peck, and T. Komentani, *Nature*, **332**, 314 (1988).
4. Z. Z. Sheng and A. M. Hermann, *Nature*, **332**, 138 (1988).
5. T. E. Sutto and B. A. Averill, *Chem. Mater.*, **4**, 1092 (1992).
6. S. Li, K. V. Ramanujachary, and M. Greenblatt; *Physica C*, **166**, 535 (1990).
7. Data were collected on a Scintag Powder Diffractometer using Cu $K_{\alpha 1}$ radiation (1.59059 Å). Reflections were indexed using a Fortran 4 X-ray fitting program.[8]
8. C. M. Clark, D. K. Smith, and G. Ohnson, Fortran 4 X-Ray fitting Program (1973). Unit-cell dimensions and standard deviations are results from three individual preparations of the material.

40. BARIUM YTTRIUM COPPER OXIDE CRYSTALS

$$2BaCuO_2 + 0.5Y_2O_3 + CuO + \left(0.25 - \frac{x}{2}\right)O_2 \rightarrow Ba_2YCu_3O_{7-x}$$

Submitted by L. F. SCHNEEMEYER,* J. V. WASZCZAK,* and R. B. van DOVER*
Checked by T. A. VANDERAH†

The high-temperature superconductors are a remarkable set of materials that present us with fascinating scientific and technological challenges. The cuprate $Ba_2YCu_3O_7$, the first material with a $T_c > 77$ K, has a tripled perovskite structure with Cu occupying the small perovskite *B* site while Ba and Y are ordered on the perovskite *A* site. Bulk crystals are key to

* AT&T Bell Laboratories, Murray Hill, NJ 07974.
† NIST, Gaithersburg, MD 20899.

unraveling the relationship between the structure and bonding and properties, especially the anisotropy in conductivity, critical current density (J_c), and critical magnetic fields. Additionally, they provide insights into the role of grain boundaries in ceramics, the nature of flux pinning sites, and the nature and importance of native defects.

A number of laboratories[1–4] have reported similar approaches to the growth of $Ba_2YCu_3O_7$ from $BaCuO_2$–CuO melts in a partially melting region of the BaO–$YO_{1.5}$–CuO phase diagram as shown in Fig. 1.[5] Here we describe a reliable approach for the growth of platelet crystals with surface areas on the order of a few square millimeters and thicknesses from tens of micrometers to a few tenths of a millimeter. A uniform oxygen content is established by a postgrowth anneal under controlled conditions of temperature and oxygen partial pressure. Crystals are characterized by X-ray diffraction and by measurements of the temperature dependence of the resistivity and ac and dc susceptibilities.

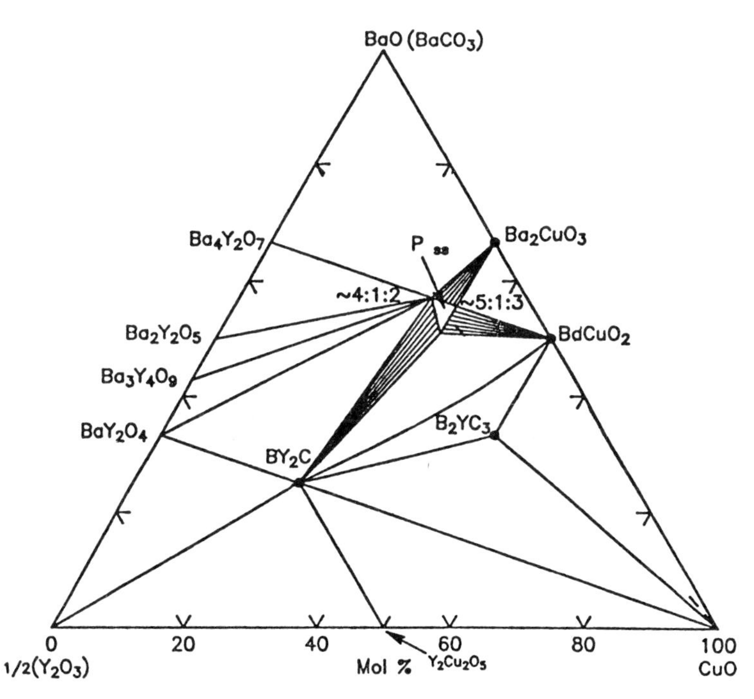

Figure 1. BaO–CuO–$YO_{1.5}$ ternary phase diagram at 975–1000°[5] with the regio of partial melting marked by the dotted line.

Procedure

Commercial high-purity CuO (typically 99.999%) (10 g, 0.1257 mol), Y_2O_3 (typically 99.999%) (1.5980 g, 7.077×10^{-3} mol), and $BaCO_3$ (typically 99.99%) (11.0247 g, 0.05586 mol), which corresponds to a metals ratio 4Ba:1Y:9Cu, are ground together and placed in half a 40 mL cylindrical ZrO_2 crucible (sliced lengthwise) as shown in Fig. 2. The crucible is tilted slightly with the closed end of the crucible down. A space approximately $\frac{3}{16}$ in. wide is left between the crucible end and the powder to allow the melt to collect and provide an area where much of the crystal growth will take place. The crucible is then placed in a temperature-controlled box furnace oriented as shown in the figure. The atmosphere is air. The furnace is heated at 200°/h to 1000°, soaked at 1000° for 3 h, then cooled from 1000 to 890° (the solidification temperature for the flux) at cooling rates as high as 15°/h. A cooling rate of 10°/h is recommended here. The cooling rate controls the thickness of the product crystals. Cooling rates less than 1°/h are needed to produce crystals with thickness approaching 1 mm.[6] However, thermal fluctuations in most furnaces are of order 1°, so $\leqslant 100$ μm is typical of the crystal thicknesses obtained. Finally, the furnace is turned off and allowed to cool to room temperature. The crystals absorb oxygen from the air as they cool and therefore, the harvested crystals are orthorhombic, but oxygen deficient, and are twinned.

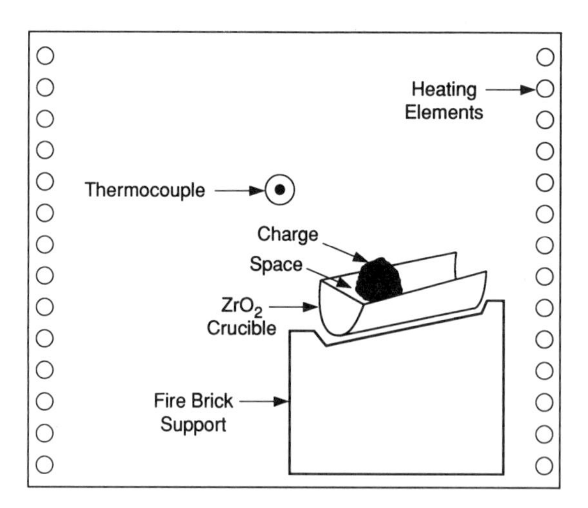

Figure 2. Schematic drawing of the furnace configuration for crystal growth.

We exploit the inherent gradient in a box furnace to render the product crystals largely flux-free. Box furnaces typically have temperature gradients that can be as large as tens of degrees. The crucible end with the growth space is oriented towards the center of the box furnace, which will be the warmer region. Because the $BaCuO_2/CuO$ based flux exhibits a marked creeping tendency, nearly flux-free crystals can be obtained as a result of preferential wetting of the crucible surface. During the last stage of the cooling, the flux creeps toward the cooler side of the crucible, which is towards the side and front of the furnace, thus away from the area where the crystals have grown. The furnace that we use has a 10-in.^{3} volume and the gradients are large. Gradients vary from furnace to furnace. It may be necessary to experiment with the placement of the crucible in the furnace in order to optimize both the growth of $Ba_2YCu_3O_7$ crystals and the advantageous flux creep.

The crucible can act as a source of impurities that can contaminate the crystals, a common problem in flux growth. The choice of a container for the growth of $Ba_2YCu_3O_7$ is complicated by the ability of this phase to accommodate a wide variety of ions as dopants and the reactivity of the $BaCuO_2$–CuO melts toward many common crucible materials. High-density, high-purity ZrO_2 crucibles (available from Deguissit, or McDanel, etc.) are recommended since incorporation of Zr^{4+} is low (<0.1 at.%). High density ThO_2 crucibles with the even larger Th^{4+} are also useful,[2] but Th is radioactive and must be handled and disposed of properly. Gold crucibles are often used and produce crystals with 1–7 at.% Au incorporated by substitution on the Cu chain site.[7] Substitution at this level has been shown to increase the T_c by up to 2 K in both polycrystals and single crystals. Relatively inexpensive and readily available high-density, high-purity alumina (Al_2O_3) crucibles are also popular. However, the resulting crystals have a typical composition $Ba_2YCu_{2.9}Al_{0.1}O_7$ and lowered T_c, ≈ 85 K, because of Al substitution on chain copper sites.[8] We note, however, that the use of alumina crucibles has been reported to favor the growth of large, thick $Ba_2YCu_3O_7$ crystals.[9] In general, the incorporation of impurities is undesirable because they add an ill-determined variation on an already complex material, $Ba_2YCu_3O_7$.

The oxygen content of $Ba_2YCuO_3O_{7-x}$ is a function of the temperature and oxygen partial pressure during annealing as shown in Fig. 3.[10] Inside the crystal, oxygen moves by a solid-state diffusion process. The diffusion constant is temperature-dependent[11,12] but is relatively slow, $D_{ab} = 3 \times 10^{-12}$ cm^2/s at 400°, and diffusion is mainly in the a–b plane. Thus, long anneals are required to obtain homogeneous, fully oxygenated crystals. For millimeter-sized crystals, as-grown crystals should be annealed at 400–450° for >4 weeks. Crystals can be sealed under oxygen in a clean, degassed

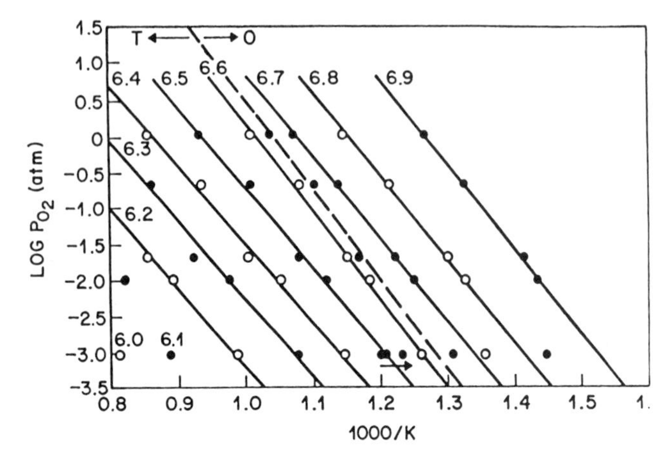

Figure 3. Oxygen pressure as a function of temperature for various values of x in $Ba_2YCu_3O_x$. The tetragonal (T)-orthorhombic (O) transition at $x \approx 6.63$ is shown.

fused-silica tube, then held at temperature in a tube furnace. Silver foil or clean, dry Al_2O_3 or ZrO_2 or $Ba_2YCu_3O_7$ ceramic should be used to insulate the crystals from the silica.

All members of the lanthanoid series $Ba_2RECu_3O_7$ with the exceptions of Ce, Tb, and Sc form tripled perovskites that are isomorphous with $Ba_2YCu_3O_7$. Crystals of these phases can also be grown from $BaCuO_2$/CuO-based melts.

Properties

Crystals of $Ba_2YCu_3O_7$ are shiny metallic gray and grow with a thin platelet shape. Fully oxygenated crystals are orthorhombic, space group *Pmmm*, with lattice constants $a = 3.8265(4)$ Å, $b = 3.8833(2)$ Å, and $c = 11.6813(10)$ Å at 293 K.[13] The X-ray powder patterns is given in Table I.

To measure the temperature dependence of the resistivity, it is necessary to make electrical contact to the crystal. The best contacts are made using thin film pads of Ag or Au deposited onto the clean crystal surface shielded with an appropriate mask. Contact to these pads may then be made using Ag-filled paste or epoxy. Figure 4 illustrates the a–b resistivity for a fully oxygenated, twinned, $Ba_2YCu_3O_7$ crystal. The resistivity decreases nearly linearly with temperature to about 120 K, at which point superconducting fluctuations begin to contribute significantly to the conductivity. The resistivity continues to decrease to the transition temperature, T_c, where the resistivity abruptly drops to zero.

TABLE I. Powder X-ray Diffraction Pattern for $Ba_2YCu_3O_7$[15]

h	k	l	d_{obs} (Å)	I/I_0 (%)
0	0	2	5.844	2
0	0	3	3.893	11
1	0	0	3.822	3
0	1	2	3.235	3
1	0	2	3.198	5
0	1	3	2.750	60
1	0	3	2.726	100
1	1	0		
1	1	1	2.653	2
1	1	2	2.469	3
0	0	5	2.336	11
1	0	4	2.321	3
1	1	3	2.232	13
0	2	0	1.946	23
0	0	6		
2	0	0	1.911	10
1	1	5	1.775	3
0	1	6	1.741	2
0	2	3		
1	0	6	1.734	2
1	2	0		
2	0	3	1.716	2
2	1	0		
1	2	1		
1	2	2	1.662	1
1	2	3	1.584	24
1	1	6		
2	1	3	1.569	11

Several features of the resistivity curve are measures of crystal quality. The room-temperature resistivity, ρ_{ab}, should be low, of order 150–250 $\mu\Omega\cdot$cm. Higher values typically indicate incomplete oxygenation or microcracking in the crystal. The resistivity between ≈ 125 and 300 K can be extrapolated to $T = 0$ to infer $\rho(T = 0)$. This value should be $\leqslant 5\ \mu\Omega\cdot$cm in relatively clean crystals. An excess of $Ba_2YCu_4O_8$ intergrowths may lead to a value for $\rho(T = 0)$ that is negative, however. The transition width is also indicative. Very pure, fully oxygenated crystals should have transition widths <0.25 K

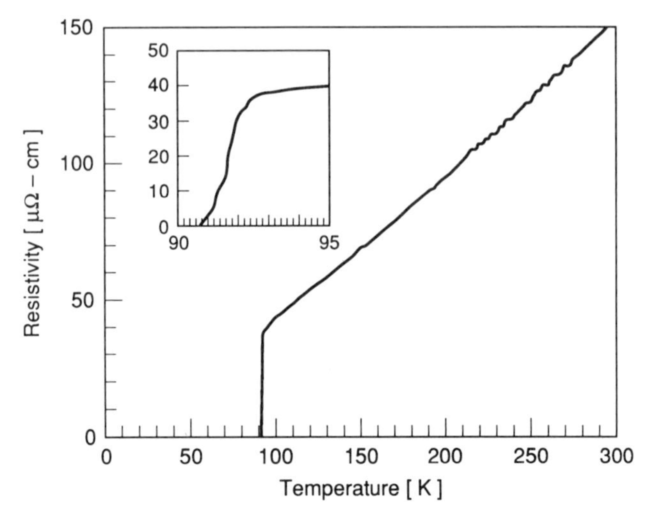

Figure 4. The temperature dependence of the resistivity of a fully oxygenated $Ba_2YCu_3O_7$ crystal. The inset shows the superconducting transition in detail.

for the 10–90% portion of the transition. The observation of a "foot" at the bottom of the transition can indicate the presence of macroscopic inhomogeneities, such as poorly oxygenated regions or microcracks. The T_c values of undoped crystals are found to vary from ≈ 89 to 93 K, where the variation is attributed to ill-understood details of ordering on the oxygen chain sublattice.

A second commonly used characterization is measurement of the dc magnetic susceptibility. Since an ideal superconductor completely excludes magnetic flux ($B = 0$), we have $0 = B = H_a + \pi M$, so $\chi_v = M/H_a = -1/4\pi$, where H_a is the applied field [in oersteds (Oe)] and M is the measured magnetization ($\text{emu} \cdot \text{cm}^{-3}$). However, in practice, we must also consider the effects of the demagnetizing field.[14] Then, $\chi_v = M(1 - N_d/4\pi)/H$, where N_d is the demagnetization coefficient, which depends on geometry. For example, for a sphere $N_d/4\pi = \frac{1}{3}$, while for a platelet with the field applied normal to the plane, $N_d/4\pi \rightarrow 1$.

The result for a typical high quality crystal is shown in Fig. 5, where the data were taken with the field parallel to the c axis, that is, normal to the plane of the platelet. Data were taken by cooling in zero field, then applying a field of 10 Oe and measuring the susceptibility on warming (denoted zfc, for zero-field-cooled), and also by cooling in an applied field (fc). The curves are often referred to as *exclusion* and *expulsion*, respectively, or *shielding* and *Meissner curves*, respectively. The ratio between the low-temperature ($T \rightarrow 0$)

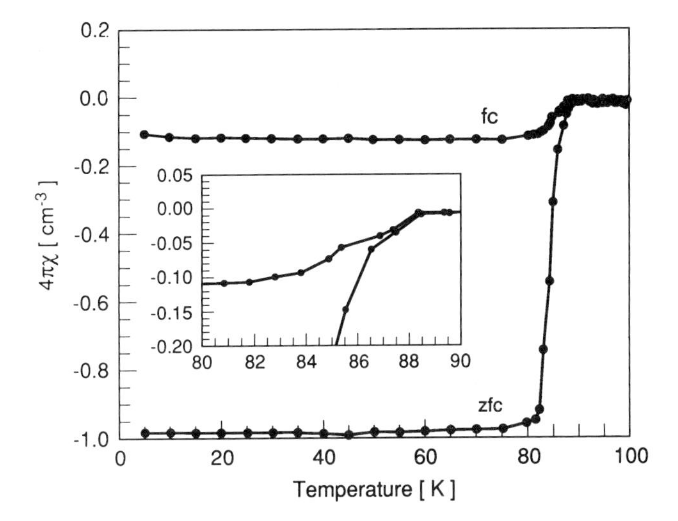

Figure 5. The temperature dependence of the dc susceptibility for a field-cooled and zero-field-cooled $Ba_2YCu_3O_7$ crystal. The region of the transition is detailed in the inset.

susceptibilities $\chi_v(\mathrm{fc})/\chi_v(\mathrm{zfc})$ is often called the *Meissner fraction* and is used as a rough indication of sample quality. However, under field-cooled conditions, it is difficult to avoid trapping flux that reduces the apparent susceptibility as seen in Fig. 5. Also, it is nearly impossible to distinguish between flux trapping and the possibility of an inhomogeneous superconductor. Thus we emphasize that the Meissner fraction is never definitive and at best is qualitative.

The issue of how to reliably and usefully characterize the quality of a single-crystal sample of $Ba_2YCu_3O_7$ is not fully resolved. Resistivity and susceptibility are relatively crude indicators. The suitability of any crystal for a given measurement and subsequent interpretation must be judged with attention to possible sample limitations.

References

1. D. L. Kaiser, F. Holtzberg, B. A. Scott, and T. R. McGuire, *Appl. Phys. Lett.*, **51**, 1040 (1987).
2. L. F. Schneemeyer, J. V. Waszczak, T. Siegrist, R. B. van Dover, L. W. Rupp, B. Batlogg, R. J. Cava, and D. W. Murphy, *Nature*, **332**, 601 (1987).
3. H. J. Scheel and F. Licci, *J. Crystal Growth*, **85**, 607 (1987).
4. T. A. Vanderah, C. K. Lowe-Ma, D. E. Bliss, M. W. Decker, M. S. Osofsky, E. F. Skelton, and M. M. Miller, *J. Crystal Growth*, **118**, 385 (1992), and references cited therein.
5. R. S. Roth, J. R. Dennis, and K. C. Davis, *Adv. Ceram. Mater.*, **2**, 303 (1987).

6. W. Sadowski and H. J. Scheel, *J. Less-Common Met.*, **150**, 219 (1989).
7. W. Wong-Ng, F. W. Gayle, D. L. Kaiser, S. F. Watkins, and F. R. Fronczek, *Phys. Rev. B*, **41**, 4220 (1990).
8. T. Siegrist, L. F. Schneemeyer, J. V. Waszczak, N. P. Singh, R. L. Opila, B. Batlogg, L. W. Rupp, and D. W. Murphy, *Phys. Rev. B*, **36**, 8365 (1987).
9. P. K. Gallagher, H. M. O'Bryan, S. A. Sunshine, and D. W. Murphy, *Mater. Res. Bull.*, **22**, 995 (1987).
10. Th. Wolf, W. Goldacker, B. Obst, G. Roth, and R. Flukiger, *J. Crystal Growth*, **96**, 1010 (1989).
11. S. I. Bredikhin, G. A. Emel'chenko, V. Sh. Shechtman, A. A. Zhokhov, S. Carter, R. J. Carter, J. A. Kilner and B. C. H. Steele, *Physica C*, **179**, 286 (1991).
12. K. Kishio, K. Suzuki, T. Hasegawa, T. Yamamoto, K. Kitazawa, and K. Fueki, J. *Solid State Chem.*, **82**, 192 (1989).
13. T. Siegrist, S. Sunshine, D. W. Murphy, R. J. Cava, and S. M. Zahurak, *Phys. Rev. B*, **35**, 7137 (1987).
14. T. van Duzer and C. W. Turner, *Principles of Superconductive Devices and Circuits*, Elsevier, New York, 1981.
15. R. J. Cava, B. Batlogg, R. B. van Dover, D. W. Murphy, S. Sunshine, T. Siegrist, J. P. Remeika, E. A. Reitman, S. Zahurak, and G. P. Espinosa, *Phys. Rev. Lett.*, **58**, 1676 (1987).

41. SYNTHESIS OF THE PEROVSKITE SERIES $LaCuO_{3-\delta}$

Submitted by JOSEPH F. BRINGLEY*† and BRUCE A. SCOTT
Checked by KENNETH R. POPPELMEIER,‡ GREGG A. TAYLOR‡ and BOGDAN DABROWSKI¶

The Cu^{3+} state is difficult to stabilize in the copper oxides. Under the conditions at which cuprates are typically synthesized (900–1100°, $10^{-5} - 1$ atm O_2), the 2 + valence state is by far the most stable and widely observed oxidation state of copper. The cuprates can be oxidized (or reduced) relative to Cu^{2+} by doping with an appropriate cation (e.g., $La_{1.85}Sr_{0.15}CuO_4$; $Cu^{2.15+}$), by annealing at a lower temperature in oxygen (oxidation), or at a higher temperature in a reducing atmosphere (reduction).[1] As a rule, higher oxidation states of copper become increasingly stable as the temperature is lowered and oxygen activity increased, while lower oxidation states are favored at higher temperatures and decreasing oxygen activity. For example, formally Cu^{3+} occurs in the alkali cuprates $MCuO_2$

* IBM Research Division, Thomas J. Watson Research Center, Yorktown Heights, NY 10598.

† Present address: Research Laboratories, Eastman Kodak Company, Rochester, NY 14650-2033.

‡ Department of Chemistry, Northwestern University, Evanston, IL 60208-3113.

¶ Department of Physics, Northern Illinois University, Dekalb, IL 60115.

(M = Na,K,Rb,Cs)[2,3] that are prepared in 1 atm O_2 at 450°, but are unstable above ~500°. For the vast majority of materials, however, kinetics are too slow at such low temperatures to provide an appreciable reaction rate. More elaborate procedures are required to stabilize Cu^{3+} at higher temperatures, such as the application of high pressure. As an example, 65 kbar of hydrostatic oxygen pressure is required to prepare rhombohedral $LaCuO_3$ (high-pressure form). The experimental procedure for preparing this material, however, requires a large press, and produces only very small amounts of material.[4]

An alternative approach to the stabilization of Cu^{3+} in oxides, presented here, involves the use of highly reactive oxide precursors coupled with their reaction in elevated oxygen pressure (≤ 1 kbar). A number of setups are commercially available to perform these syntheses, and only moderate temperatures and oxygen pressures are required. The present syntheses demonstrate the stabilization of Cu^{3+} in the perovskite series $LaCuO_{3-\delta}$.[5,6]

Materials and Methods

High-pressure oxygen reactions were performed in an externally heated René alloy vessel. An illustration of the high-pressure setup is shown in Fig. 1. The reaction chamber was pressurized at or below room temperature with high-purity oxygen. Reaction temperatures were measured at a shallow recess in the reactor wall at the sample position, and pressure measured with an uncalibrated gauge. High-purity deionized water was used in all procedures.

A. SYNTHESIS OF TETRAGONAL $LaCuO_{3-\delta}$ ($0.0 \leq \delta \leq 0.2$)

$$\tfrac{1}{2}La_2O_3 + CuO \rightarrow LaCuO_{3-\delta}$$

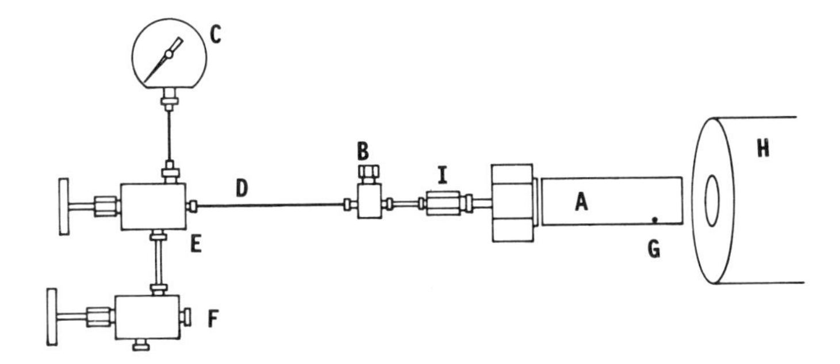

Figure 1. High-pressure oxygen apparatus: *A* Renè alloy reaction vessel; *B* safety rupture disk; *C* pressure gauge; *D* high-pressure capillary tubing; *E* two-way valve; *F* gas inlet; *G* thermocouple recess; *H* tube furnace; *I* plunger.

Procedure

Lanthanum oxide (.336 g, 0.001 mol) and CuO (0.164 g, 0.002 mol) are weighed and placed in a 250 mL Erlenmeyer flask containing a Teflon-coated magnetic stir bar. Deionized H_2O (50.0 mL) is added to the flask, and 10.0 mL of concentrated HNO_3 is then added slowly with stirring. The suspension is gently heated until all of the solid is dissolved, the solution diluted to 100.0 mL with H_2O and the contents cooled to 5° in an ice bath. A solution of freshly prepared 6*N* aqueous NaOH is then added at a rate of approximately 5 mL/min until a deep blue precipitate appears (pH $\simeq$ 10). Care should be taken not to add too much base, since CuO will precipitate from highly basic solutions. The hydroxide precipitate is then collected by vacuum filtration in air on a coarse glass frit and rinsed with 100 mL H_2O. The precipitate is then washed free of sodium ion by stirring it vigorously in 300 mL H_2O, allowing the precipitate to settle and recollecting it on a glass frit. This procedure is repeated twice to ensure that the precipitate is completely free of sodium ion. The precipitate should be exposed to air as little as possible because the hydroxides will slowly absorb CO_2. Exposure to organic solvents should also be avoided, because the oxycarbonates of lanthanum that form on heating the precipitate in an oxygen atmosphere inhibit the oxidation of the perovskite. After being allowed to partially dry in air, the precipitate is placed in a platinum crucible and heated at 100°/h to 600°, held there for 8 h, and cooled to 100°, all under flowing O_2. The resulting black solid is then ground in a mortar and pestle and pressed into a pellet of 1.25 cm diameter in a piston cylinder die. The pellet is wrapped in Pt foil and placed in the high-pressure oxygen reactor. The reactor is then sealed gas tight and cooled to −80° in a liquid nitrogen bath. The reactor is mounted horizontally in a tube furnace and quickly connected to an external oxygen source from which it is pressurized to 2000 psi (140 bar). The reactor is sealed only after the temperature has increased to −50° and is then allowed to warm to room temperature, at which point the pressure gradually increased to 200 bar. The reactor is then heated at 100°/h to 900° where the pressure reaches a maximum of 600 bar O_2. The reactor is held at this temperature for 2*d* and then cooled 100°/h to 100°. The contents are then removed from the furnace.

Properties

The product is analyzed by powder X-ray diffraction; all diffraction lines indexable in tetragonal symmetry, space group $P4/m$, $a = 3.8187(1)$ Å, $c = 3.9727(1)$ Å. The oxygen content is determined by thermogravimetric analysis (TGA) by virtue of the fact that $LaCuO_{3-\delta}$ is cleanly decomposed to

La_2CuO_4 and Cu_2O at 1000° in inert gas. The tetragonal phase of $LaCuO_{3-\delta}$ is stable over the range $0.0 \leq \delta \leq 0.2$. Samples for which $\delta = 0$ are obtained for $P_{O_2} \geq 500$ bar, but in general, oxygen stoichiometries are dependent on several factors, including the homogeneity and purity of the starting materials, and the diffusion of oxygen through powdered reactants at reaction temperature. For this reason, large, densely packed pellets should be avoided.

B. SYNTHESIS OF MONOCLINIC $LaCuO_{3-\delta}$ ($0.2 \leq \delta \leq 0.4$)

$$LaCuO_{3-\delta} \rightarrow LaCuO_{2.67} + \delta/2O_2$$

Procedure

The monoclinic phase of $LaCuO_{3-\delta}$ ($0.2 \leq \delta \leq 0.4$) may be prepared by performing the above synthesis at a lower oxygen pressure, ~100–300 bar. This method has the advantage that the full range of oxygen stoichiometries (i.e., $0.2 \leq \delta \leq 0.4$) can be obtained. However, the compound is more easily prepared by reducing the tetragonal perovskite ($0.0 \leq \delta \leq 0.2$) in 0–1 atm O_2. A sample of tetragonal $LaCuO_{3-\delta}$ contained in a Pt crucible is placed into a tube furnace. The material is heated slowly to 500° held there for 12 h, and then cooled to room temperature, all under oxygen flow. The sample is then removed from the furnace.

Properties

The product is characterized by X-ray powder diffraction and thermogravimetric analysis. All diffraction lines could be indexed using a monoclinic cell, space group $P2/m$, $a = 8.6288(1)$ Å, $b = 3.8308(1)$ Å, $c = 8.6515(1)$ Å, $\beta = 90.214°$. TGA is consistent with the stoichiometry $LaCuO_{2.67}$.

C. SYNTHESIS OF ORTHORHOMBIC $LaCuO_{3-\delta}$ ($0.43 \leq \delta \leq 0.5$)

$$LaCuO_{3-\delta} \rightarrow LaCuO_{2.5} + \delta/2O_2$$

Procedure

The orthorhombic phase of $LaCuO_{3-\delta}$ ($0.43 \leq \delta \leq 0.5$) is prepared in a manner similar to that for the monoclinic phase (above), but is obtained under more reducing conditions. Various oxygen stoichiometries can be obtained by quenching the materials from different temperatures or by reoxidizing the material at lower temperature. A sample of tetragonal or monoclinic $LaCuO_{3-\delta}$ is added to a Pt crucible and placed into a tube furnace. The material is then heated slowly to 725° held there for 12 h, and cooled to room temperature under a flow of nitrogen. The sample was then removed from the furnace.

Properties

The product is characterized by X-ray powder diffraction and thermogravimetric analysis. All diffraction lines could be indexed in orthorhombic symmetry, space group *Pbam*, $a = 5.5491(1)$ Å, $b = 10.4782(2)$ Å, $c = 3.8796(1)$ Å. TGA is consistent with the stoichiometry $LaCuO_{2.50}$.

References

1. J. B. Goodenough, *Supercond. Sci. Technol.*, **3**, 26 (1990).
2. K. Hestermann and R. Hoppe, *Z. Anorg. Allg. Chem.*, **367**, 249 (1969).
3. N. E. Brese, M. O'Keefe, R. B. Von Dreele, and V. G. Young, *J. Solid State Chem.*, **83**, 1 (1989).
4. G. Demazeau, C. Parent, M. Pouchard, and P. Hagenmuller, *Mater. Res. Bull.*, **7**, 913 (1972).
5. J. F. Bringley, B. A. Scott, S. J. La Placa, R. F. Boehme, T. M. Shaw, M. W. McElfresh, S. S. Trail, and D. E. Cox, *Nature*, **347**, 263 (1990).
6. S. J. La Placa, J. F. Bringley, B. A. Scott, and D. E. Cox, *Acta Cryst.*, C**49**, 1415 (1993).

42. LITHIUM NIOBIUM OXIDE: $LiNbO_2$ AND SUPERCONDUCTING Li_xNbO_2

$$Li_3NbO_4 + 2NbO \rightarrow 3LiNbO_2$$

$$LiNbO_2 + 0.25Br_2 \rightarrow Li_{0.5}NbO_2 + 0.5LiBr$$

$$LiNbO_2 + 0.55NOBF_4 \rightarrow Li_{0.45}NbO_2 + 0.55NO + 0.55LiBF_4$$

$$LiNbO_2 + 0.4(NH_4)_2Ce(NO_3)_6 \rightarrow Li_{0.6}NbO_2 + 0.4Li(NH_4)_2Ce(NO_3)_6$$

Submitted by MARGRET J. GESELBRACHT* and ANGELICA M. STACY*
Checked by MATTHEW ROSSEINSKY†

The partially deintercalated material, Li_xNbO_2, is the first example of a layered oxide containing an early transition metal that exhibits superconductivity.[1] We report here the synthesis of superconducting samples of

* Department of Chemistry, University of California at Berkeley, Berkeley, CA 94720.
† Inorganic Chemistry Laboratory, University of Oxford, South Parks Road, Oxford OX1 3QR, UK.

Li_xNbO_2 by partial deintercalation of the fully lithiated parent compound, $LiNbO_2$. The parent compound was first synthesized by Meyer and Hoppe by the reaction of Li_2O with NbO and NbO_2:[2]

$$Li_2O + NbO + NbO_2 \xrightarrow[\text{Ni bomb, Ar}]{700^\circ} 2LiNbO_2$$

A common impurity of this route is the lithium niobium(V) oxide, $LiNbO_3$, as can be identified by powder X-ray diffraction. Kumada et al. optimized the synthesis of $LiNbO_2$ by employing Li_3NbO_4 as the lithium source.[3]

$$Li_3NbO_4 + 2NbO \xrightarrow[\text{Silica tube, vacuum}]{1050^\circ} 3LiNbO_2$$

Chang et al. reported the preparation of $LiNbO_2$ by either reaction of LiH with NbO_2 under hydrogen or by ion exchange of Na^+ with Li^+ in $NaNbO_2$.[4] We have made further refinements on the method of Kumada et al. and have obtained reproducibly large quantities of pure polycrystalline $LiNbO_2$. The superconducting material, Li_xNbO_2, is obtained by deintercalation reactions with chemical oxidizing agents at room temperature under an inert atmosphere.

A. LITHIUM NIOBIUM OXIDE: $LiNbO_2$

Procedure

$LiNbO_2$ is prepared by the reaction of Li_3NbO_4 and NbO at 1050° as in Kumada et al.[3] Unless otherwise specified, the reactants are heated to the reaction temperature in 1–2 h. The lithium niobium(V) oxide, Li_3NbO_4, is made as a white powder by heating a 3:1 molar ratio of Li_2CO_3 (6.236 g, 84.41 mmol, Cerac, 99.8%) and Nb_2O_5 (7.478 g, 28.13 mmol, Cerac, 99.95%) in an alumina crucible at 900° for 3 days. Niobium monoxide, NbO, is obtained by heating a 3:1 molar ratio of Nb powder (5.118 g, 55.09 mmol, Cerac, 99.8%) and Nb_2O_5 (4.882 g, 18.37 mmol) in a fused-silica tube, sealed under vacuum, at 1100° for 3 days.

In a typical preparation of $LiNbO_2$, pressed pellets (9.5 mm diameter) of a mixture of Li_3NbO_4 (4.493 g, 25.28 mmol) and NbO (5.507 g, 50.57 mmol) in a 1:2 molar ratio are placed in an alumina boat, and the boat is sealed in a fused-silica tube to minimize reaction with the silica. It has been found that even with this precaution, the integrity of the silica tube is weakened during the reaction. Frequently, the tube fractures on cooling from 1050°, causing the pellets to oxidize to $LiNbO_3$. To prevent this, the reaction tube is sealed in a larger fused-silica tube (22 mm i.d., 25 mm o.d.) that provides an outer shield should the inner tube (17 mm i.d., 19 mm o.d.) fracture at any time. The set of tubes is heated over a period of 10 h to 1050°, held at that temperature for 96 h, and then cooled to room temperature over a period of 10 h. At the end of the firing period, it is common to find that the inner tube is substantially discolored with evidence of tube attack. Alternatively, we have been able to make large quantities (25-mm-diameter pellets, ~20 g) by running this reaction under a dynamic flow of forming gas (5% H_2 in Ar). The product of this reaction has a surface layer of $LiNbO_3$ that can be physically removed from the pellet, leaving a pure core of $LiNbO_2$.

Properties

The $LiNbO_2$ phase is obtained in the form of a burgundy-red powder that is reasonably stable to atmospheric moisture. Samples can be worked with on the benchtop, but should be stored in a desiccator to minimize decomposition.

The X-ray powder diffraction pattern can be indexed to the hexagonal cell reported by Meyer and Hoppe; the refined lattice parameters are $a = 2.9063(6)$ Å and $c = 10.447(2)$ Å. Weight gain on oxidation in air is 11.8%; this is consistent with a stoichiometry of $Li_{0.96}NbO_2$, assuming that the Nb atoms are all oxidized to Nb^{5+}. Atomic absorption spectroscopy yields $Li_{0.93}NbO_2$. $LiNbO_2$ is diamagnetic at all temperatures from 2 to 300 K in applied magnetic fields from 5 to 40 kG. Since a diamagnetic signal is difficult to measure because of the detrimental effects of even the smallest levels of paramagnetic impurities, magnetic susceptibility is a sensitive probe of the purity of individual samples. A more detailed discussion of the properties of $LiNbO_2$ can be found in Geselbracht et al.[5]

B. LITHIUM NIOBIUM OXIDE: Li_xNbO_2 ($x < 1$)

Procedure

Deintercalation reactions that utilize chemical oxidizing agents result in the partial removal of lithium ions from the parent compound, $LiNbO_2$. Both Br_2 and $NOBF_4$ have been used to prepare Li_xNbO_2 where $x = 0.5$ and 0.45 for Br_2 and $NOBF_4$, respectively.[1] All deintercalation reactions have been carried out under argon by using standard Schlenk line techniques. $Li_{0.50}NbO_2$ is prepared by heating to reflux 1.7 g (12.89 mmol) of $LiNbO_2$ powder in 10 mL of an acetonitrile solution that is 0.3*M* in Br_2 for 3 days. $Li_{0.45}NbO_2$ is prepared by stirring a 70-mL slurry of 0.97 g (7.36 mmol) of

$LiNbO_2$ powder and 1.7 g (14.55 mmol) of $NOBF_4$ in acetonitrile for 10 days. For both reactions, the resultant gray powder is filtered, washed with acetonitrile and ethanol, and dried *in vacuo*.

Stoichiometric amounts of $(NH_4)_2Ce(NO_3)_6$ dissolved in acetonitrile also have been used to synthesize Li_xNbO_2. In a typical reaction, $Li_{0.6}NbO_2$ is prepared by adding 1.0 g (7.58 mmol) of $LiNbO_2$ to 60 mL of an acetonitrile solution that was 0.13*M* in $(NH_4)_2Ce(NO_3)_6$ (7.80 mmol). The burgundy-red powder turned grayish-black on contact with the solution and gave the slurry an olive green appearance. After 14 days of stirring under argon, the reaction mixture was filtered, and the gray powder was washed with acetonitrile and ethanol and dried *in vacuo*.

Properties

The partially deintercalated Li_xNbO_2 is also relatively stable to atmospheric moisture and can be worked with on the benchtop for short periods of time, although storage in a desiccator is recommended. Lithium content is determined by atomic absorption spectroscopy or inductively coupled plasma emission spectroscopy, by using 3:1 $HF:HNO_3$ to dissolve the samples. X-ray powder diffraction reveals little change from the starting material. The refined lattice parameters for $Li_{0.50}NbO_2$ are $a = 2.919(1)$ Å and $c = 10.453(4)$ Å, for $Li_{0.45}NbO_2$ are $a = 2.9227(9)$ Å and $c = 10.455(5)$ Å, and for $Li_{0.6}NbO_2$ are $a = 2.9229(6)$ Å and $c = 10.460(3)$ Å.

The partially deintercalated materials show Pauli paramagnetism at all temperatures above 5.5 K. However, when cooled in low applied magnetic fields ($H \sim 25$ G), the susceptibility drops to large diamagnetic values below 5.5 K; this transition is indicative of the onset of superconductivity. The depth of the transition in the magnetic susceptibility–temperature plot, in this case, the value of the diamagnetic susceptibility at the lowest temperature measured (2 K), is related to the volume fraction of superconducting material present in the sample. This value differs for Li_xNbO_2 depending on the synthesis conditions. The absolute value of the diamagnetic susceptibility at 2 K is much smaller for materials synthesized in refluxing acetonitrile ($T = 80°$) than for materials synthesized at room temperature, indicating a smaller fraction of superconducting material in the samples prepared at 80°. Furthermore, we have found that the superconductivity of materials that are prepared at room temperature can be destroyed by annealing the samples under an argon atmosphere at 80° for 5 days. In light of these findings, we recommend carrying out all deintercalation reaction at room temperature in order to prepare the best superconducting samples.

References

1. M. J. Geselbracht, T. J. Richardson, and A. M. Stacy, *Nature*, **345**, 324–326 (1990).
2. G. Meyer and R. Hoppe, *J. Less-Common Met.*, **46**, 55–65 (1976).
3. N. Kumada, S. Muramatu, F. Muto, N. Kinomura, S. Kikkawa, and M. Koizumi, *J. Solid State Chem.*, **73**, 33–39 (1988).
4. S-H. Chang, H-H. Park, A. Maazaz, and C. Delmas, *Compt. Acad. Sci. Paris, Ser. II*, **308**, 475–478 (1989).
5. M. J. Geselbracht, A. M. Stacy, A. R. Garcia, B. G. Silbernagel, and G. C. Kwei, *J. Phys. Chem.*, **97**, 7102–7107 (1993).

Chapter Six

MISCELLANEOUS SOLID-STATE COMPOUNDS

43. ZEOLITE MOLECULAR SIEVES

Submitted by L. D. ROLLMANN* and E. W. VALYOCSIK*
Checked by R. D. SHANNON†

Reprinted from *Inorg. Synth.*, **22**, 61 (1983)

Zeolites are three-dimensional, crystalline networks of AlO_4 and SiO_4 tetrahedra; a unit negative charge is associated with each AlO_4 tetrahedron in the framework. Their crystallization is often a nucleation-controlled process occurring from molecularly inhomogeneous, aqueous gels, and the particular framework structure that crystallizes can be strongly dependent on the cations present in these gels.[1]

Synthesis methods are described below for four very different but important zeolite structures, A,[2] Y,[3] tetramethylammonium (TMA) offretite,[4] and tetrapropylammonium (TPA) ZSM-5.[5,6] These four were selected because they span the composition range from 1:1 Si:Al to a potentially aluminum-free zeolite structure (A to ZSM-5). In addition, these syntheses provide examples of fundamental concepts in crystallization such as templating (TMA offretite and ZSM-5), low-temperature nucleation (Y), and variable reactant (silica) sources.

* Mobil Research and Development Corp., Central Research Division, P.O. Box 1025, Princeton, NJ 08540.

† Central Research and Development Dept., E. I. du Pont, Wilmington, DE 19898.

The chemical description of a zeolite synthesis mixture requires special comment. It is conventional to present reaction mixtures as mole ratios of added ingredients:

$$\frac{SiO_2}{Al_2O_3}, \quad \frac{H_2O}{SiO_2}, \quad \frac{OH^-}{SiO_2}, \quad \frac{Na^+}{SiO_2}, \quad \frac{R}{SiO_2}$$

wherein R (if present) may be a component such as a quaternary ammonium cation or a potassium ion. By convention, moles of hydroxide are calculated by assuming, for example, that sodium silicate is a mixture of silica, sodium hydroxide, and water; that sodium aluminate is an analogous mixture but with alumina in place of silica; and that aluminum sulfate is a mixture of alumina, water, and sulfuric acid. In the descriptions given below, it is also recognized that alumina consumes 2 mol of hydroxide (i.e., acts as 2 mol of acid) on its incorporation into a zeolite framework as aluminate ion. Hydroxide: silica ratios are then calculated by subtracting moles of acid added from moles of hydroxide and then dividing by moles of silica present. Organics such as amines are never included in calculating $OH:SiO_2$ ratios.

Determination of purity in zeolites is a second area of concern. Elemental analysis is generally not a satisfactory criterion since almost all zeolite structures can exist in a range of compositions (i.e., of $SiO_2:Al_2O_3$ ratio). For example, the A structure has been crystallized with a $SiO_2:Al_2O_3$ ratio from 2 to 6;[7] at the other extreme, ZSM-5 has even been synthesized with essentially no aluminum.[6].

X-ray powder diffraction patterns are the most common measure of purity in zeolite samples. If the diffraction pattern shows no evidence for crystalline (or amorphous) contaminants, purity is then estimated by comparing intensities of reflections (at *d* spacings smaller than about 6 Å) with those of an authentic sample of the same composition and crystal size. Except for such large-scale commercial products as NaA and NaY, "authentic" samples are normally obtained by repeated and varied crystallization experiments.

A. ZEOLITE A[2]

$$2NaAlO_2 + 2(Na_2SiO_3 \cdot 9H_2O) \rightarrow$$
$$Na_2O \cdot Al_2O_3 \cdot 2SiO_2 \cdot 4.5H_2O + 4NaOH + 11.5H_2O$$

Procedure

Sodium aluminate (13.5 g, approximately 0.05 mol alumina and 0.07 mol Na_2O; commercial sodium aluminate contains about 40% Al_2O_3, 33% Na_2O and 27% H_2O) and sodium hydroxide (25 g, 0.62 mol) are dissolved in

300 mL of water in a magnetically stirred 600-mL beaker and brought to a boil. The aluminate solution is added, with vigorous stirring, to a hot solution of sodium metasilicate, $Na_2SiO_3 \cdot 9H_2O$ (14.2 g, 0.05 mol), in 200 mL water in a 1-L beaker, also equipped with a Teflon-coated magnetic stirrer. The entire mixture is heated with stirring at about 90° until the suspension will settle quickly when stirring is stopped (2–5 h). The suspension is then filtered hot and the solid washed repeatedly with water (four 100-mL portions) and dried in an oven at 110° to give about 7–8 g (80–90% yield based on SiO_2) of $Na_2O \cdot Al_2O_3 \cdot 2SiO_2 \cdot 4.1H_2O$.

Anal. Calcd.: Na_2O, 17.3; Al_2O_3, 28.5; SiO_2, 33.5. Found: Na_2O, 16.1; Al_2O_3, 28.8; SiO_2, 34.2. The purity of the sample is determined by inspection of its X-ray diffraction pattern.

Properties[7]

The product, NaA, is a white crystalline solid with a crystal density of 1.27 g/cm³. Its crystals are normally 1-2μm in diameter and have cubic symmetry. A typically unit cell formula is $Na_{12}[(AlO_2)_{12}(SiO_2)_{12}] \cdot 27H_2O$; the sodium ions are readily exchanged in aqueous solution by cations such as calcium or potassium. In the sodium form, after dehydration at 350–400° *in vacuo*, zeolite A will sorb about 0.25 g H_2O per gram of ash (at room temperature, 4 torr), but it will not sorb hexane. In the dehydrated calcium form, 0.27 g H_2O per gram (4 torr) and 0.145 g hexane (10 torr) per gram are sorbed, but benzene is not. The X-ray diffraction pattern of NaA is as follows:

hkl	*d* (Å)	*I*/*I* (0)
100	12.29	100
110	8.71	69
111	7.11	35
210	5.51	35
211	5.03	2
220	4.36	6
221, 300	4.107	36
311	3.714	53
320	3.417	16
321	3.293	47
410	2.987	55

B. ZEOLITE Y[3]

$$2NaAlO_2 + 5SiO_2 + xH_2O \rightarrow Na_2O \cdot Al_2O_3 \cdot 5SiO_2 \cdot xH_2O$$

Procedure

A solution of 13.5 g sodium aluminate (0.05 mol alumina and 0.07 mol Na_2O) and 10 g sodium hydroxide (0.25 mol) in 70 g water is prepared in a 200-mL beaker and added, with vigorous magnetic stirring, to 100 g 30% silica sol (0.5 mole SiO_2; commercial colloidal silica suspensions typically contain 30% SiO_2 together with 0.1–0.5% Na_2O, as stabilizer) in a 250-mL polypropylene bottle. The reaction mixture, which is defined by the following mole ratios of components:

$$\frac{SiO_2}{Al_2O_3} = 10 \qquad \frac{H_2O}{SiO_2} = 16 \qquad \frac{OH^-}{SiO_2} = 0.6 \qquad \frac{Na^+}{SiO_2} = 0.8$$

is then set aside to age. After 24–48 h at room temperature, the bottle is placed in a steam chest at about 95°. After 48–72 h, daily samples of the solid in the bottom of the bottle are taken, filtered, washed, dried, and analyzed by X-ray diffraction for crystallinity. Special care should be taken during sampling to avoid mixing the sample or inadvertently seeding the mixture. When the diffraction pattern reaches a limiting intensity, the hot mixture is removed from the steam chest and filtered and the solid washed with water (four times 100 mL) and dried at 100°. About 30 g of NaY (50–60% based on SiO_2) is obtained, with an approximate molar composition of $Na_2O \cdot Al_2O_3 \cdot 5.3SiO_2 \cdot 5H_2O$.

Anal. Calcd.: Na_2O, 10.5; Al_2O_3, 17.3; SiO_2, 53.9. Found: Na_2O, 10.9; Al_2O_3, 17.2; SiO_2, 53.4. A final X-ray diffraction pattern is taken for inspection and for comparison with that of an authentic sample of NaY.

Properties[7]

The NaY produced is a white crystalline solid with a crystal density of 1.27 g/cm^3. Its crystals are usually smaller than 1 μm and have cubic symmetry. A typical unit-cell formula would be $Na_{56}[(AlO_2)_{56}(SiO_2)_{136}] \cdot 250H_2O$, about 70% of the sodium ions being readily exchanged by cations such as ammonium ion. The remainder can be exchanged with persistence. Ammonia can be removed from the resultant NH_4Y by heating to about 450°. In the sodium form, after dehydration at 350–400° *in vacuo*, zeolite Y will sorb about 25% of its weight in water (1 torr, 25°), about 19% in hexane (10 torr), and about 25% in benzene (10 torr). The X-ray diffraction pattern of NaY is as follows:

hkl	*d* (Å)	*I*/*I* (0)
111	14.29	100
220	8.75	9
311	7.46	24
331	5.68	44
333, 511	4.76	23
440	4.38	35
620	3.91	12
533	3.775	47
444	3.573	4
711, 551	3.466	9
642	3.308	37
731, 553	3.222	8
733	3.024	16

C. TMA OFFRETITE[4]

Procedure

A solution of 5.2 g sodium aluminate (0.02 mol alumina and 0.03 Na_2O), 14.6 g NaOH (0.36 mol), and 8.2 g KOH (0.15 mol) in 76 g water is prepared in a 200-mL beaker equipped with a magnetic stirrer. To this is added 11.0 g 50% tetramethylammonium chloride (0.05 mol), and the resultant solution is poured quickly into 112 g 30% silica sol (0.56 mol SiO_2) in a 250-mL polypropylene bottle. The mixture, which has the following composition

$$\frac{SiO_2}{Al_2O_3} = 27 \qquad \frac{H_2O}{SiO_2} = 16 \qquad \frac{OH^-}{SiO_2} = 1.0 \qquad \frac{Na^-}{SiO_2} = 0.75$$

$$\frac{K^+}{SiO_2} = 0.27 \qquad \frac{TMA^+}{SiO_2} = 0.09$$

is vigorously shaken and placed in a steam chest at 95° to crystallize. After 48–72 h, daily samples of the solid in the bottom of the bottle are taken, filtered out, washed, dried, and analyzed by X-ray diffraction for crystallinity. When the diffraction pattern reached a limiting intensity, the mixture is removed from the steam chest and filtered, and the solid washed with water (four times 100 mL) and dried at 100°. About 16 g of TMA offretite (25–30% based on SiO_2) is obtained, with an approximate molar composition of 0.3 $(TMA)_2O \cdot 0.5K_2O \cdot 0.4Na_2O \cdot Al_2O_2 \cdot 7.7SiO_2 \cdot 7.1H_2O$.

Anal. Calcd.: Na_2O, 3.1; N, 1.0; C, 3.5; K_2O, 5.8; Al_2O_3, 12.5; SiO_2, 56.9. Found: Na_2O, 2.9; N, 1.0; C, 3.6; K_2O, 5.9; Al_2O_3, 12.6; SiO_2, 56.9. A final X-ray diffraction pattern is taken for inspection and for comparison with that of an authentic sample of TMA offretite.

Properties[7]

Tetramethylammonium offretite has a crystal density of 1.55 g/cm^3. Its crystals are usually oval agglomerates, about 0.3×1.5 μm in size, with hexagonal symmetry. A typical unit-cell formula would be $TMA \cdot 2K \cdot Na[(AlO_2)_4(SiO_2)_{14}] \cdot 7H_2O$, the TMA (and a portion of the potassium ions) being trapped within gmelinite-and ε-cages, respectively, and therefore not readily exchanged by sodium or ammonium ions. In the as-synthesized form, after calcination to 500° in air, this zeolite will sorb about 12% of its weight in butane and 7% isobutane (100 torr). Its X-ray diffraction pattern is as follows:

hkl	*d* (Å)	*I*/*I*(0)
100	11.45	100
001	7.54	16.5
110	6.63	55.2
101	6.30	9.9
200	5.74	15.0
201	4.57	26.5
210	4.34	43.3
211, 002	3.76	89.2
102	3.59	43.0
220	3.31	18.6
202	3.15	17.4

In particular, a pure sample of TMA offretite will not show reflections at $d = 9.2$, 5.34, and 4.16 Å, which are the "odd-l" lines of erionite.

D. ZSM-5

Procedure

A solution of 0.9 g $NaAlO_2$ (0.0035 mol alumina, 0.01 mol NaOH) and 5.9 g NaOH (0.15 mol) in 50 g water is prepared in a 200-mL beaker equipped with magnetic stirrer and labeled "solution A." In a second beaker, a solution ("B") is prepared by adding 8.0 g tetrapropylammonium bromide (0.03 mol) to a stirred mixture of 6.2 g of 96% H_2SO_4 (0.12 mol) and 100 g H_2O. Solutions A and B are poured simultaneously into a solution of 60 g 30% silica sol (0.3 mol SiO_2, 0.003 mol Na_2O, and 50 g H_2O in a 250-mL polypropylene bottle). The bottle is immediately capped and vigorously shaken to form a gel with the composition

$$\frac{SiO_2}{Al_2O_3} = 85 \qquad \frac{H_2O}{SiO_2} = 45 \qquad \frac{OH^-}{SiO_2} = 0.1 \qquad \frac{Na^+}{SiO_2} = 0.5 \qquad \frac{TPA^+}{SiO_2} = 0.1$$

It is then placed in a steam chest at 95° to crystallize. After 10–14 days, periodic samples of the solid in the bottom of the bottle are taken, filtered out, washed, dried, and analyzed by X-ray diffraction for crystallinity. When the diffraction pattern reaches a limiting intensity, the mixture is removed from the steam chest and filtered, and the solid washed with water (four times 100 mL) and dried at 110°. (Crystallization times can be reduced to 1 day or less when conducted in a stirred autoclave at higher temperatures; at 140–180°, for example.) About 19 g of ZSM-5 (85% yield based on SiO_2) is obtained with a molar composition of $1.8(TPA)_2O \cdot 1.2Na_2O \cdot 1.3Al_2O_3 \cdot OSi_2 \cdot 7H_2O$.

Anal. Calcd.: Na_2O, 1.1; N, 0.7; C, 7.4; Al_2O_3, 1.9; SiO_2, 85.3. Found: Na_2O, 1.1; N, 0.7; C, 8.0; Al_2O_3, 1.9; SiO_2, 84.6 A final X-ray diffraction pattern is taken for inspection and for comparison with that of an authentic sample of ZSM-5.

Properties[5,6,8,9]

ZSM-5 has a crystal density of 1.77 g/cm^3. Its crystals have orthorhombic symmetry, as synthesized, and can vary widely in size. Compositionally ZSM-5 is unusual, in comparison with the examples given above, in that it can be prepared in the absence of aluminum. Organics can be removed from ZSM-5 samples by careful oxidative calcination at about 500°. Alkali-metal cations, if present, can be exchanged by ammonium ion, for example, to produce NH_4ZSM-5. Calcined samples of NH_4ZSM-5 will sorb about 11%hexane (25°, 20 torr). The X-ray diffraction pattern of ZSM-5 is characterized by the following significant lines:

d (Å)	Relative Intensity
11.1	S
10.0	S
7.4	W
7.1	W
6.3	W
6.04 / 5.97	W
5.56	W
5.01	W
4.60	W
4.25	W
3.85	VS
3.71	S
3.04	W
2.99	W

References

1. L. D. Rollmann, *Adv. Chem. Series*, **173**, 387 (1979).
2. J. Ciric, *J. Colloid Interface Sci.*, **28**, 315 (1968).
3. D. W. Breck, U.S. Patent 3,130,007 to Union Carbide Corp., 1964.
4. E. E. Jenkins, U.S. Patent 3,578,398 to Mobil Oil Corp., 1971.
5. R. J. Argauer and G. R. Landolt, U.S. Patent 3,702,886 to Mobil Oil Corp., 1972.
6. F. G. Dwyer and E. E. Jenkins, U.S. Patent 3,941,871 to Mobil Oil Corp., 1976.
7. D. W. Breck, *Zeolite Molecular Sieves*, Wiley, New York, 1974.
8. G. T. Kokotailo, S. L. Lawton, D. H. Olson, and W. M. Meier, *Nature*, **272**, 437 (1978).
9. E. L. Wu, S. L. Lawton, D. H. Olson, A. C. Rohrman, Jr., and G. T. Kokotailo, *J. Phys. Chem.*, **83**, 2777 (1979).

44. SUBSTITUTED β-ALUMINAS

Submitted by J. T. KUMMER*
Checked by M. STANLEY WHITTINGHAM†

Reprinted From *Inorg. Synth.*, **19**, 51 (1979)

β-Alumina has the empirical formula $Na_2O \cdot 11Al_2O_3$. In reality the compound is massively defective and contains considerably more (~ 25%) Na_2O

* Ford Motor Company, Research Staff, Dearborn, MI 48121.
† Corporate Research Laboratories, Exxon Research and Engineering Co., P.O. Box 45, Linden, NJ 07036.

than indicated by the empirical formula. β-Alumina has a hexagonal layer structure[1] of the space group $P6_3/mmc$ with the lattice constants $a = 5.59$ Å and $c = 22.53$ Å. The sodium ions are situated exclusively in planes perpendicular to the c axis that contain, in a loose packing, an equal number of sodium and oxygen ions. This unusual structure results in the sodium ion possessing a high mobility in this plane, and the resulting high ionic conductivity is the prime reason for the recent interest in this compound.[1]

Substituted β-alumina can be made by an ion exchange procedure in a molten salt using single crystals of sodium β-alumina as starting material.[2] Small single crystals of β-alumina, if not available, can be obtained from fusion-cast bricks of β-alumina (Monofrax H, 14 kg each).‡ These bricks fracture easily, yielding single crystals of β-alumina that are very thin in the c direction ($\leqslant 0.03$ cm) and up to 1 cm in diameter. If there is a difference in size between the ion introduced into the β-alumina and the ion removed in the ion-exchange process there is a very much larger change in c-axis dimension of the crystal than in the a-axis dimension. In general, for crystals 0.03 cm thick and 1 cm in diameter, the physical integrity of sodium β-alumina crystals is preserved during ion exchange with other monovalent cations. The author does not know if very large single crystals will preserve their physical integrity during ion exchange. Polycrystalline ceramic material will, in general, fracture during the ion-exchange process as a result of the unequal change in the a and c dimensions. An exception is the exchange of polycrystalline Na^+ β-alumina by Ag^+ ion, where the change along the c axis is small.

The ion exchange can be done either by direct exchange of the Na^+ ion in β-alumina with the desired cation in a molten salt medium using an appropriate anion to improve the exchange equilibria[2] or by first exchanging the Na^+ in β-alumina with Ag^+ ion in molten silver nitrate and then exchanging the Ag^+ ion in this material with the desired cation in a molten salt or other media. In general, the latter procedure is preferred. The use of silver β-alumina as an intermediate has two advantages: (1) by employing a metal chloride melt, the exchange reaction with silver β-alumina can be driven to completion as a result of the formation of silver chloride in the melt; and (2) complete exchange is determined by the absence of silver ion in the exchanged material — there should be no fear of contamination error in the analytical procedure since silver is not a common contaminant in the laboratory. Because of the large atomic weight of silver, the silver β-alumina contains approximately 19 wt. % silver, and the exchange can be monitored be weight changes in many cases.

‡ Available from Monofrax Div., Carborundum Co., P.O.Box A, Falconer, NY 14733.

The analysis of β-aluminas for stable cations can be made by a fusion process. The β-alumina can be dissolved in molten Li_2CO_3 (or K_2CO_3), the resulting glass dissolved in dilute nitric acid, and the solution analyzed by atomic adsorption. Activation analysis or X-ray fluorescence analysis can also be used.

A. SILVER β-ALUMINA (ALUMINUM SILVER(I) OXIDE)

$$Na_2O \cdot 11Al_2O_3 + 2AgNO_{3(1)} \rightarrow Ag_2O \cdot 11Al_2O_3 + 2NaNO_{3(1)}$$

Procedure

Sodium β-alumina single crystals are dried at 500° for 3 h and cooled in a desiccator. The exchange of the sodium ion by the silver ion is carried out using molten silver nitrate at 300° contained in a Vycor or fused-quartz vessel. Pyrex should not be used because of the presence of potassium and sodium in the glass. One gram of dried crystals of sodium β-alumina is placed in a Vycor test tube approximately 2 cm in diameter and about 14 cm long. Ten grams of reagent grade silver nitrate is added, and the mixture is heated to 300° in a furnace. The time required for exchange equilibrium increases as the square of the diameter of the largest crystals used. For crystals of 2-mm diameter, the time to reach 99% equilibrium is 3 h. For crystals of 1-cm diameter the time is 75 h. The crystals float at first and then sink to the bottom of the test tube. Stirring is not necessary as sodium nitrate is less dense than silver nitrate. At the end of the time allowed for attainment of 99% equilibrium, the molten silver nitrate is decanted from the crystals into a porcelain crucible and the Vycor test tube containing the exchanged crystals is cooled to room temperature. The crystals are washed with water to dissolve the residual silver nitrate, and then with alcohol or acetone, and are then dried at 200°. The crystals contain less than 0.1% sodium.

Properties

Silver β-alumina crystals are colorless. They remain clear and transparent on heating to 1000° and do not react when contacted with chlorine for several hours at 560°. The lattice constants are $a = 5.594$ Å, $c = 22.498$ Å. The crystals are not hygroscopic.

B. THALLIUM(I) β-ALUMINA (ALUMINUM THALLIUM(I) OXIDE)

$$Na_2O \cdot 11Al_2O_3 + 2TlNO_{3(1)} \rightarrow Tl_2O \cdot 11Al_2O_3 + 2NaNO_{3(1)}$$

Procedure

■ **Caution.** *Thallium salts, particularly their vapors, are extremely toxic. Melts should be prepared in a hood.*

The procedure for producing thallium β-alumina from sodium β-alumina is the same as that used above for silver β-alumina. Thallium (I) nitrate is used in place of silver nitrate. For crystals of 1-mm in diameter the time to reach 99% equilibrium is about 75 h. The exchanged crystals contain less than 0.1% sodium.

Properties

Thallium (I) β-alumina crystals are colorless and do not react with chlorine at 700°. The lattice constants are $a = 5.597$ Å, $c = 22.883$ Å. The crystals are not hygroscopic.

C. LITHIUM β-ALUMINA (ALUMINUM LITHIUM OXIDE)

$$Ag_2O \cdot 11Al_2O_3 + 2LiCl_{(1)} \rightarrow Li_2O \cdot 11Al_2O_3 + 2AgCl_{(1)}$$

Procedure

One gram of silver β-alumina (see above) is placed into a fused quartz test tube about 2 cm in diameter and about 14 cm long. Five grams of lithium chloride is added. It is important that the lithium chloride used have a very low content of other alkali-metal impurities, except Cs, since the ion-exchange equilibria greatly favor the presence of the other alkali metals in the β-alumina crystals over lithium. Essentially all the impurity ends up in the crystals. The fused-quartz test tube is heated to 650° in a furnace. For crystals 1-cm in diameter the time to reach 99% equilibrium is approximately 16 h. The molten salt is decanted and the crystals are allowed to cool to room temperature. Methyl alcohol containing about 10% propylamine or ethylenediamine is used to wash the product and thereby remove the silver chloride and residual lithium salts. The sample is dried at 400° and stored in a desiccator. The lithium β-alumina crystals contain less than 0.05% Ag. If the lithium chloride used contains a trace of sodium or potassium, it can be prepurified by treatment with silver β-alumina at 650°. Each gram of silver β-alumina will remove about 30 mg of sodium from the melt. The molten lithium chloride, after decantation from the pretreatment silver β-alumina, can be used to prepare the product, lithium β-alumina.

Properties

Lithium β-alumina crystals are colorless and hygroscopic. They should be kept in a desiccator and dried at 400° before use. The lattice constants are $a = 5.596$ Å, $c = 22.570$ Å.

D. POTASSIUM β-ALUMINA (ALUMINUM POTASSIUM OXIDE)

$$Ag_2O \cdot 11Al_2O_3 + 2KCl_{(1)} \rightarrow K_2O \cdot 11Al_2O_3 + 2AgCl_{(1)}$$

The preparation of this compound from silver β-alumina is similar to the preparation of lithium β-alumina. The melt consists of 10 g of potassium chloride. The exchange temperature is 800°. For crystals with diameters of 1 cm it takes about 16 h to reach 99% of equilibrium. The potassium salts used should contain less than 0.1 wt.% sodium. After decantation of the melt the crystals are washed with water containing 2% propylamine or ethylēnediamine to remove residual potassium salts and silver chloride. The sample is dried at 200°. The potassium β-alumina contains less than 0.05% silver.

Properties

Potassium β-alumina crystals are colorless and are not hygroscopic. The lattice constants are $a = 5.596$ Å, $c = 22.729$ Å.

E. RUBIDIUM β-ALUMINA (ALUMINUM RUBIDIUM OXIDE)

$$Ag_2O \cdot 11Al_2O_3 + 2RbCl_{(1)} \rightarrow Rb_2O \cdot 11Al_2O_3 + 2AgCl_{(l)}$$

The preparation of this substance from silver β-alumina is similar to the preparation of lithium β-alumina. The melt consists of 10 g of rubidium chloride. The exchange temperature is 800°. For crystals 2 mm in diameter it takes about 16 h to reach 99% of equilibrium. The rubidium salts used should contain less than 0.02% potassium and less than 0.1% sodium. After decantation of the melt the crystals are washed with water containing 2% propylamine of ethylenediamine to remove residual potassium salts and silver chloride. They are dried at 200°. The rubidium β-alumina crystals contain less than 0.05 wt.% silver.

Properties

Rubidium β-alumina crystals are colorless and are not hygroscopic. The lattice constants are $a = 5.597$ Å, $c = 22.877$ Å.

F. AMMONIUM β-ALUMINA (ALUMINUM AMMONIUM OXIDE)

$$Na_2O \cdot 11Al_2O_3 + 2NH_4NO_{3(l)} \rightarrow (NH_4)_2O \cdot 11Al_2O_3 + 2NaNO_{3(l)}$$

Preparation

■ **Caution.** *Ammonium nitrate is hazardous in large amounts and can be detonated with shock waves. There have been no problems when using 10 g of material, but it should be handled with care.*

Because ammonium salts decompose at high temperature, the exchange must be carried out at temperatures below 180°. At this temperature the rate of exchange is low and the time for complete exchange is high. One gram of sodium β-alumina is placed in a Vycor test tube about 2 cm in diameter and 14 cm long along with 10 g of ammonium nitrate. This test tube is placed in a furnace at 170–180° as close to 170° (the melting point) as is feasible. The time to reach 99% equilibrium is approximately 140 h for crystals $\frac{1}{2}$ mm in diameter. At the end of the exchange process the molten salt is decanted, and the test tube with the crystals is cooled to room temperature. The crystals are washed with water and dried at 200°. These dried crystals are placed in the Vycor test tube along with a fresh 10 g of ammonium nitrate, and the above procedure is repeated. The crystals are dried at 200°. The ammonium β-alumina crystals contain less than 0.3% sodium.

Properties

Ammonium β--alumina crystals are colorless and are not hygroscopic. The lattice parameters are $a = 5.596$ Å, $c = 22.888$ Å. The IR spectra show characteristic absorption bands at 3180, 3070, and 1430 cm^{-1}.

G. GALLIUM(I) β-ALUMINA (ALUMINUM GALLIUM(I) OXIDE)

$$Ag_2O \cdot 11Al_2O_3 + 2Ga \xrightarrow[GaI_{(l)}]{} Ga_2O \cdot 11Al_2O_3 + 2Ag$$

Preparation

One gram of silver β-alumina crystals is placed in a Vycor tube with a 1-cm i.d. and 20 cm length and closed at one end. Three grams of gallium and 4 g of iodine are added, and the Vycor tube is necked down near the middle in preparation for sealing off. The tube is evacuated with a mechanical pump to

< 1 torr pressure and sealed. The sealed tube is placed in a cold furnace and then heated to about 290°. For crystals 1 mm diameter the time of heating is 48 h. The furnace is cooled to room temperature and the ampule is removed. The ampule is broken open and the melt is dissolved in 10% HCl solution, leaving a pool of gallium–silver metal and gallium (I) β-alumina crystals.

Properties

Gallium(I) β-alumina crystals are transparent and reddish brown.[3] They become colorless when heated 16 h at 540° in air. The lattice constants are $a = 5.600$ Å, $c = 22.718$ Å. The density is 3.51 g/cm^3, and the compound is stable up to 750° in dry air. The material is not hygroscopic.

H. NITROSYL β-ALUMINA (ALUMINUM NITROSYL OXIDE)

$$NOCl + AlCl_3 \rightarrow [NO][AlCl_4]$$

$$Ag_2O \cdot 11Al_2O_3 + 2[NO][AlCl_4] \rightarrow [NO]_2O \cdot 11Al_2O_3 + 2Ag\,AlCl_4$$

Preparation

■ **Caution.** *This preparation must be done in a hood because NOCl is very toxic.*

This material is prepared by allowing silver β-alumina to exchange with molten [NO] [$AlCl_4$].[4]

Using a long-stem funnel, 4 g of anhydrous $AlCl_3$ is placed in the bottom of a Vycor test tube 5 cm in diameter and 60 cm long that has previously been flushed with dry argon. The tube is then necked down about 20 cm from the bottom in preparation for sealing off. With argon flowing across the open mouth of the tube, the $AlCl_3$ is sublimed from the tube bottom to above the necked-down section and the tube is sealed off at the neck. Nitrosyl chloride gas* is bled from the lecture bottle through Teflon and glass tubing into a small cold trap in dry ice where about 2.5 g (approximately 1.7 mL) of the NOCl is condensed. The Vycor tube containing the $AlCl_3$ and with argon flowing across the open mouth is immersed in dry ice and the NOCl is distilled through a long glass tube from the cold trap (boiling point −5°) and condensed in the bottom of the Vycor tube to contact the $AlCl_3$. The Vycor tube is placed in an ice-salt bath and allowed to stand 1 h with occasional

* Available from Matheson Gas Products, P.O.Box 85, East Rutherford, NJ 07073.

agitation to allow reaction, after which it is heated to 200° to melt the [NO] [$AlCl_4$].

The tube is cooled to −78° and 1.5 g of 0.1 -mm-diameter dried (400°) silver β-alumina crystals are added to the $NOAlCl_4$ through a long-stemmed funnel. Additional NOCl, about 0.5 g (0.35 mL), is added to suppress the decomposition of [NO] [$AlCl_4$], and the top of the Vycor tube is sealed off. The tube is heated to 200° for 24 h. At the end of this time the tube is cooled and broken open in the hood and the melt is dissolved in distilled water containing ethylenediamine to dissolve the AgCl. (■ **Caution.** *NO_2 fumes are given off.*) The crystals of nitrosyl β-alumina contain less than 0.5 wt% Ag.

Properties

Nitrosyl β-alumina crystals are colorless and nonhygroscopic. The lattice constants are $a = 5.597$ Å, $c = 22.711$ Å. The IR spectrum contains a strong absorption band at 2245 cm^{-1}, a frequency indicative of the N–O stretching motion of the nitrosonium ion. The material is thermally unstable. Particles 150–250 μm in size decompose above 400°. Particles smaller than 45 μm decompose above 150°. The density of the material is 3.22 g/cm^3.

References

1. J. T. Kummer, *Progr. Solid State Chem.*, **7**, 141 (1972).
2. Y. Y. Yao, J. T. Kummer, *J. Inorg. Nucl. Chem.*, **29**, 2453 (1967).
3. R. H. Radzilowski, *Inorg. Chem.*, **8**, 994 (1969).
4. R. H. Radzilowski, and J. T. Kummer, *Inorg. Chem.*, **8**, 2531 (1969).

45. VANADYL PHOSPHATES AND ORGANYLPHOSPHONATES

Submitted by JOHN F. BRODY* and JACK W. JOHNSON*
Checked by JACK VAUGHEY†

Because of their layered structures, vanadyl phosphates and organylphosphonates undergo many reactions in the solid phase at temperatures much lower than those typical for classical solid-state reactions. The relatively weak bonding that holds the layers together allows access of molecular reagents to the interlayer region, initiating intercalation or oxidation–reduction reactions under relatively mild conditions. Vanadyl phosphates

* Exxon Research and Engineering, Annandale, NJ 08801.
† Department of Chemistry, University of Houston, Houston, TX 77204.

and organylphosphonates have been used as precursors for oxidation catalysts, as hosts for the selective recognition of alcohol isomers, and as general substrates for both coordinative and redox intercalation reactions.

The preparation of $(VO)PO_4 \cdot 2H_2O$ follows that originally reported by Ladwig.[1] Many preparations of $(VO)(HPO_4) \cdot 0.5H_2O$ have been reported in the literature related to the catalysis of the oxidation of butane to maleic anhydride. They differ primarily in the method of reduction of the vanadium from the pentavalent to the tetravalent state. HCl, isobutyl alcohol, and benzyl alcohol have all been employed as reducing agents.[2] Following are two preparations of $(VO)(HPO_4) \cdot 0.5H_2O$.[3] The first starts from $(VO)PO_4 \cdot 2H_2O$ and uses *s*-butanol as the reducing agent. It produces a slightly more crystalline product than the second procedure that starts more conveniently from V_2O_5 and H_3PO_4, using ethanol as the solvent and reducing agent. The preparation of the vanadyl organylphosphonates is derived from the second method of preparing $(VO)(HPO_4) \cdot 0.5H_2O$, with organylphosphonic acid substituted for phosphoric acid.[4,5]

A. VANADYL(V) PHOSPHATE, $(VO)PO_4 \cdot 2H_2O$

$$V_2O_5 + 2H_3PO_4 + H_2O \rightarrow 2(VO)PO_4 \cdot 2H_2O$$

Procedure

Vanadium pentoxide (V_2O_5), 24.00 g (0.132 mol), a rust-colored solid, was finely ground with a mortar and pestle in a hood and added through a funnel into a 1-L, three-necked round-bottomed flask containing 300 mL distilled water. The mixture was stirred at room temperature using a mechanical paddle stirrer fitted through the center neck of the flask. Then 132 mL (1.93 mol) phosphoric acid (85% H_3PO_4) was added to the reaction flask, followed by another 277 mL water, and 2 mL concentrated nitric acid (70% HNO_3). The reaction flask was fitted with a reflux condenser in the second neck and a thermometer/temperature controller in the remaining neck. The rust-colored mixture was then heated to reflux and stirred over night. Soon after refluxing started, the mixture became a bright yellow color. Refluxing was discontinued after 16 hr and the bright yellow mixture was allowed to cool. The contents of the reaction flask were filtered through a 500 mL medium-porosity fritted glass funnel to obtain a clear yellow filtrate and an yellow solid. The solid was washed on top of the fritted funnel with 4 × 100-mL aliquots of distilled water, followed by vacuum filtration. The filtrates from the water washing were deep red-brown in color. The solid was then washed on top of the fritt with 3 × 25-mL aliquots of acetone. The filtrate from the first acetone wash was light brown, and the successive acetone wash filtrates were clear and colorless. The yellow solid was suction-dried on top of the fritt for $\frac{1}{2}$ h and then spread out on top of a large watch glass and allowed to air-dry to yield 25.91 g (0.131 mol, 49.6%) $(VO)PO_4 \cdot 2H_2O$.

Properties

$(VO)PO_4 \cdot 2H_2O$ is a platy, bright yellow microcrystalline solid. When exposed to organic vapors in the air, it takes on a greenish tint as a result of surface reduction. The amount of V^{4+} present can be measured by EPR (electron paramagnetic resonance) spectroscopy and varies from 1 to 2%.[6] It is most conveniently characterized by its X-ray powder diffraction pattern, which can be indexed on a tetragonal unit cell with $a = 6.21$ Å and $c = 7.41$ Å. The strongest lines are the (00l) series corresponding to the layer repeat distance 7.41 Å. The water content is variable, depending on the ambient humidity. The dihydrate can be obtained by storing the compound over water. $(VO)PO_4 \cdot 2H_2O$ can be converted topotactically to anhydrous α-$(VO)PO_4$ by heating at 250° in flowing air. Alcohols,[7] pyridine and its 4-substituted derivatives,[6] and amides[8] can be intercalated into $(VO)PO_4 \cdot 2H_2O$. In the presence of reducing agents, cations A^+ can be reductively intercalated to form $A_x(VO)PO_4 \cdot nH_2O$.[9]

B. VANADYL(IV) HYDROGEN PHOSPHATE HEMIHYDRATE, $(VO)(HPO_4) \cdot 0.5H_2O$: $(VO)PO_4 \cdot 2H_2O$/*s*-BuOH METHOD

$$2(VO)PO_4 \cdot 2H_2O + CH_3CH_2CHOHCH_3 \rightarrow 2(VO)(HPO_4) \cdot 0.5H_2O + CH_3CH_2COCH_3 + 3H_2O$$

Procedure

Vanadyl(V) phosphate ($(VO)PO_4 \cdot 2H_2O$), 8.00 g (40.4 mmol), a bright yellow solid, was added through a funnel into a 250-mL, three-necked, round-bottomed flask containing 160 mL *s*-butanol and stirred at room temperature using a magnetic stir bar. The reaction flask was fitted with a reflux condenser in one neck and a thermometer–temperature controller in another neck. The yellow-colored mixture was then heated to reflux for 21 h. During the reaction the mixture changed from yellow to light blue color. Refluxing was discontinued and the mixture was allowed to cool. The contents of the reaction flask were filtered through a 250-mL medium- porosity fritted glass funnel to obtain a clear yellow filtrate and a light blue solid. 2-Butanone can be observed by GLC (gas-liquid chromatography) in the filtrate. The solid was washed on top of the fritted funnel with 4 × 50-mL aliquots of acetone, followed by vacuum filtration. Filtrates from the first three acetone washes ranged from clear orange to very light tan, and the filtrate from the final wash was clear and colorless. The fritted funnel containing the solid product was dried in vacuum at room temperature for 8 h to yield 4.74 g of light blue $(VO)(HPO_4)\cdot 0.5H_2O$ (27.6 mmol, 68.3%).

Anal. Found (calcd.): 28.59% V (29.63), 18.19% P (18.02), 1.23% H (1.17).

Properties

$(VO)(HPO_4)\cdot 0.5H_2O$ prepared by this method is a light blue microcrystalline solid formed of platy crystallites of size ~ 5 μm. Its X-ray powder diffraction pattern can be indexed on an orthorhombic unit cell, space group *Pmmn* with $a = 7.434$ Å, $b = 9.620$ Å, and $c = 5.699$ Å. The strongest line is (001), corresponding to the layer repeat distance 5.699 Å. Redox titration confirms the vanadium oxidation state as 4.00. The water of hydration is lost and the hydrogen phosphate groups condense on heating at 400° under inert gas to yield vanadyl diphosphate, $(VO)_2(P_2O_7)$, through a topotactic mechanism.[10] $(VO)_2(P_2O_7)$ is an active, selective catalyst for the oxidation of butane to maleic anhydride.[11].

C. VANADYL(IV) HYDROGEN PHOSPHATE HEMIHYDRATE, $(VO)(HPO_4)\cdot 0.5H_2O$: V_2O_5/H_3PO_4/EtOH METHOD

$$V_2O_5 + 2H_3PO_4 + 3\,CH_3CH_2OH \rightarrow 2(VO)(HPO_4)\cdot 0.5H_2O + CH_3CH(OCH_2CH_3)_2 + 3H_2O$$

Procedure

Vanadium pentoxide (V_2O_5), 15.00 g (0.082 mol), was finely ground with a mortar and pestle in a hood and added through a funnel to a 2-L, three-necked, round-bottomed flask containing 500 mL ethanol (95%). The mixture was stirred at room temperature using a mechanical paddle stirrer fitted through the center neck of the flask. Then 22.6 mL (0.330 mol) phosphoric acid (85% H_3PO_4) was added to the reaction flask, followed by another 400 mL ethanol (95%). The reaction flask was fitted with a reflux condenser in the second neck and a thermometer–temperature controller in the remaining neck. The rust-colored mixture was then heated to reflux and stirred for 11 days. During the reaction the mixture changed from orange to olive-green to a pale blue-green color. Refluxing was discontinued and the blue-green mixture was allowed to cool. The contents of the reaction flask were filtered through a 500-mL medium porosity fritted glass funnel to obtain a clear very pale green filtrate and a light green-blue solid. Acetaldehyde diethyl acetal can be observed by GLC in the filtrate. The solid was washed on top of the fritted funnel with 3 × 50-mL aliquots of acetone, followed by vacuum filtration. The filtrates from the acetone washing were clear and colorless. The fritted funnel containing the solid product was dried in vacuum at room temperature for 16 h to yield 28.46 g (0.166 mol, 100%) of $(VO)(HPO_4)\cdot 0.5H_2O$.

Anal. Found (cald.): 29.39% V (29.63), 17.79% P (18.02), 1.31% H (1.17).

Properties

$(VO)(HPO_4)\cdot 0.5H_2O$ prepared by this method is a blue-green microcrystalline solid formed of 10-μm clumps of platy crystallites of size the order of 0.5 μm. Its X-ray powder diffraction pattern is very similar to that of the material produced by the $(VO)PO_4\cdot 2H_2O$/*s*-BuOH method, but the lines are somewhat broader, in keeping with the smaller crystallite size. The two materials are identical as judged by IR spectroscopy, elemental analysis, and oxidation-state titration.

D. VANADYL(IV) PHENYLPHOSPHONATE, $(VO)(C_6H_5PO_3)\cdot H_2O\cdot C_2H_5OH$

$$V_2O_5 + 2C_6H_5PO_3H_2 + 5CH_3CH_2OH \rightarrow 2(VO)(C_6H_5PO_3)\cdot C_2H_5OH + CH_3CH(OC_2H_5)_2 + 2H_2O$$

Procedure

Phenylphosphonic acid ($C_6H_5PO_3H_2$), 10.43 g (66.0 mmol), a white solid, was added through a funnel to a 500 mL, two-necked, round-bottomed flask containing 350 mL ethanol (95%) and stirred at room temperature with a magnetic stir bar. The phenylphosphonic acid is very soluble in the ethanol and within minutes a clear colorless solution is obtained. Vanadium pentoxide (V_2O_5), 6.00 g (33.0 mmol), a rust-colored solid, was finely ground with a mortar and pestle in a hood and added to the reaction flask. The reaction flask was outfitted with a reflux condenser and a thermometer–temperature controller. The orange-colored mixture was then heated to reflux and stirred for 6 days. During the reaction the mixture changed from orange to mint green to a pale blue-green color. Refluxing was discontinued and the blue-green mixture was allowed to cool. The contents of the reaction flask were filtered through a 250-mL medium-porosity fritted glass funnel to obtain a clear gold filtrate and a light blue solid. Acetaldehyde diethyl acetal can be observed by GLC in the filtrate. The solid was washed on top of the fritted funnel with 7×50-mL aliquots of chilled acetone. The filtrates from the first five acetone washes ranged from deep gold to very light gold and the filtrates from the final two washes were clear and colorless. The solid product was suction-dried on top of the fritted funnel at room temperature for 3 h to yield 17.96 g $(VO)(C_6H_5PO_3) \cdot H_2O \cdot EtOH$ (62.57 mmol, 94.8%).

Anal. Found (cald.): 32.39%C (33.47), 4.22% H (4.56), 11.2% P (10.79), 18.5% V (17.74).

Properties

$(VO)(C_6H_5PO_3) \cdot H_2O \cdot EtOH$ is a light blue, fluffy microcrystalline powder. Its X-ray powder diffraction pattern can be indexed on an orthorhombic unit cell with $a = 9.96$ Å, $b = 12.06$ Å, and $c = 9.69$ Å. The strongest line is (010), corresponding to the layer repeat distance 12.06 Å.

E. VANADYL(IV) PHENYLPHOSPHONATE, $(VO)(C_6H_5PO_3) \cdot 2H_2O$

$$(VO)(C_6H_5PO_3) \cdot H_2O \cdot C_2H_5OH + H_2O \rightarrow (VO)(C_6H_5PO_3) \cdot 2H_2O + C_2H_5OH$$

Procedure

The ethanol can be removed from $(VO)(C_6H_5PO_3) \cdot H_2O \cdot EtOH$ by washing the solid with water at 60°, followed by vacuum drying at 65°. Then 17.9 g (62.3 mmol) $(VO)(C_6H_5PO_3) \cdot H_2O \cdot EtOH$ was added through a funnel to a 1-L Erlenmeyer flask containing 350 mL distilled water and a magnetic stir bar. The flask was stirred in a 56° oil bath for 17 h. A persistent froth was present, which was removed from the surface and discarded. The medium green mixture was filtered through a 250-mL medium-porosity fritted glass funnel to obtain a clear light green filtrate and a medium green solid. The solid was washed on top of the fritted funnel with 4×100-mL aliquots of distilled water and vacuum-filtered until the washing filtrates were clear and colorless. The filtrates from the washing ranged in color from light green to light yellow, with the final filtrates clear and colorless. The fritted funnel containing the solid product was placed in a vacuum oven and evacuated at 65° for 16 h to yield 8.09 g (31.2 mmol, 50.1%) $(VO)(C_6H_5PO_3) \cdot 2H_2O$.

Anal. Found (calcd.): 27.81% C (27.82), 3.45% H (3.50), 11.6% P (11.96), 19.9% V (19.66).

Properties

$(VO)(C_6H_5PO_3)\cdot 2H_2O$ is a blue-green microcrystalline solid. Its X-ray powder diffraction pattern can be indexed on an orthorhombic unit cell with $a = 10.03$ Å, $b = 9.69$ Å, and $c = 9.77$ Å. The strongest line is (010) corresponding to the layer repeat distance 9.69 Å. Because of the near-equality of the axis lengths, there are many overlapping peaks in pattern. Straight-chain alcohols are intercalated into $(VO)(C_6H_5PO_3)\cdot 2H_2O$ at slightly elevated temperatures to form $(VO)(C_6H_5PO_3)\cdot H_2O\cdot ROH$.[4]

F. VANADYL(IV) HEXYLPHOSPHONATE, $(VO)(C_6H_{13}PO_3)\cdot H_2O\cdot C_6H_5CH_2OH$

$$V_2O_5 + 2C_6H_{13}PO_3H_2 + 3C_6H_5CH_2OH \rightarrow 2(VO)(C_6H_{13}PO_3)\cdot H_2O\cdot C_6H_5CH_2OH + C_6H_5CHO + H_2O$$

Procedure

Benzyl alcohol ($C_6H_5CH_2OH$), 60.0 mL, was added to a 100 mL, two-necked, round-bottomed flask along with 3 mL distilled water. The mixture was stirred at room temperature with a magnetic stir bar. Then 1.00 g (6.02 mmol) *n*-hexylphosphonic acid ($C_6H_{13}PO_3H_2$) was added to the reaction flask and within minutes the white flakes dissolved. Then 0.498 g (2.736 mmol) vanadium pentoxide (V_2O_5) was finely ground with a mortar and pestle in a hood and added to the reaction flask. The reaction flask was fitted with a reflux condenser and a thermometer–temperature controller. The orange-colored mixture was then heated with stirring to 85° for 3 days. During the reaction the mixture changed from orange to leaf green to a pale blue-green color. Refluxing was discontinued and the blue-green mixture was allowed to cool. The contents of the reaction flask were filtered through a 100-mL medium-porosity fritted glass funnel to obtain a pale yellow filtrate and a light blue-green solid. Benzaldehyde can be observed by IR spectroscopy in the filtrate. The solid was washed on top of the fritted funnel with 2 × 25-mL aliquots of diethyl ether (Et_2O) and suction-dried for $\frac{1}{2}$ h to yield 1.75 g (4.91 mmol, 89.7%) $(VO)(C_6H_{13}PO_3)\cdot H_2O\cdot C_6H_5CH_2OH$.

Anal. Found (calcd.): 41.71% C (43.71), 6.30% H (6.49), 9.19% P (8.67), 14.5% V (14.26).

Properties

$(VO)(C_6H_{13}PO_3)\cdot H_2O\cdot C_6H_5CH_2OH$ is a light blue microcrystalline solid. Its X-ray powder diffraction pattern can be indexed on an orthorhombic unit cell with $a = 10.06$ Å, $b = 18.81$ Å, and $c = 9.84$ Å. The strongest line is (010), corresponding to the layer repeat distance 18.81 Å. The benzyl alcohol can be removed from the compound by evacuation at 140°, producing $(VO)(C_6H_{13}PO_3)\cdot H_2O$, which can selectively intercalate alcohols at ambient temperature to form compounds of the formula $(VO)(C_6H_{13}PO_3)\cdot H_2O\cdot ROH$.[5]

References

1. G. Ladwig, *Anorg. Allg. Chem.*, **338,** 266 (1965).
2. G. Busca, F. Cavani, G. Centi, and F. Trifiro, *J. Catal.*, **99,** 400 (1986).
3. J. W. Johnson, D. C. Johnston, A. J. Jacobson, and J. F. Brody, *J. Am. Chem. Soc.*, **106,** 8123 (1984).
4. J. W. Johnson, A. J. Jacobson, J. F. Brody, and J. T. Lewandowski, *Inorg. Chem.*, **23,** 3842 (1984).
5. J. W. Johnson, A. J. Jacobson, W. M. Butler, S. E. Rosenthal, J. F.Brody, and J. T. Lewandowski, *J. Am. Chem. Soc.*, **111,** 381 (1989).
6. J. W. Johnson, A. J. Jacobson, J. F. Brody, and S. M. Rich, *Inorg. Chem.*, **21,** 3820 (1982). D. Ballutaud, E. Bordes, P. Courtine, *Mater. Res. Bull.*, **17,** 519 (1982).
7. K. Beneke and G. Lagaly, *Inorg. Chem.*, **22,** 1503 (1983); L. Benes, J. Votinsky, J. Kalousova, and J. Klikorka, *Inorg. Chim. Acta*, **114,** (1986).
8. M. Martinez-Lara, L. Moreno-Real, A. Jiminez-Lopez, and A. Rodriguez-Garcia, *Mater. Res. Bull.*, **21,** 13 (1986).
9. J. W. Johnson and A. J. Jacobson, *Angew. Chem.*, **95,** 422 (1983); *Angew. Chem., Int. Ed. Engl.*, **22,** 412 (1983); A. J. Jacobson, J. W. Johnson, J. F. Brody, J. C. Scanlon, and J. T. Lewandowski, *Inorg. Chem.*, **24,** 1782 (1985); M. R. Antonio, R. L. Barbour, and P. R. Blum, *Inorg. Chem.*, **26,** 1235, (1987).
10. E. Bordes, P. Courtine, and J. W. Johnson, *J. Solid State Chem.*, **55,** 270 (1984).
11. E. Bordes and P. Courtine, *J. Catal.*, **57,** 236 (1977); B. K. Hodnett, *Catal. Rev. Sci. Eng.*, **27,** 373 (1985); G. Centi, F. Trifiro, J. R. Ebner, and V. M. Franchetti, *Chem. Rev.*, **88,** 55 (1988).

46. QUATERNARY CHLORIDES AND BROMIDES OF THE RARE-EARTH ELEMENTS: ELPASOLITES $A^I_2B^IRE^{III}X_6$ [r (A^I)>r (B^I)]

Submitted by GERD MEYER*
Checked by S.-J. HWU† and J. D. CORBETT†

Reprinted from *Inorg. Synth.*, **22,** 10 (1983)

A. CESIUM LITHIUM THULIUM CHLORIDE, $Cs_2LiTmCl_6$

$$4CsCl + Li_2CO_3 + Tm_2O_3 \xrightarrow{HCl(aq)} (2Cs_2LiTmCl_6)\ (hyd)$$

$$\xrightarrow[-H_2O]{HCl(g)500^\circ} 2Cs_2LiTmCl_6$$

Quaternary chlorides and bromides of the rare-earth elements (RE = Sc, Y, La, lanthanides), the so-called elpasolites, $A^I_2B^IRE^{III}X_6$ [$r(A^I) > r(B^I)$, (A^I = Cs, Rb,Tl,K; B^I = Li,Ag,Na,K; X = Cl,Br] are known in many of the theoretically possible cases.[1-7] They have been proved to be useful tools in investigating the properties of the trivalent rare-earth ions because of the often ideal octahedral surrounding of RE^{3+} by X^- (site symmetry O_h). Elpasolites have been and will be widely used for optical measurements and Raman and ESR (electron spin resonance) spectra. Susceptibility data have also been collected and interpreted to evaluate the behavior of ions with 4f electrons in ligand fields of different strength. Distances between RE^{3+} ions are relatively large; the (REX_6) octahedra are "isolated" so that no magnetic interactions between the paramagnetic ions occur.

Procedure

Cesium chloride, 673.4 mg (4 mmol), 73.9 mg (1 mmol) Li_2CO_3, and 385.9 mg (1 mmol) Tm_2O_3, all available as reagent-grade or ultrapure from various sources, are used as starting materials. Approximately 200 mL concentrated hydrochloric acid is added to the reactants as in Chapter 2 Section 15.A of this volume, and the same further procedure followed as described there.

* Institut für Anorganische und Analytische Chemie I, Justus-Liebig-Universitaet Giessen, Heinrich-Buff-Ring 58, 6300 Giessen, Germany.

† Department of Chemistry and AmesLaboratory, Iowa State University, Ames, IA 50011.
The Ames Laboratory is operated for the U.S. Department of Energy by Iowa State University under Contract No. W-7405-ENG-82.

$Cs_2LiTmCl_6$ is obtained as a slightly sintered powder. Because of its sensitivity to moisture, it should be stored and handled in a dry box and for longer periods of time sealed under inert gas (N_2 or Ar) in Pyrex tubes.

The yield is approaching the theoretical value (2 mmol $Cs_2LiTmCl_6$:1.309 g).

Properties

$Cs_2LiTmCl_6$ is colorless and moisture-sensitive. For its characterization, common X-ray (Debye–Scherrer or, better, Guinier patterns) and/or spectroscopic techniques are used. The first nine d values of cubic-face-centered $Cs_2LiTmCl_6$ (a = 1043.9 pm) are: 6.022(54), 3.688(100), 3.145(25), 3.011(30), 2.608(64), 2.393(9), 2.129(43), 2.007(11), 1.844(39) Å; numbers in parentheses are calculated intensities on a 1–100 scale.

Figure 1 shows the Raman spectra of $Cs_2LiTmCl_6$ and $Cs_2LiTmBr_6$.

B. OTHER CHLORO AND BROMO ELPASOLITES

Other chloro elpasolites are obtained by the same preparative route by selecting the appropriate starting materials (carbonates and chlorides as well; Ag_2O is also useful). For the preparation of bromo elpasolites, chlorides as

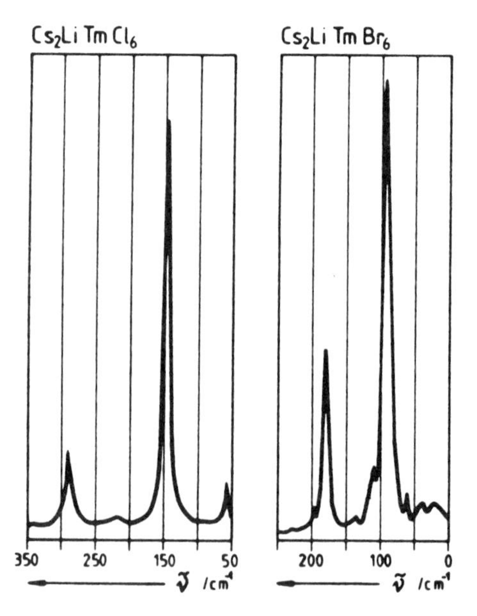

Figure 1. Raman spectra of $Cs_2LiTmCl_6$ and $Cs_2LiTmBr_6$.

starting materials of course have to be replaced by bromides, hydrochloric acid by hydrobromic acid, and HCl gas by HBr gas, which is also obtainable in steel cylinders or lecture bottles.

When Pr or Tb compounds are to be prepared, the use of "Pr_6O_{11}" or "Tb_4O_7" (the sesquioxides are not available commercially) requires more time to obtain a clear solution (usually 15 min or more), and additional HCl solution may be needed, for example, to dissolve precipitated binary halides such as AgCl and TlCl.

Larger-scale syntheses should be possible with appropriate sizes of the crucible, quartz tube, and tubular furnace.

In some cases (elpasolites $Cs_2NaRECl_6$[1]) the desired products crystallize from aqueous HCl solution so that drying the residue in the HCl gas stream is not necessary. However, to make sure that the expected compound forms, the drying step should be carried out. In general, melting in the open-flow system described above should be avoided to prevent undesirable effects such as incongruent melting, sublimation of binary components, or creeping of the material over the edges of the crucible. Therefore, the temperature should be maintained at 400–450° in the case of the rubidium chlorides and compounds of the $Cs_2LiREBr_6$ type.

Large single crystals of $Cs_2NaRECl_6$-and $Cs_2NaREBr_6$-type compounds can be grown by a Bridgman technique.[8–11]

Four different crystal structure types are known for elpasolites depending on the composition and the intensive variables (temperature, pressure). *K* is a cubic face-centered K_2NaAlF_6 type,[12] *the* elpasolite (there is also a lower symmetry variety, *T*, which can be indexed tetragonally[5]) with $[BX_6]$ and $[REX_6]$ octahedra sharing common corners, and B^I and RE^{III} occupying the octahedral interstices alternately between layers of composition AX_3 with the stacking sequence *ABCABC* . . . .

The 2*L* form has the trigonal Cs_2LiGaF_6 structure type,[13] with a stacking sequence *AB* . . . . The $[BX_6]$ and $[REX_6]$ octahedra all share common faces. Both 6*L* (HT-K_2LiAlF_6-type,[14] stacking sequence: *ABCBAC* . . .) and 12*L* (rhombohedral Cs_2NaCrF_6 type,[13] stacking sequence *ABABCACABCBC* . . .) are polytypes between the polymorphs *K* and 2*L* with confacial bioctahedra (6*L*) and triple octahedra (12*L*), respectively, connected via common corners with single octahedra. Figure 2 shows the characteristic structural features.

Thermally induced phase transitions are found for $Rb_2NaTmCl_6$ ($T \rightarrow K$)[5] and $Cs_2LiLuCl_6$ ($6L \rightarrow K \rightarrow 6L$)[15] as well as for others. In general, higher temperatures favor structures with more face-sharing octahedra, that is, with larger molar volumes.

Known chloro and bromo elpasolites are summarized in Tables I and II together with their lattice constants. Mixed crystals with their particular properties have also been prepared.[5]

TABLE I. Lattice Constants (in pm) of Cubic and Tetragonal Chloro and Bromo Elpasolites

RE	$Rb_2LiRECl_6$	$Rb_2NaRECl_6$	$Cs_2LiRECl_6$	$Cs_2AgRECl_6$	$Cs_2NaRECl_6$	Cs_2KRECl_6	$Cs_2LiREBr_6$	$Cs_2NaREBr_6$
La			1072.4		1099.2	(1137.9)[a]	1127.9	1151.9[b]
								1158.5
Ce			1067.5	1090.7	1094.6	1128.1	1123.6	1150.3
						1135.1		
Pr			1065.1	1087.8	1091.2	1130.9	1120.7	1146.8
Nd			1061.9	1085.2	1088.9	1126.8	1118.2	1144.6
Sm	1042.7[b]		1057.7	1080.0	1083.4	1121.3	1113.4	1139.3
	1050.0							
Eu	1040.4[b]	1064.1	1055.4	1078.0	1081.0	1118.0	1111.4	1137.8
	1048.2	1069.9						
Gd	1038.8	1061.3	1053.4	1075.8	1079.2	1116.4	1109.7	1136.1
	1043.8	1067.6						
Tb	1037.8		1050.4	1073.5	1076.7	1116.0	1107.3	1133.4
Dy	1035.4	1057.6	1048.7	1071.2	1074.3	1115.9	1105.7	1131.4
		1065.2						

Y	1034.4	1056.6 1063.3	1047.9	1069.8	1073.2	1112.8	1104.7	1130.1
Ho	1033.9	1056.2 1062.5	1047.0	1069.5	1072.9	1111.6	1103.4	1129.3
Er	1030.7	1053.9 1060.5	1045.1	1068.2	1070.4	1109.9	1102.1	1127.9
Tm	1029.6	1053.9 1059.9	1043.9	1065.4	1068.6	1107.2		1125.3
Yb	1028.3	1050.9 1057.6	1041.8	1064.4	1067.7	1107.4		1124.5
Lu	1026.9	1049.4 1057.0		1062.6	1065.5	1102.7		1123.3
Sc	1010.7	1036.4		1045.2	1048.8	1088.0		1107.0
Ref.	5	5	4	3	1	2[c]	7	6,7

[a] Presumably tetragonal.
[b] Tetragonal with *a* above and *c* below.
[c] Ce, Lu and Sc compounds are author's own observations.

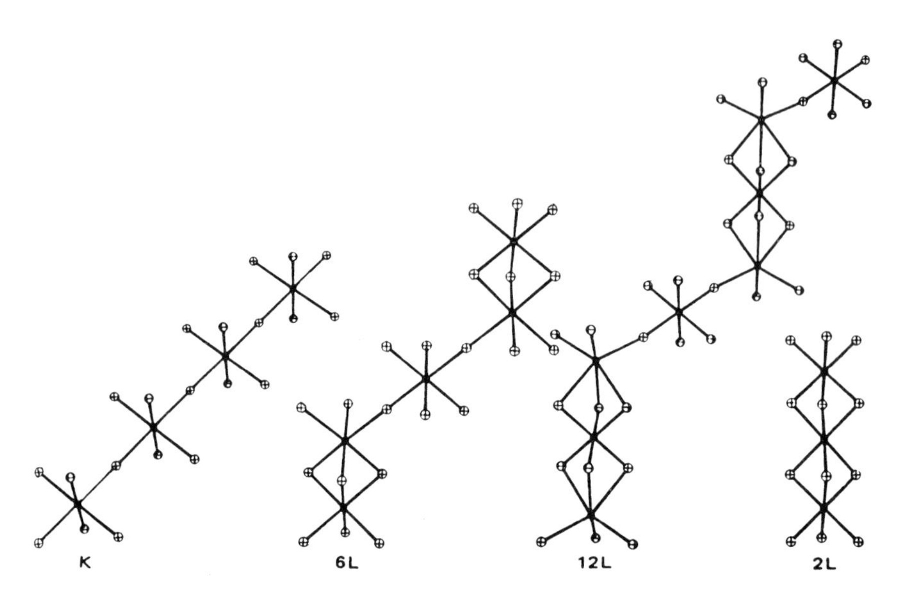

Figure 2. Characteristic structural features in $A^{I}_{2}B^{I}RE^{III}X_6$-type compounds ("elpasolites"). Sections in [00.1] direction show the principal connection of $[BX_6]$ and $[REX_6]$ octahedra (see text).

TABLE II. Trigonal and Hexagonal Chloro and Bromo Elpasolites: Crystal Structures and Lattice Constants[4,7]

	Structure Type	*a*/pm	*c*/pm
$Cs_2LiLuCl_6$	6*L*	738.6	1821
$Cs_2LiScCl_6$	12*L*	730.0	3601
$Cs_2LiTmBr_6$	6*L*	777.4	1925.2
$Cs_2LiYbBr_6$	6*L*	777.0	1924.1
$Cs_2LiLuBr_6$	6*L*	775.9	1919.6
$Cs_2LiScBr_6$	2*L*	761.8	651.3

All these compounds are most conveniently characterized by powder X-ray diffraction methods. To calculate line intensities, computer programs such as LAZY-PULVERIX[16] are readily available; the required crystallographic data may be found in References 12–14.

References

1. L. R. Morss, M. Siegal, L. Stenger, and N. Edelstein, *Inorg. Chem.*, **9**, 1771 (1970).
2. G. Baud, L. Baraduc, P. Gaille, and J.-C. Cousseins, *C. R. Hebd. Seances Acad. Sci. C*, 272, 1328 (1971).
3. G. Meyer and P. Linzmeier, *Rev. Chim. Miner.*, **14**, 52 (1977).
4. G. Meyer and H.-C. Gaebell, *Z. Anorg. Allg. Chem.*, **445**, 147 (1978).
5. G. Meyer and E.Dietzel, *Rev. Chim. Miner.*, **16**, 189 (1979).
6. G. Meyer and P. Linzmeier, *Z. Naturforsch.*, **32b**, 594 (1977).
7. G. Meyer and H.-C. Gaebell, *Z.Naturforsch.*, **33b**, 1476 (1978).
8. R. W. Schwartz, T. R. Faulkner, and F. S. Richardson, *Mol. Phys.*, **38**, 1767 (1979).
9. R. W. Schwartz and P. N. Schatz, *Phys. Rev. B*, 8, 3229 (1973).
10. G. Mermant and J. Primot, *Mat. Res. Bull.*,**14**, 45 (1979).
11. H.-D. Amberger, *Z. Naturforsch.*, **35b**, 507 (1980).
12. L. R. Morss, *J. Inorg. Nucl. Chem.*, **36**, 3876 (1974).
13. D. Babel and R. Haegele, *J. Solid State Chem.*, **18**, 39 (1976).
14. H. G. F. Winkler, *Acta Cryst.*, **7**, 33 (1954).
15. G. Meyer and W. Duesmann, *Z. Anorg. Allg. Chem.*, **485**, 133 (1982).
16. K. Yvon, E. Parthe, and W. Jeitschko, *J. Appl. Cryst.*, **10**, 73 (1977).

47. PREPARATION OF Ta_2S_2C

$$2Ta + 2S + C \xrightarrow{C_6Cl_6} Ta_2S_2C$$

Submitted by R. P. ZIEBARTH* and F. J. DiSALVO†
Checked by LAVRE MONCONDUIT† and RAYMOND BREC‡

Ta_2S_2C is a layered compound[1] composed of tightly bound two-dimentional slabs containing five close-packed layers of atoms sequenced S–Ta–C–Ta–S. These slabs are in turn held together by van der Waals interactions that give Ta_2S_2C a graphitic character. In spite of its apparent low-dimensional character, the compound exhibits metallic behavior down to 4.2 K. A number of intercalated derivatives containing alkali metals[2], transition metals[3], and organic compounds[4] have also been prepared.

Ta_2S_2C was first prepared as one of the products of the reaction of a 1 : 1 : 1 mixture of tantalum, sulfur, and carbon at 1200°.[1] Later reports[2] indicated that the compound could be prepared in pure form by the reaction of stoichiometric quantities of tantalum, sulfur, and carbon in a graphite crucible (sealed in evacuated silica jacket) at 1200°. Recent work[5] has shown

* Department of Chemistry, The Ohio State University, Columbus, OH 43221.
† Department of Chemistry, Cornell University, Ithaca, NY 14853.
‡ Institut des Matériaux de Nantes, 2, rve de la Houssiniére, 44072 Nantes Cédex 03 France.

that catalytic amounts of halogens are essential to the preparation of the compound at 1150–1200° and Ta_2S_2C can be prepared in evacuated fused-silica ampules by the addition of trace quantities of C_6Cl_6, C_6Br_6, or I_2.

Procedure

■ **Caution.** *The presence of adsorbed water in graphite powders not recently baked out under vacuum may lead to excessive pressure in and the subsequent explosion of the sealed silica reaction ampule. Extensive devitrification of the silica reaction ampules may occur during the high-temperature reaction, which can also result in explosion of the reaction vessel. A face shield, and gloves should be worn and care should be exercised when removing the reaction ampules from the furnace.*

A 1.00 g sample of Ta_2S_2C was prepared from a stoichiometric mixture of tantalum powder (Wah Chang, 60 mesh, 99.9%), sulfur powder (Atomergic Chemicals Corporation, 99.9999%), and spectroscopic-grade carbon powder (National Carbon Co., Division of Union Carbide) that had been degassed by heating to 1000° under vacuum ($\sim 10^{-4}$ torr) for 12 h; 10 mg of C_6Cl_6 (Aldrich, 97%) was also included in the reaction ampule as a source of halogen. Reactants were placed in a 12-mm-o.d. (10-mm i.d.) silica tube that was evacuated and sealed off with a hot gas/oxygen flame. The 12-mm reaction tube was rinsed with acetone, dried, and sealed in an evacuated 16-mm-o.d. silica tube. The outer tube reduces devitrification of the inner reaction tube and thus reduces the risk of reaction tube breakage on cooling.

The "double-sealed" reaction tube was placed in a SiC furnace and heated at 100°/h to 400° then ramped at 10°/h to 600°. The slow increase in temperature between 400 and 600° provides time for the sulfur to react to form tantalum sulfides and thus prevents the buildup of an excessive sulfur pressure in the reaction tube. At 600° the vapor pressure of sulfur over liquid sulfur is approximately 10 atmospheres. Larger scale reactions may require the reaction to be held at 600° to provide additional time for the reaction of Ta and S. From 600° the reaction was heated at 50°/h to 1180°. After 5 days the furnace was cooled to 850° and the silica reaction vessel was carefully removed (note safety precautions above) and cooled on a firebrick.

Properties

The product is a microcrystalline black powder with a graphitic character. The powder diffraction pattern has been reported by Beckmann et al. Grinding tends to introduce some stacking disorder into the material and broadens reflections associated with the *hkl* ($h + k \neq 0$) lines in the powder diffraction pattern. In addition, the intensities of the 001 X-ray reflections are often "enhanced" when using a powder diffractometer, due to the nonrandom orientation of the platelike crystallites in the sample holder. In some cases, depending on the quality of the silica tube and the exact reaction temperature, reaction with the silica tube may result in some impurity phases. In one reaction we found a moderately strong peak at $d = 4.03$ A (intensity 18% of strongest peak). The samples are usually a mixture of stacking variants dominated by one- and three-layer repeat sequences. The ratio of such variants may depend on the exact reaction temperature and/or the degree of grinding.

References

1. O. Beckmann, H. Boller, and H. Nowotny, *Montashefte für Chemie*, **101**, 61(1970).
2. R. Brec, J. Ritsma, G. Ouvrard, and J. Rouxel, *Inorg. Chem.*,**16**, 660 (1977).
3. H. Boller and R. Sobczak, *Montashefte für Chemie*, **102**, 1226 (1971).
4. R. Schöllhorn and A. Weiss, *Z. Naturforsch.*, **28b**, 716 (1973).
5. R. P. Ziebarth, J. K. Vassiliou, and F. J. DiSalvo, *J. Less-Common Met.*, **156**, 207 (1989).

48. SCANDIUM STRONTIUM BORATE, ALUMINUM STRONTIUM YTTRIUM BORATE, AND LANTHANUM MAGNESIUM STRONTIUM BORATE

$$\tfrac{1}{2}Sc_2O_3 + 3Sr(NO_3)_2 + \tfrac{3}{2}B_2O_3 \rightarrow Sr_3Sc(BO_3)_3 + 3\text{“}N_2O_5\text{”}$$

$$0.465\,Al_2O_3 + 6Sr(NO_3)_2 + 0.535\,Y_2O_3 + 3B_2O_3 \rightarrow Sr_6YAl_{1.07}Al_{0.93}(BO_3)_6 + 6\text{“}N_2O_5\text{”}$$

$$2La(NO_3)_3 \cdot 6H_2O + Mg(NO_3)_2 \cdot 2H_2O + 5Sr(NO_3)_2 + 3B_2O_3 \rightarrow La_2Sr_5Mg(BO_3)_6 + 8H_2O + 9\text{“}N_2O_5\text{”}$$

$$2\text{“}N_2O_5\text{”} \rightarrow 4NO_2 + O_2$$

Submitted by KATHLEEN I. SCHAFFERS and DOUGLAS A. KESZLER*
Checked by L. F. SCHNEEMEYER†

Borates have been extensively used for some time in the glass, ceramic, and porcelain industries to produce materials with small coefficients of thermal

* Center for Advanced Materials Research and Department of Chemistry, Oregon State University, Corvallis, OR 97331–4003.
† AT&T Bell Laboratories, 600 Mountain Ave., Murray Hill, NJ 07974.

expansion. More recently, they have proved to be useful as optical materials and heterogeneous catalysts. The compounds BaB_2O_4[1] and LiB_3O_5[2] are exceptional materials for frequency conversion of laser light at high powers to short wavelengths, and the compound Cr^{3+}:$ScBO_3$[3] functions as a broadly tunable solid-state laser material. The compound $Cu_{2-x}Zn_xAl_6B_4O_{17}$[4] is one example of a borate that serves as a heterogeneous catalyst for the production of methanol.

We recently reported the existence of the largest structural family of borates discovered to date.[5] The family has the nominal composition $A_6MM'(BO_3)_6$, where A = Sr, Ba, or large lanthanide and the elements M and M′ are any of those cations having a + 2, + 3, or + 4 formal oxidation state and a preference for octahedral coordination. The structure of the materials contains chains (Fig. 1) consisting of octahedra alternately occupied by atoms M or M′ and separated by triangular planar BO_3 groups. The distorted octahedron occupied by atom M′ is smaller than that occupied by atom M. The sizes of the octahedra are dictated primarily by the interactions of the chain with the A atoms that link the chains into the three-dimensional structure.

We have prepared 135 compounds that crystallize in this structural type; the large number of derivatives results from the variety of elements that will occupy the A, M, and M′ sites. We describe here the preparation of three types of derivatives. The compound $Sr_3Sc(BO_3)_3$ contains A = Sr and M = M′ = Sc, a formally + 3 cation. The compound $Sr_6Y_{1.07}Al_{0.93}(BO_3)_6$ contains M = Y and M′ = Al, two cations of formal charge + 3 but of disparate sizes. The compound $La_2Sr_5Mg(BO_3)_6$ contains M = Sr and M′ = Mg, cations of formal charge + 2, and the larger La atom distributed on the A site.

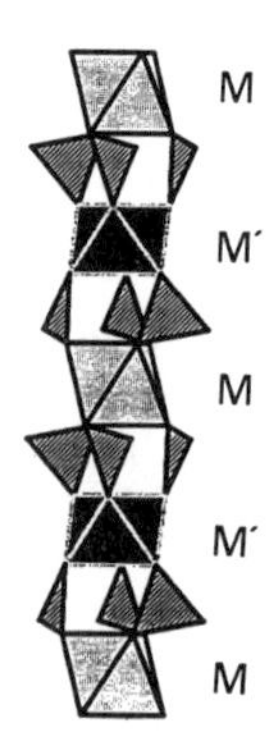

Figure 1. A single chain in the structure of the family of borates $A_6MM'(BO_3)_6$.

Procedure

Standard high-temperature techniques are efficient means for the synthesis of the materials. These techniques have proved to be superior to heating precipitates from aqueous solutions formed with an assortment of reagents at various pH levels. The reagents used in the syntheses are $Sr(NO_3)_2$ (reagent-grade or Puratronic, Johnson-Matthey), Sc_2O_3 (99.0%, Boulder Scientific), Y_2O_3 (99.99%, Rare Earth Products), Al_2O_3 (99.99%, CERAC), $La(NO_3)_{3.6}H_2O$ (99.9%, AESAR), $Mg(NO_3)_2.2H_2O$ (Puratronic, Johnson-Matthey), and B_2O_3 (99.99%, ALFA). To prepare the compounds, stoichiometric quantities of the reagents are mixed and ground with a 3 mol% excess of B_2O_3. The use of the reagent B_2O_3 is preferred to boric acid, H_3BO_3, as the acid tends to froth on dehydration, subsequently wetting the walls of the crucible. The reagent B_2O_3, however, is difficult to retain in an anhydrous condition.[6] We dry B_2O_3 by heating it at 393 K for 3 days followed by melting in a Pt crucible at 1073 K for 3.5 h. A 3 mol% excess of the reagent is used in each synthesis to compensate for residual H_2O and minor volatilization losses during heating. Prior to use, $Sr(NO_3)_2$ is heated in a Pt crucible at 573 K for 3 days, and Sc_2O_3 is heated at 1173 K for 5 h in a gold crucible under flowing O_2. Drying $Sr(NO_3)_2$ and Sc_2O_3 in this way is not critical for a successful synthesis. We observe a mass loss of 1% or less following this procedure.

The compounds are readily prepared in a Pt crucible by flash heating (as described below) ground quantities of the starting materials: 784 mg (3.65 mmol) $Sr(NO_3)_2$, 85 mg (0.62 mmol) Sc_2O_3, and 131 mg (1.88 mmol) B_2O_3 for $Sr_3Sc(BO_3)_3$; 772 mg (3.65 mmol) $Sr(NO_3)_2$, 69 mg (0.30 mmol) Y_2O_3, 31 mg (0.30 mmol) Al_2O_3, and 128 mg (1.84 mmol) B_2O_3 for $Sr_6Y_{1.07}Al_{0.93}(BO_3)_6$; and 373 mg (0.86 mmol) $La(NO_3)_3.6H_2O$, 456 mg (2.15 mmol) $Sr(NO_3)_2$, 80 mg (0.43 mmol) $Mg(NO_3)_2.2H_2O$, and 91 mg (1.31 mmol) B_2O_3 for $La_2Sr_5Mg(BO_3)_6$. The furnace is set to the desired temperature followed by introduction of the samples. The initial heating is performed at 923 K for 30 min and the second at 1023 K for 20 min. As oxides of nitrogen are released in these initial heatings, the furnace is placed in a fume hood. The sample is then reground and flash-heated at 1303 K for 3 h. The crystallinity of the sample $Sr_6Y_{1.07}Al_{0.93}(BO_3)_6$ is improved by annealing at 1373 K for 24 h.

Another method may be employed to reduce the particle sizes of the starting materials, thereby increasing their reactivity. Stoichiometric quantities of the reagents are first heated at 923 K for 1 h to decompose the nitrates. The resulting mixture, corresponding to 1 mmol of product, and 1.5 mL of 20% NH_4OH(aq) (Baker, ultrapure, ULTREX) are placed in a 23-mL Teflon container and heated in a Parr 4745 digestion bomb at 393 K for 16 h. The lid

TABLE I. Diffraction Data

2θ(°)*	*hkl*	I/I_{100}
	$Sr_3Sc(BO_3)_3$	
19.43	021	8
21.10	012	6
24.37	211	30
25.72	202	18
29.19/29.37/29.71	$003/220/21\bar{2}$	100
32.17	$31\bar{1}$	36
32.76	$11\bar{3}$	14
35.47	401	11
36.46	312	29
38.52	$23\bar{1}$	33
39.19	140	28
41.89/42.27	$223/32\bar{2}$	28
45.62	214	27
49.20/49.60	$15\bar{1}/413$	48
50.56	$31\bar{4}$	19
52.32/52.81	152/205	32
53.90/54.42	431/520	32
56.85	$43\bar{2}$	12
	$Sr_6Y_{1.07}Al_{0.93}(BO_3)_6$	
21.23	012	7
24.34	211	39
25.84	202	13
29.32/29.75	$003/21\bar{2}$	100
32.08	$31\bar{1}$	34
32.95	$11\bar{3}$	11
35.38	401	5
36.44	312	26
38.37	$23\bar{1}$	22
39.08	140	24
42.01	223	34
45.89	214	23
49.03/49.67	$15\bar{1}/413$	53
50.81	$31\bar{4}$	20
52.24	152	17
53.20/53.75/54.25	205/431/520	33
56.73	$43\bar{2}$	12

*λ = 1.54184 Å.

of the Teflon container is removed after cooling, and the liquid is evaporated at 393 K. The sample is then flash-heated as detailed above or according to the following schedule: 923 K, 1 h; 998 K, 4 h; 1098 K, 12 h; 1198 K, 12 h; and 1273 K, 24 h.

Properties

X-ray powder diffraction data for the compounds $Sr_3Sc(BO_3)_3$ and $Sr_6Y_{1.07}Al_{0.93}(BO_3)_6$ are given in Table I. Rhombohedral unit-cell parameters for each compound prepared by the two different synthetic methods are listed in Table II. Cell parameters were determined by least-squares analysis of 10 reflections collected by using an automated Philips powder diffractometer and corrected with NIST Si standard 640b. The nonstoichiometry of the Y–Al phase has been verified by single-crystal X-ray analysis. Powder diffraction measurements indicate that it is a line phase at this stoichiometry.

Differential thermal analysis of $Sr_3Sc(BO_3)_3$ prepared by the flash-heating method indicates a melting point of 1523 K. The melting point of Au was used as a standard to correct the data.

The derivatives $Sr_3Sc(BO_3)_3$ and $Sr_6Y_{1.07}Al_{0.93}(BO_3)_6$ are particularly interesting as possible new laser materials when doped with the ion Cr^{3+}. Fluorescence spectra from samples of $Sr_3Sc_{0.98}Cr_{0.02}(BO_3)_3$ and $Sr_6YAl_{0.98}Cr_{0.02}(BO_3)_6$ prepared by flash heating exhibit maxima in their broad emission bands at 745 nm.

TABLE II. Unit-Cell Data

Compound	*a* (Å)	*c* (Å)	*V* (Å^3)
	Flash Synthesis		
$Sr_3Sc(BO_3)_3$	12.151(1)	9.178(1)	1173.6(1)
$Sr_6Y_{1.07}Al_{0.93}(BO_3)_6$	12.194(1)	9.102(1)	1172.1(2)
$LaSr_2Mg(BO_3)_3$	12.309(1)	9.259(2)	1214.9(3)
	NH_4OH Synthesis		
$Sr_3Sc(BO_3)_3$	12.144(1)	9.184(1)	1173.0(1)
$Sr_6Y_{1.07}Al_{0.93}(BO_3)_6$	12.179(1)	9.096(1)	1168.4(1)
$LaSr_2Mg(BO_3)_3$	12.306(1)	9.261(1)	1214.6(2)

References

1. D. Eimerl, L. Davis, S. Velsko, E. K. Graham, and A. Zalkin, *J. Appl. Phys*, **62**, 1968 (1987).
2. C. Chen, Y. Wu, A. Jiang, B. Wu, and G. You, *J. Opt. Soc. Am B*, **6**, 616 (1989).
3. S. T. Lai, B. H. T. Chai, M. Long, and R. C. Morris *IEEE J. Quant. Electron*, **QE-22**, 1931 (1986).
4. K. I. Schaffers, T. Alekel III, P. D. Thompson, J. R. Cox, and D. A. Keszler, *J. Am. Chem. Soc.*, **112**, 7068 (1990).
5. A. Zletz, U.S. Patent 709,790, 11 March 1985.
6. J. D. MacKenzie, *J. Phys. Chem*, **63**, 1875 (1959).

49. PREPARATION OF ZrO_2, ZnO, AND ZnS THIN FILMS

Submitted by Y-M. GAO*, R. KERSHAW*, and A. WOLD*
Checked by D. M. SCHLEICH,†

A. GROWTH AND CHARACTERIZATION OF ZrO_2 FILMS

$$Zr(C_5H_7O_2)_4 + 24O_2 \rightarrow ZrO_2 + 20CO_2 + 14\ H_2O$$

The high melting point, chemical inertness, high refractive index, wide bandgap, high dielectric constant and electrical resistivity have made ZrO_2 an important refractory material.[1] Zirconium dioxide thin films can be used for optical coatings, protective coatings, and insulating layers.[2,3] Various fabrication techniques have been employed for the preparation of high-quality films of ZrO_2, such as pulsed-laser evaporation,[4] ion-assisted deposition,[5] chemical-vapor deposition,[6] and metalorganic chemical-vapor deposition.[7] Recently a novel ultrasonic nebulization and pyrolysis technique has been developed for the preparation of ZrO_2 films.[8]

Procedure

Zirconium oxide films have been prepared by an ultrasonic nebulization and pyrolysis technique.[9] A solution (0.02*M*) of zirconium acetylacetonate in ethanol was nebulized by a commercial ultrasonic humidifier (Holmes Air transducer operates at 1.63 MHz) and was carried into a horizontal reactor by argon (Fig. 1). The reactor was heated by a two-zone mirror furnace (Transtemp Co., Chelsea, MA).

* Brown University, Providence, RI 02912.

† ISITEM, Universite de Nantes, La Chantretie, Rue Christian Pauc, CP 3023, 44087 Nantes, France.

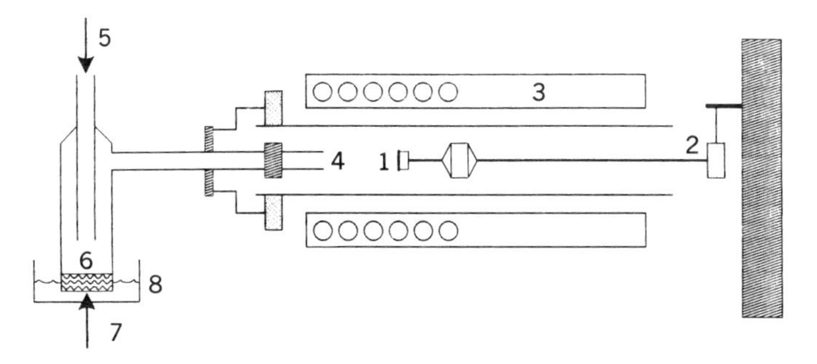

Figure 1. Ultrasonic humidifier–horizontal reactor setup: (1) substrate; (2) substrate rotation motor; (3) furnace; (4) spray nozzle; (5) carrier gas; (6) solution; (7) membrane; (8) ultrasonic humidifier.

The substrate was held perpendicular to the gas flow in the furnace by a silica holder that was rotated by a low-speed motor to achieve best uniformity. Both the efficiency of the deposition and the uniformity of the films were affected by the deposition parameters used. To obtain the best uniformity, the tip of the nozzle, which was 2.5 cm long, was bent upward at an angle of about $15 \pm 5°$. Typical reaction parameters were: furnace temperature, 485°; argon flow rate, 4.5 L/min; distance between nozzle and substrate, 10 cm. Under such conditions, a film of 2000 Å could be grown in 15 min using 15 mL of solution. After deposition, films were annealed in a flowing oxygen atmosphere at the same temperature for 30 min. Both (100) silicon and silica wafers were used as substrates. Cleaning of the silicon substrates was carried out just prior to the deposition according to the procedure described by Fournier et al.[10]The procedure was to clean the silicon substrates with (1:1) hydrochloric acid for 10 min, then with (1:2) hydrofluoric acid for 0.5 min. The substrates were rinsed with distilled water and finally semiconductor-grade acetone prior to deposition of a film. Silica substrates were cleaned with hydrochloric acid, distilled water, and semiconductor-grade acetone prior to the deposition.

Characterization

The thickness of the films on silicon substrates was determined by ellipsometry using a Rudolph Research Auto EL-II ellipsometer,[11] whereas those on silica substrates were measured by interference fringes in their UV–visible spectra.

X-ray diffraction patterns of the zirconium oxide films were obtained using a Philips diffractometer and monochromated high-intensity Cu K_{α_1} radiation

($\lambda = 1.5405$ Å). Diffraction patterns were taken with a scan rate of 1° 2θ/min over the range $15° < 2\theta < 80°$. The two principal diffraction peaks were found at 35.4° and 31.3° and corresponded to the (200) and (111) reflections.

Characterization of the surface topography of films deposited on silicon was achieved by examining replicas of the surface in a transmission electron microscope. A replication solution (Ladd) was applied to the surface of the film and allowed to dry. The plastic formed a negative image on the surface. A carbon–platinum film was shadow-evaporated from a source pellet onto the stripped plastic at a glancing angle followed by a uniform layer of carbon. Each carbon replica was liberated using acetone, mounted on 200-mesh copper grids and examined in a Philips 420 scanning transmission electron microscope (STEM) operating at 60 kV. The grain sizes were measured on a zirconium oxide film to be 0.3 μm in thickness.

Optical measurements of the films on silica substrates were performed using a Cary model 17 dual-beam ratio recording spectrophotometer in the range 200–1600 nm. Measurements were made in the transmission mode. The optical bandgap was deduced from the transmittance near the absorption edge. The thickness of each film was obtained from the fringes in the spectra.

Current–voltage measurements were performed to determine the dC breakdown voltage of the films. The measurements were carried out with samples in an $Au/ZrO_2/Si$ configuration.[9] Gold was evaporated through a mask to obtain an array of circular electrodes of 1-mm^2 area on the surface of the zirconium oxide film. Contact to a gold dot electrode was obtained by touching a gold-tipped micromanipulator to the surface. Contact to the silicon substrates (*n*-type with a resistivity of 0.01 Ω·cm) consisted of indium alloy (Indalloy 9, Indium Corporation of America) ultrasonically bonded to the back surface. The indium alloy contacts were checked for ohmic behavior before the current–voltage response was measured (with the gold dot made negative with respect to the substrate). A 1-mm boundary along the edge of the substrate was ignored.

References

1. D. C. Bradley and P. Thornton, in *Comprehensive Inorganic Chemistry*, Vol. 3, A. F. Trotman-Dickenson (ed.), Pergamon Press, Oxford, 1973; 3, p. 426.
2. K. V. S. R. Apparao, N. K. Sahoo, and T. C. Bagchi, *Thin Solid Films*, **129**, L71 (1985).
3. M. Balog, M. Schieber, S. Patai, and M. Michman, *J. Cryst. Growth*, **17**, 298 (1972).
4. H. Sankur, J. DeNatale, w. Gunning, and J. G. Nelson, *J. Vac. Sci. Technol.*, **A5** (5), 2869 (1987).
5. O. R. McKenzie, D. J. H. Cockayne, M. G. Sceats, P. J. Martin, W. G. Sainty, and R. P. Netterfield, *J. Mater. Sci.*, **22** (10), 3725 (1987).
6. R. N.Tauber, A. C. Dumbri, and R. E. Caffrey, *J. Electrochem. Soc., Solid State Sci.* **118**, 747 (1971).

7. M. Balog, M. Schieber, M. Michman, and S. Patai, *Thin Solid Films*, **47**, 109 (1977).
8. Y-M. Gao, P. Wu, R. Kershaw, K. Dwight, and A. Wold, *Mater. Res. Bull.*, in press.
9. P. Wu, Y-M. Gao, R. Kershaw, K. Dwight, and A. Wold, *Mater. Res. Bull.*, **25** (3), 357 (1990).
10. J. Fournier, W. DeSisto, R. Brusasco, M. Sosnowski, R. Kershaw, J. Baglio, K. Dwight, and A. Wold, *Mater. Res. Bull.*, **23**, 31 (1988).
11. R. Brusasco, R. Kershaw, J. Baglio, K. Dwight, and A. Wold, *Mater. Res. Bull.*, **21**, 301 (1986).

B. GROWTH OF ZnO FILMS

$$Zn(C_2H_3O_2)_2 + 4O_2 \rightarrow ZnO + 4CO_2 + 3H_2O$$

Zinc oxide with wurtzite structure is a semiconductor with a large bandgap. Under certain conditions, it possesses piezoelectric and photoconductive properties, and its electrical conductivity varies over a large range. Therefore, zinc oxide thin films can be used for transparent and conductive coatings, IR reflective coatings, and piezoelectric and guided optical wave devices. Various fabrication techniques for obtaining good ZnO thin films have been utilized, such as sputtering,[1] ionized-cluster beam deposition,[2] spray pyrolysis,[3] and chemical-vapor deposition (CVD).[4] The technique of spray pyrolysis can be used to prepare zinc oxide thin films.

Procedure

Zinc oxide films can be prepared by a modification of the procedure used for the preparation of ZrO_2 films. A mixed solution of zinc acetate ("Baker Analyzed" Reagent, J. C. Baker Chemical Co.) and acetic acid (Reagent A.C.S., Fisher Scientific Co.) was used as source material. A stream of oxygen gas carried the nebulized solution into the reaction chamber. Both the efficiency of the deposition and the uniformity of the films were affected by the deposition parameters used. Table I shows a set of typical reaction

TABLE I. Reaction Parameters

Solution	0.1 *M* in zinc acetate and 0.05 *M* acetic acid
Substrate temperature	380°
Substrate to nozzle distance	12.7 cm
I.D. of nozzle	9.5 mm
I.D. of reactor	36 mm
Oxygen flow rate	5 L/min

conditions in the study. Under such conditions, a film of 2500 Å could be grown in 30 min using 10 mL of solution. Both silica and silicon wafers were used as substrates.

Characterization

Films of zinc oxide were characterized by the same procedures used for ZrO_2. The three strongest X-ray diffraction peaks were found to be at 36.3°, 31.8°, and 34.5°, and these corresponded to the (101), (100), and (002) reflections of zinc oxide. The measured grain sizes for a 0.3 μm thick zinc oxide film was 0.1 μm and approximately 0.3 μm for a film of 1 μm thickness. The resistivity of zinc oxide, in the dark, was 1×10^3 Ω·cm.

References

1. T. Minami, H. Nanto, and S. Takata, *Thin Solid Films*, **124**, 43 (1985).
2. T. Takagi, I. Yamada, K. Matsubara, and H. Takaoka, *J. Cryst. Growth*, **45**, 318 (1978).
3. L. Bahadur, M. Hamdani, J. F. Koenig, and P. Chartier, *Solar Energy Mater*, **14**, 107 (1986).
4. M. Shimizu, T. Horii, T. Shiosaki, and A. Kawabata, *Thin Solid Films*, **96**, 149 (1982).

C. GROWTH OF ZnS FILMS BY CONVERSION OF ZnO FILMS WITH H_2S

$$ZnO + H_2S \rightarrow ZnS + H_2O$$

Zinc sulfide is an IR window material and a transparent semiconductor with a large direct bandgap. It also possesses piesoelectric, photoconductive, and electroluminescent properties. Thin films of zinc sulfide can be utilized for infrared antireflection coatings, light-emitting diodes (LEDS), electroluminescent (EL) displays, multilayer dielectric filters, optical phase modulation, and light guiding in integrated optics. In recent years, there has been a large amount of effort directed at the growth of high-quality films of ZnS. Various fabrication techniques have been employed, such as ion-beam sputtering,[1] atomic-layer epitaxy (ALE),[2] molecular-beam epitaxy (MBE)[3], metal organic chemical-vapor deposition (MOCVD),[4] and spray pyrolysis.[5]

Recently a novel ultrasonic nebulization and pyrolysis method has been developed in this laboratory.[6] High quality films of zinc oxide can be grown using this simple technique. Maruska[7] reported the conversion of CdO films to CdS films by an ion-exchange process. It therefore appeared feasible to convert ZnO films with H_2S into ZnS films.

Procedure

■ **Caution.** *All experiments carried out under flowing H_2S should be performed in a well-ventilated hood. Excess H_2S can be trapped by the use of gas traps containing solid sodium hydroxide pellets.*

Zinc sulfide films were prepared by sulfurization of zinc oxide films with hydrogen sulfide.

Zinc oxide films on silicon and silica substrates were prepared by ultrasonic nebulization of zinc acetate solutions and thermal conversion of the zinc acetate to zinc oxide films. The procedure has been described in Section B. Zinc oxide films as deposited were annealed in a mixture of hydrogen sulfide and argon (H_2S:Ar = 1:1) in a horizontal tube furnace. The flow rate of the gas mixture was 50 cm^3/min. The temperature of the furnace was raised gradually from room temperature to 500° in 4 h. The furnace was maintained at 500° for another 3 h and then cooled down slowly in the H_2S/Ar atmosphere. Completion of the conversion was verified by X-ray diffraction analysis and infrared spectroscopy of the films.

Characterization

X-ray diffraction analysis was performed on the zinc sulfide films as deposited on the substrates. The diffraction patterns showed that the films consisted of ZnS crystallized as a mixture of cubic and hexagonal phases. Scanning electron microscopy studies showed a submicrometer grain texture with the size of most particles being about 0.3 μm. The measured direct optical gap was found to be 3.65 eV.

References

1. T. E. Varitomos and R. W. Tustison, *Thin Solid Films*, **151**, 27 (1987).
2. A. Hunger and A. H. Kitai, *J. Cryst. Growth*, **91**, 111 (1978).
3. M. Kitagawa, Y. Tomomura, A. Suzuki, and S. Nakajima, *J. Cryst. Growth*, **95**, 509 (1989).
4. S. Yamaga, A. Yoshikawa, and H. Kasai, *J. Cryst. Growth*, **86**, 252 (1988).
5. H. L. Kwok and Y. C. Chau, *Thin Solid Films*, **66**, 303 (1980).
6. W. DeSisto, M. Sosnowski, F. Smith, J. DeLuca, R. Kershaw, K. Dwight, and A. Wold, *Mater. Res. Bull.*, **24**, 753 (1989).
7. H. P. Maruska, A. R. Young, II, and C. R. Wronski, *Proceedings of* 15*th IEEE Photovoltaics Specialists Conference*, (1981), p.1030.

CONTRIBUTOR INDEX

Volume 30

SUBJECT INDEX

Prepared by THOMAS E. SLOAN

The Subject Index for *Inorganic Syntheses*, Volume 30 is based on the Chemical Abstracts Service (CAS) Registry nomenclature. Each entry consists of the CAS Registry name, the CAS Registry Number, and the page reference. The inverted form of the CAS Registry Name (parent index heading) is used in the alphabetically ordered index. Generally one index entry is given for each CAS Registry Number. Some less common ligands and organic rings may have a separate alphabetical listing with the same CAS Registry Number as given for the index compound, e.g., *1,4,8,11-tetrazacyclotetradecane*, nickel(2+) deriv. [770776-74-5]. Simple salts, binary compounds, and ionic lattice compounds, including nonstoichiometric compounds, are entered in the usual uninverted way, e.g., *Chromium chloride*, *($CrCl_2$)* [10049-05-5]. Salts of oxo acids are entered at the acid name, e.g., *Sulfuric acid*, disodium salt [7757-82-6].

FORMULA INDEX

Prepared by THOMAS E. SLOAN

The formulas in the *Inorganic Syntheses, Volume 30* Formula Index are for the total composition of the entered compound. In many cases, especially ionic complexes, there are significant differences between the Inorganic Syntheses formula entry and the CAS Registry formula, e.g., Sodium tetrahydroborate(1−) [16940-66-2], the Inorganic Syntheses formula index entry formula is BH_4Na, whereas, the CAS Registry formula is $BH_4 \cdot Na$ and in CAS formula indexes the entry is at the BH_4 formula for the tetrahydroborate(1−) ion. The formulas consist solely of the atomic symbols (abbreviations for atomic groupings or ligands are not used) and are arranged in alphabetical order with carbon and hydrogen always given last, e.g., $Br_3CoN_4C_4H_{16}$. To enhance the utility of the formula index, all formulas are permuted on the atomic symbols for all atom symbols. $FeO_{13}Ru_3C_{13}H_3$ is also listed at $O_{13}FeRu_3C_{13}H_3$, $C_{13}FeO_{13}Ru_3H_3$, $H_3FeO_{13}Ru_3C_{13}$ and $Ru_3FeO_{13}C_{13}H_3$. Ligands are not given separate formula entries in this I.S. Formula Index.

Water of hydration, when so identified, and other components of clathrates and addition compounds are not added into the formulas of the constituent compound. Components of addition compounds (other than water of hydration) are entered at the formulas of both components.

CHEMICAL ABSTRACTS SERVICE REGISTRY NUMBER INDEX